Record Cultures

Record Cultures

The Transformation of the U.S. Recording Industry

KYLE BARNETT

UNIVERSITY OF MICHIGAN PRESS • ANN ARBOR

First paperback edition 2021

Published in the United States of America by the
University of Michigan Press
Manufactured in the United States of America
Printed on acid-free paper

A CIP catalog record for this book is available from the British Library.

Library of Congress Cataloging-in-Publication data has been applied for.

First published in paperback July 2021.

Library of Congress Cataloging-in-Publication Data

Names: Barnett, Kyle, author.
Title: Record cultures : the transformation of the U.S. recording industry / Kyle Barnett.
Description: Ann Arbor : University of Michigan Press, 2020. | Includes bibliographical references and index.
Identifiers: LCCN 2019039017 (print) | LCCN 2019039018 (ebook) | ISBN 9780472131037 (hardcover) | ISBN 9780472124312 (ebook)
Subjects: LCSH: Sound recording industry—United States—History. | Popular music—Social aspects—United States—History.
Classification: LCC ML3790 .B27 2020 (print) | LCC ML3790 (ebook) | DDC 384—dc23
LC record available at https://lccn.loc.gov/2019039017
LC ebook record available at https://lccn.loc.gov/2019039018

ISBN: 978-0-472-03877-0 (pbk : alk paper)

For Lisa

Acknowledgments

I lived with this book a long while and I am deeply grateful for those that helped along the way. This project began while I was a graduate student at the University of Texas Department of Radio-Television-Film. When I brought Tom Schatz my proposal to frame the story of the recording industry as a media industry during a crucial, transitional era, he immediately embraced the idea. Without his support and friendship, this book would likely not exist. At Texas, I was fortunate to find a committee of brilliant and supportive scholars: Jim Buhler, John Downing, James Hay, Mary Kearney, and Craig Watkins. I am so thankful for the friendship, camaraderie, and critical eye of my friends and colleagues at Texas, who read various versions of what appears here: Christopher Lucas, Afsheen Nomai, Allison Perlman, Jennifer Petersen, and Avi Santo.

At Bellarmine University, I found lasting friendship and support from Ruth Wagoner and Gail Henson, who supported my research and assisted me in navigating the learning curve endemic to life as a junior professor. My fellow faculty in Bellarmine's Department of Communication have helped this come together, among a blur of classes, meetings, and campus changes big and small. I am especially grateful to the friendship and collegiality of Moira O'Keeffe, who offered writing advice and support with great patience and kindness. I benefited in no small way from the assistance of Sue Mauldin, our administrative assistant in Bellarmine's Department of Communication, who fuses optimism and care for students, faculty, and staff, with a practical disposition perfect for the role. Thanks to numerous undergraduate work-study students who assisted me, under Sue's tutelage, over the course of this book project. I am indebted for the help of Bellarmine's first-rate reference librarians: Daniel Bays, John Boyd,

Kevin Peers, and Martha Perry-Lundgren for their tireless work in tracking down hard-to-find sources and answering my arcane questions. Thanks to my undergraduate and graduate students over the years, at Bellarmine and elsewhere, whose discussions with me about music, media, and history are reflected here in meaningful ways. Those students could attest to my writing time in the coffee houses of Louisville, and I am accordingly thankful for the (past and present) baristas of Sunergos Coffee on Norris Place and Quills Coffee on Baxter Avenue.

At the University of Michigan Press, Mary Francis proved to be the ideal editor for a book that stands at the intersection of music, media, and culture. Mary assisted and supported this project at various stages with insight and patience. Thanks to all at the University of Michigan Press who helped me along: Sara Cohen, Susan Cronin, Mary Hashman, Carol Reed, and Samuel Killian.

I am thankful to the archivists and librarians who assisted me in no small way. Their assistance in this detective work was and is a pleasure. I am grateful for the help of archivists and librarians at the Starr-Gennett Foundation (then curated by Elizabeth Surles) and the Wayne County Historical Society in Richmond, Indiana. Thanks to Deborah Gillaspie of the Chicago Jazz Archive at the University of Chicago, as well as Matt Holdzkom, Jordan Ryan, and the staff of the Indiana Historical Society (IHS) in Indianapolis. My deepest thanks to the family of the late John MacKenzie, who made sure his Gennett Records research was made available to researchers at the IHS. In Nashville, I benefited from the advice of senior historian John Rumble of the Country Music Hall of Fame. Thanks to a tip from the late author and collector Raymond Wile and the help of archivists at the New York Public Library of the Performing Arts' Rodgers and Hammerstein Archives of Recorded Sound, I researched OKeh and Columbia Records materials that to my knowledge have yet to be cataloged. I'm also thankful to the staff of the Camden County Historical Society in Camden, New Jersey, where I researched RCA-Victor history and the Port Washington Historical Society and Ozaukee County Historical Society in Wisconsin, where I researched Paramount Records. The holdings of such city and county historical societies hold untold treasures, of which we remain largely unaware. At Kentucky's Berea College, I was greatly assisted by archivist Harry Rice in researching the Southern Appalachian Collection in the Hutchins Library Special Collections and Archives. Various trips to the Library of Congress' Recorded Sound Reference Center were crucial for this research, where I benefited from the expertise of Bryan Cornell, Karen

Fishman, and David Sager. I was granted access to archival audio interviews of musicians and talent scouts in the Southern Folklife Collection thanks to Steven Weiss and staff at the University of North Carolina's Wilson Library, and from Martin Fisher and Greg Reish at Middle Tennessee State University's Center for Popular Music. I am also deeply grateful to the collectors, fans, and amateur researchers from whom I have learned much over the years. Their work has been crucial to media research and deserves to be recognized.

I am thankful for the anthology editors that commented on several book chapters that were precursors to what I present here. Some of my work on Gennett in chapter 1 first appeared in "The Recording Industry's Role in Media History," in *Convergence Media History*, edited by Janet Staiger and Sabine Hake (Routledge, 2009). Some passages focused on scouts as cultural intermediaries in chapter 2 were first published in "Record Men: Talent Scouts in the U.S. Recording Industry, 1920–1935," in *Making Media Work: Cultures of Management in the Entertainment Industries*, edited by Derek Johnson, Derek Kompare, and Avi Santo (New York University Press, 2014). Thanks to the Berea College Special Collections and Archives, the Library of Congress, the Country Music Hall of Fame, and the Billy Rose Theatre Division of the New York Public Library of the Performing Arts for images reproduced here. I am fortunate to have Chicago's Halloween Martin gracing the cover of this book, who many believe to be the first female radio disc jockey in Chicago, and quite possibly the United States. (Her life and career deserve more attention.) The image first appeared (without a photographer's credit) in the November 1932 issue of *The Chicagoan* and appears here thanks to Quigley Publishing Company, a division of QP Media, Inc.

I received several fellowships that assisted in the research and writing of this book. At the University of Texas, I received a Jesse Jones Fellowship that afforded me resources and time to work. Thanks to Bellarmine University for several Faculty Development Fellowships that supported this research, which allowed funding to visit archives. Thanks to Bellarmine for the sabbatical in which I was able to focus on research without the teaching and administrative responsibilities of a college professor at a liberal arts college. Finally, I am grateful for the generous support of the American Musicological Society for a grant in support of this research.

This book was given new purpose by the late David Sanjek's presentation, "First I Look at the Purse: The Contamination of Popular Music Studies by Agoraphobia," at the 2011 Experience Music Project pop music conference at UCLA. In his presentation, Sanjek called for popular music scholars to write the kind of books more common in film studies, in which detailed analyses of a given period in film history was given book-length treatment.* Sanjek wondered aloud why popular music histories were envisioned as epic narratives, meant to define (or redefine) the field into a less precarious existence. I do not claim this is the book Sanjek had in mind, but I hope that it contributes to such a mission. Thanks to Ned Raggett and Eric Weisbard, who helped confirm my memory of Sanjek's presentation.

Thanks to all who facilitated this writing in various ways: Chris Chinchar, Charlie Dahan, Adele Deaton, Leonard DeGraaf, Christine Ehrick, Evan Finch, Hollis Griffin, Eric Harvey, Michele Hilmes, Paul Israel, Chris Jennings, Rick Kennedy, Gary Koehl, Dale Lawrence, Pam Lindroos, David and Sue Long, Karl Hagstrom Miller, Michael Newman, Mike O'Connor, Alisa Perren, Diane Pecknold, Nolan Porterfield, Malcolm Rockwell, Rene Rodgers, Tony Russell, Alex Russo, Nathan Salsburg, Jeff Schwartz, Yan Shuang (闫爽), Heather Murphy Sloane, Rob Sloane, Jacob Smith, Janet Staiger, Jonathan Sterne, Kevin Trogden, Jennifer Waits, and Eric Weisbard. I am especially thankful for the suggestions of Kathryn Fuller-Seeley and Charles McGovern, whose careful reading of this book in its later stages greatly improved the final result.

I am fortunate beyond measure to be spending my life with Lisa Barnett, who has supported this project in every possible way. She read various drafts of this manuscript, helped me deal with conceptual and logistical issues, and provided me with time to research and write. I am grateful beyond words for the life we have built together, first as a duo and later as a trio with our daughter Leila (龚乐洋). I am so grateful to have them both at the very center of my life.

*David Sanjek, "First I Look at the Purse: The Contamination of Popular Music Studies by Agoraphobia," presented at the Experience Music Project Pop Conference, UCLA, February 26, 2011.

Contents

Digital materials related to this title can be found on the Fulcrum platform via the following citable URL: https://doi.org/10.3998/mpub.9901441

INTRODUCTION

Phonograph Boom

The Expansion of the U.S. Recording Industry

The unprecedented expansion of the U.S. recording industry that began almost a century ago fundamentally changed the shape and scope of the country's popular music. This expansion was sparked by smaller labels' inability to compete with the industry's most established labels and ended with the first major consolidation of media entertainment industries at the close of the decade. It all felt so speculative and provisional; how long would the public find this or that sound appealing? Recording directors and talent scouts did the best they could to keep up. As freelance talent scout Harry Charles noted, "you had to grab what you could get."[1] This book frames U.S. recording companies *as* media industries, with deep and abiding ties to other such industries. The recording companies' ties to radio would increase so dramatically by the end of the 1920s that it reshaped both industries, in what was the first major consolidation between entertainment industries in the twentieth century. With radio as the "new national fad," recording companies large and small sought to experiment in the depth and breadth of what they recorded, the technologies through which they recorded, and in connection to the other key entertainment media of the era, most importantly radio and film, in what would be the recording industry's "most turbulent and productive" era to date.[2]

While all sources agree that the recording industry expanded dramatically during the 1910s, estimates vary given gaps in the historical record as well as the recording industry's traditionally low barrier to entry when compared to other media industries. Roland Gelatt's foundational book *The Fabulous Phonograph, 1877–1977* notes just three phonograph-producing companies in 1913 and 73 by 1916.[3] More recently, Rick Kennedy has noted a similar expansion of recording companies between 1914 and 1917, as the number of recording companies went from a few dozen to more than 150.[4] The 1919 *Census of Manufactures* (published in the 1920 Census) listed the "phonograph and graphophones" businesses among its "principal industries" (ranked fiftieth overall) making "products valued at $100,000,000 but less than $500,000,000."[5] Comparatively, motion pictures appeared in the table with "photograph materials," ranked eightieth overall, while "musical instruments, pianos," ranked ninetieth."[6] The absence of radio from this list was understandable, given its nascent status as well as President Woodrow Wilson's banning of U.S. amateur radio stations from transmitting for the course of the war unless they served military purposes (the ban was lifted in 1919).[7] While the pages of *Talking Machine World* often exhibited a booster's enthusiasm for the phonograph trade, the paper's enthusiasm during the late 1910s was well founded.

While the Big Three had asserted their dominance a decade before, this did not prevent competition from new companies entering the market. By the 1910s, companies were making their own phonographs (often discount models) and recordings, modeled after the "vertical cut" disc records process established in France by Páthe Frères in 1909.[8] The problem for consumers—a perennial one in media—was that consumers had to choose between the lateral-cut discs playable on the Big Three's machines or choose the vertical-cut format offered up by newer companies, including Starr (later Gennett), OKeh, Paramount, Aeolian-Vocalion, and Pathé's American division.[9] By the end of World War I, many of the key companies here—big and small—were already in existence, but new companies began to emerge, while existing companies created new labels. The impetus was a 1918 battle between Victor and the Starr Piano Company, an unlikely challenger that would battle Victor's proprietary right to lateral-cut recordings, and win. As Sutton and Nauck suggest, it took a 1922 Supreme Court decision to settle the matter, but lower court decisions emboldened companies to develop lateral-cut disc lines, an "industry-wide" media transition in 1919 and 1920, that allowed consumers to play a much larger variety of discs on their previously purchased talking machines.[10]

Sales of sound recordings surged with "returning troops, wartime workers flush with cash, and music fans whose collections had been depleted by contributions to the Slacker Record Drives," which gave records to soldiers to listen to in sometimes far-flung locales, a kind of aural trip back home.[11] The wartime slowdown was replaced with a boom, putting those companies most involved in war-effort production (especially Victor, with its impressive Camden works) at a temporary disadvantage.[12] Companies old and new began making discs for what was a new, broader market, creating a boom that would lead to a dramatic expansion of recording culture in the United States. Roughly two months after the November 11, 1918, armistice that marked the close of the Great War, a January 15, 1919, *Talking Machine World* headline read: "Our National Victory Should Make Business Men Optimistic."[13] Both the recording and playback process remained mechanical, but the phonograph boom came as old patents expired, court decisions opened up the industry, and new companies sought to compete against the industry's established leaders. A year later, industry predictions were coming true. The trade journal's January 15, 1920, issue was full of advertisements from companies like Aeolian-Vocalion, Brunswick-Balke-Collender, Gennett, OKeh, Pathé and others, labels that, as Allan Sutton has suggested, were all producing their own lateral-cut recordings in what was a new competitive context.[14] An April 1919 advertisement declared, "There Has Never Been Such a Boom in the Record Business," telling retailers that it wasn't a matter of how many records that could be sold, but a question of "Where Can you Get Enough Popular Records Right Away?"[15] While the Victor phonograph works was still being restored from its wartime role as arms manufacturer, the domestic U.S. market for sound recordings was so great that the company couldn't keep up. And in the month of August 1919, no new Victor recordings were released so that the company could fill its many back orders.[16] Edison's ability to respond to recording industry changes was already a problem before World War I, when Edison himself professed technological quality and clarity over the use of celebrity, an increasingly important dynamic in both sound recordings and cinema in the 1910s.[17] To make matters worse, Edison himself became the company's de facto music director and found himself alienated by the emerging music of the World War I era, especially jazz.[18] Increasing attention was given to both rural—and more specifically southern—consumers at a time when the recording industry was expanding from its traditional center in the American Northeast.[19] Industry trade papers offered tips for attracting consumers while keeping up with demand, through novel and eye-catching window displays or having just the right mix of newly released recordings. The June 14, 1919, issue of *Music Trade Review*, written by OKeh Records' advertising manager, announced that "The Phonograph Industry Has Become an Ideal Musical-Commercial Power." The article noted that "the phonograph and record industry is decidedly not a fad industry" but one imbued with a "lasting power . . . as vital to the progress of mankind as the telephone, telegraph, the printing press, and several other epoch making inventions of man."[20] On the same page, it was noted that a Pathé Records advance list used an increasingly common approach by companies to fill orders accurately and quickly: retail dealers and wholesalers used telegraph codes and technology to ensure speed and accuracy in filling rush orders.[21]

It was the emerging recording companies that first began recording a wider swath of vernacular American musical and spoken-word performances in an attempt to identify niches through which they could compete with the Big Three oligopoly: Victor, Columbia, and Edison. It was through this search that the naming, defining, and categorizing of jazz, race records, and old-time music impacted the industrial practices of record companies who defined those genres for the buying public. This isn't to say all the music represented in these genres became central to popular music culture. Each genre's music existed both at the center and along the margins of the decade's popular music. Countless other recordings, some of which constituted vital genres and subgenres all their own, continued despite changes that would slow or stop the expansion of the industry's recordings by decade's end.

It is not that recording companies weren't already busy recording various aspects of the American vernacular at the start of the 1920s. It's that the breadth and depth of that activity expanded greatly during the decade, through industrial organization and cultural circulation of sound media, in constant negotiations between musicians and audiences, scouts, wholesalers and retailers, recording engineers, and recording directors. These employees worked in several registers at once: (1) they managed the selection, manufacturing, circulation, and sale of sound recordings for the companies with which they were affiliated, (2) they managed audiences by gauging tastes and participating in the construction of genres, and (3) they necessarily managed a range of social and cultural realities across huge swaths of American society as part of their everyday work.[22]

But even larger dynamics were at work: The convergence of U.S. media industries that began in the late 1920s meant that the recording industry, largely its own industrial entity since its origins in the 1890s, would cease to stand alone as a discrete entity. They would become part of larger radio (and, to a lesser extent, film) companies in what was the first major consolidation of entertainment media industries in the twentieth century. It is not that recording companies were enveloped altogether, though it may have seemed that way to some in the early 1930s. In subsequent years—even during the Great Depression—new labels (or revamped ones) would continue to exist as stand-alone entities. The recording business's low barrier to entry helped ensure that. But it would no longer be the same as it had been when the leading recording companies stood alone. As is often the case, the changes were not always welcome by those most invested in how companies had operated, nor were the changes always fully spelled out in its trade papers. Many of those who had been financially and emotionally invested in the U.S. recording industry in its heyday did not come to terms with its shifting fortunes until the realities could no longer be ignored. And while the recording industry declined in many ways amidst the rise of radio, it also expanded its breadth and scope in such a way that would help ensure its long-term economic and cultural impact. This change came as the foundational genres of U.S. popular music were shaped and reshaped, before and after many of the recording industry's labels became part of entertainment conglomerates in the first major of wave of entertainment media consolidation in the twentieth century.

That these myriad changes took hold in the 1920s does not seem accidental. The broader context was a roiling set of contradictions, with new freedoms coming up against repressive and violent backlashes. As Kathleen Drowne and Patrick Huber wrote:

> A newly urbanized population clamored for the modern technologies of automobiles, refrigerators, radio, and electric appliances yet at the same time longed for the simpler lifestyles of rural America. Most Americans' standards of living rose, but so, too, did the number of violent, repressive clashes during the decade. Americans reveled in the newfound freedoms afforded them by the automobile, suffrage, consumer culture, yet simultaneously they sought to repress other freedoms through Prohibition legislation, Klan night riding, and religious Fundamentalism. The liberated "flapper" was lauded in music, movies, but . . . was nonetheless expected to settle down and adopt the lifestyle of a traditional married woman. All in all, the 1920s signified a profound shift in behaviors, attitudes, and lifestyles

of ordinary Americans, who found the modern world exciting but also extraordinarily complicated.[23]

Media in the 1920s, T. M. Kando wrote decades later, offered a "hedonistic, glamorous, and make-believe world," while cultural critic Gilbert Seldes suggested (in 1923) that the "seven lively arts" were at work creating a new American culture, in which he included popular song but only obliquely referenced sound recordings. He noted that tunes of the day expressed a broad array of current cultural attitudes, presenting "rather like contemporary fiction in giving form to social phenomena without expressing approval or disapproval."[24] The "Jazz Age" may have been given resonance via F. Scott Fitzgerald's *Tales of the Jazz Age*, but the terms had already been in circulation as changes (musical and otherwise) seemed to speed up everyday life.[25] Robert and Helen Lynd noted that residents of Muncie, Indiana—dubbed "Middletown" in their unprecedented 1929 sociological study—noted that "Mechanical inventions such as the phonograph and radio are further bringing . . . more contacts with music than ever before."[26]

By the start of the 1920s, the recording industry's focus on popular music (as opposed to spoken word recordings, re-enactments, skits, sermons, and so on) served to obscure for some its role as a *media industry*. There are many reasons for this, including the cultural slippage between our conceptions of musical recordings and musical performances, also a problem for cinema, for instance, in distancing its relationship to theatrical performance. By the end of the 1920s, it was no longer possible to ignore the recording industry's role as media industry, due to the complete recalibration of its relationships with other media industries, most notably radio. Radio's emergence during the 1920s is breathtaking in scope, even from the vantage point of commonly known touchstones: Frank Conrad's launch of Pittsburgh's KDKA in 1920s; the 1921 broadcast of the Jack Dempsey vs. Georges Carpentier boxing match (matches also appeared on 78 rpm records); bandleader Vincent Lopez's debut on WJZ in 1921, with Paul Whiteman following a year later, along with the debut of Bertha Brainard's *Broadcasting Broadway*; the first coast-to-coast radio broadcast in 1924; the rise of fictionalized radio narrative serials like *Rise of the Goldbergs* in 1929; and the transformation of radio news via the Lindbergh baby kidnapping case in 1932.[27] Radio's newly found cultural power and reach led the recording industry to first resist and then to collaborate in varying ways, in hopes of benefiting from radio's vast simultaneous reach, especially after the tentative emergence of network radio in November 1926. Radio needed the recording industry after its conceptual tendency toward liveness became increasingly untenable as programming grew. Meanwhile, after years of experiments involving film and sound combinations, the motion picture industry began its sound-on-film transition with a focus on movies that sang as much as they talked. The first half-decade or so of talking motion pictures were defined by musicals, in the form of backstage musicals and revue films of the late 1920s and early 1930s.

The convergence was not only based on content. Technological change played a part in the new closeness, determined by various cultural, industrial, and legal forces. The shifting needs of the three industries—the recording, radio, and film industries—came together in new technological practices. Electrical recording would reshape the role of the recording industry as an increasingly important supplier of aural raw material in the face of radio and cinema's demands. All three industries were deeply impacted by the adoption of electrical recording, which brought the radio and film industries together with the recording industry like never before. Both radio and film increasingly relied on the recording industry, while the recording industry itself would become increasingly linked to those industries in turn. Having these companies more closely related did not change the regulatory structures that had emerged around discrete industries, as radio was above all seen as a *communicative* medium and sound recordings and film were more commonly understood discursively as *entertainment* media, a distinction at work in Starr's analysis as well as in the academic institutionalization of media and music studies in the United States and elsewhere. But as I hope this research shows, such distinctions points more to the limits of our own understandings of these media industries rather than the beliefs and dispositions held by those most closely involved.

Between the close of the First World War and the mid-1930s, the central organizing categories of twentieth-century American popular music solidified: jazz (in its first incarnation), "race records" (later blues), and "old-time music" (later "hillbilly," then "country"). Each developed numerous subgenres and styles. Many of the subsequent genres that have emerged relate back directly or indirectly to these three categories, particularly in the social and cultural contexts and values assigned to each. To understand these sociocultural aspects of the industry is to have a good grasp of the historical arc of the recording industry's organizational schema, in which social issues play out through aesthetic categories and questions of taste.

This book maps the recording industry's historic role in the categorization of popular music through one of the industry's central means of organizing music: genre. Genre as I use the term here is understood in its industrial function, a marketing category, shaped and reshaped along the cultural circuit of production, distribution, and consumption. I have no interest here in parsing out the contours of genre in terms of stylistic approaches and lineage. Genres most commonly use aesthetic categories to mask, however thinly, social ones, though those doing so in their everyday work may not have always been aware they were doing so, given deeply ingrained beliefs and dispositions. Any brief aesthetic conversations herein should be read accordingly. That recording companies were unable (or unwilling) to parse out the differences between their aesthetic delineations with a constellation of social and cultural ones should not encourage us to do the same. There is at least some evidence in what follows that scouts were aware that their aesthetic differentiations were born out of social structures and realities more than anything else, and since they were in the business of selling sound recording, they put those understandings and misunderstandings to use. Genre categories, and their social and cultural distinctions, were (and are) created and operationalized by recording companies to sell recorded music as commodities. As such, genres are imbued with a constellation of social, cultural, and aesthetic designations with their own set of specific values, attitudes, beliefs, and dispositions.

By the 1920s, established major labels like Victor and Columbia were outpaced by smaller companies like Gennett, OKeh, and Paramount in their ability to delineate new talent and define new taste cultures. These genre categories functioned as a means for the industry to structure these "new" types of music for specific audiences through linking specific aesthetic styles with social attitudes and beliefs connected to the music. Recording companies commodified the music and artists down through genre categories, the narratives about the music, the performers, and even the details around a given performance or recording session. The key figures in developing these genre categories were the recording directors and talent scouts that functioned as intermediaries between artists, audiences, and the companies for which they worked, either as staff employees or freelancers. Talent scouts, along with recording engineers and some company executives, coordinated the day-to-day business of record making, distribution, and sales, while simultaneously discerning listeners' tastes for this or that niche.[28] These labels traversed rigid and often dangerous social and cultural divides to record ignored or disparaged music by marginalized and often vilified people.[29] The music and the people associated with it (listeners as well as performers) were represented in catalogs, advertisements, and the trade press, all of which reified these genres in various ways.

The attention to genre's importance here is ultimately in service of this book's larger focus: to explore recording companies' institutional cultures and their role in reshaping American popular music. This reshaping was due in large part to industry expansion in search of market niches, followed by recording and radio company mergers in the century's first wave of entertainment industry conglomeration. A cultural circuit of producers, consumers, and intermediaries defined genres and audiences in various ways.[30] Though profit was the first motive, recording companies were also in the business of making culture, while being cultural institutions all their own. Those institutional cultures were also reproduced in the recordings they made, in the advertisements that sold the recordings, and in the stories they told and retold for decades after the fact.

Social Contexts: Genre Formations and Technological Origins

The narratives represented here weave in and out of the arc of U.S. recording companies' fortunes during the 1920s and 1930s. In the macro narrative, the U.S. recording industry sales peaked in 1921 at $106 million, a pinnacle not reached again until 1945.[31] An early 1920s recession kept sales flat for many companies; it was followed by a recovery of sorts after 1925, then by the stock market crash and coming of the Great Depression in the early 1930s. Though some companies found their place, profiting amidst various social, industrial, and technological changes in the 1920s and 1930s, overall sales ebbed throughout this period, finding bottom in 1933, after which the industry's fortunes began to slowly climb once again.[32] But this book attempts to provide contexts for what was happening within these industrial and economic realities, as recording companies began to experiment in unprecedented ways in recording a broader range of American music, while also interacting with other media industries in new ways.

The rise of the U.S. recording industry and its subsequent consolidation into the radio and film industries in the late 1920s and early 1930s represents the earliest consolidation of entertainment media (predated only by the telecommunication consolidation of the telegraph and telephone industries). Both the dramatic expansion into new niches of the recording industry through the 1920s and its consolidation and reconstitution in the face of radio's rise in the mid-to-late 1920s and the Great Depression of the early 1930s, is reminiscent of what Paul Starr has called "constitutive choices" in the history of media "that create the material and institutional framework of fields of human activity."[33]

In Starr's use of the term, he means to map the initial emergences of new media, moments in which "ideas and culture come into play, as do constellations of power, pre-existing institutional legacies, and models from other countries,"[34] Here I extend the concept in two ways: First, Starr largely limits media to "the press, postal and telecommunications networks, cinema and broadcasting," while phonography gets scant mention in the book's almost five hundred pages.[35] Second, Starr limits his research to initial moments in which a media is created, defined, regulated, and circulated through a given society.

Media studies continues to have an odd relationship with sound recordings as media. Despite the *de rigueur* sound recording and popular music chapter in introductory media studies texts, the subject is relegated to cameo appearances in subsequent books and courses. Is it that phonographic media's traditional marginalization in mass communication research was due to deep links between the rise of broadcasting and the emergence of mass communication research? At that point, phonography's revolutionary impact had faded with its ubiquity, especially given radio's unprecedented scope and reach? Sound recordings have long been both the raw material used by and for other media of the nineteenth and twentieth centuries, and an industry all its own, a dual, defining role of lasting importance. But one of the most important contributions of media industries research has been its ability to define intermedial relationships across contexts. These intermedial connections extend well beyond early cinema and early phonography to the present day. In addition to the focus on the recording industry's transformations of the early 1920s, intermedial concerns accelerate via consolidations and collaborations across sound recording, radio, and film cultures in the late 1920s and through the 1930s. The transition from acoustic to electric recording and playback technologies was key in this regard, a technological shift that was negotiated through a comprehensive restructuring across media industries. But the impacts go far beyond this, intersecting industrial ownership, intertextuality, and the strategies and tactics of those participating along media's cultural circuit. A strength of media industries histories has been in bringing these intermedial relationships into view, as I do here, through integrating the recording industry into larger media histories, first as a discrete industry in the early decades of the twentieth century to one inexorably tied to the radio and, to a lesser extent, film industries thereafter. Part of this task is the constant reminder that the recording industry was connected to other media industries, in varying ways, all along.

Sound recording entered a society in which foundational social and economic changes were taking place. "The development of sound reproduction into recognizable media occupies a place among a whole range of social transformations in turn-of-the-century America," Jonathan Sterne writes in *The Audible Past*. "The very possibility of sound media was structured by the changing economics and social organization of invention, the growth of corporate-managerial capitalism, and the concurrent move from Victorian to consumerist forms of middle-class everyday life."[36] The fundamental changes in American business in the latter half of the nineteenth century set the stage for the recording industries that expanded in the first decades of the twentieth century. The railroad and the telegraph both were key in the development of modern approaches to production, distribution, and consumption, as well as the rise of marketing as a key element in an increasingly national (and shortly after, transnational) network.[37] The traditional "single-unit business enterprise," in which one owner or a handful of co-owners ran a given company (a shop, factory, etc.) did not disappear.[38] But it was superseded in various ways by highly organized companies that moved from strictly local or regional concerns to places far beyond. Such operations required a much broader bureaucratic structure in which all-important middle managers as well as an emerging executive elite increasingly defined what "business" would mean moving forward.[39] None of this was instant or automatic. The new managerial structures of American business came about within a specific social and historical series of contexts. Changes to these new administrative and managerial structures included greater productivity and profits at lower cost, a single unit under which a variety of complicated enterprises functioned, and a hierarchy that propelled a given company overseeing it all.[40] "The coming of the large vertically integrated, centralized, functionally departmentalized industrial organization altered the internal and external situations in which business decisions were made."[41] The upstart labels of the 1910s and 1920s found a place in the gaps of what the Big Three had recorded, focused on "handling local and more specialized aspects of the business."[42] "The new speed, regularity, and dependability of transportation" meant that wholesalers became "jobbers," who purchased goods from manufacturers outright (or sold on commission), while tying into a given company's marketing brand and building a network of retailers and filling the racks of music stores in far-flung locales.[43]

It was the "record men" of the era—sometimes talent scouts, jobbers, and salesmen all in one—that were so vital to recording companies, first located in a small portion of the American Northeast. As new companies started up in the Great Lakes Midwest, they mimicked the Big Three in their corporate structures, though they relied even more heavily on the knowledge of their employees (broadly defined) in the field. "They reported on changing demand, items particularly desired, the general economic conditions of different sections, and, above all, the credit ratings of local storekeepers and merchants. The salesmen also assisted the storekeepers in keeping a stable inventory, in improving their accounting, and even in enhancing their merchandising displays."[44] As I have argued elsewhere, those who recorded and scouted and sold this music, on behalf of companies in search of viable market niches, crossed vast cultural divides of race, class, gender, and geography, producing sounds by and for often ignored, ridiculed, and marginalized people. This came in the midst of convulsive changes in American society, as the Victorian era gave way to an increasingly liberatory and reactionary age all at once.[45] The way releases were organized—however tentatively—had, in some cases, deep impact over time. The era's cycles of production, categorization, distribution, sale, and consumption of sound recordings had lasting impacts: the reification of existing social and cultural beliefs, attitudes, and prejudices, as niche styles and tastes were shaped into genres.

U.S. recording companies were already busy recording a broad pop mélange, indebted to what William Howland Kenney calls the "Coney Island crowd" tradition, or songs, monologues, and dialogic skits that would have played in the phonograph parlors and penny arcades that would have been part of amusement park culture up to World War I.[46] Some of what was popular in the post-World War I and early jazz age years included a broad mix of "light" or popular takes on classical music via Alma Gluck, who incongruously made her name with a hugely popular version of "Carry Me Back to Old Virginny"; novelty tunes by vaudevillian singer Billy Murray; mainstream dance band music by Paul Whiteman, Isham Jones, and Ben Selvin; popular adaptations of religious songs like "The Great Judgment Morning," via Homer Rodeheaver; Arthur Collins' vaudevillian trade in racist minstrel-era "coon songs" and Al Jolson's update of minstrelsy with "Swanee"; and Tin Pan Alley numbers like Nora Bayes' take on Richard

Whiting and Raymond Egan's Orientalist tune "Japanese Sandman." For better or for worse, U.S. recording companies already included a wide variety of musical approaches and cultural portrayals.

Genres, like classifications in general, are used in a variety of ad hoc ways, on a daily basis. "We all spend large parts of our days doing classification work," Geoffrey Bowker and Susan Leigh Star write, "often tacitly, and we make up and use a range of ad hoc classifications to do so."[47] Those who produce, distribute, and consume music—musicians, fans, and the recording companies that are the focus of this book—shape and reshape borders, often reinforced when we cross them. We participate in the shaping and reshaping of these classifications and conventions (or in this study, genre categories) and they shape our everyday lives in often invisible ways, until they "break down or become objects of contention."[48] This makes the work of excavating genre work of the past easier, given that the ideologies have decayed in certain ways while remaining doggedly steadfast in others. It is nearly impossible for us to think outside these categories, which we reify even while we critique them in various ways.

Genre is a central component of this book in that it opens up to a number of important social and cultural functions of media industries' role along the cultural circuit. Like David Brackett (by way of Franco Fabbri), I see genre as fundamentally *relational.* This notion is based in a "system of difference," in which genres function within ever-changing contexts of cultural production as opposed to the mapping of "a trait-based approach."[49] Genres circulate as ever-changing "articulations" and "assemblages" in which popular music genres become recognizable in their dialogic relationship to other genres.[50] As genres emerge, they become iterative, extending genre formations through the citing of musical (and social and cultural) conventions. This approach rightly emphasizes the deeply dialogic nature of genres, not as "quasi-mechanistic" formations in their own "hermetically sealed world," but in a world in which such formations rise, shift, and change.[51]

Genres are not devised by any one actor. As Jennifer Lena has suggested, "music is in need of thick histories," in which genre formations are not solely attributed to "individual genius or serendipity" but formed through "collaborative links with skilled practitioners" of varying sorts, in which music and its circulation is part of a "participatory, community-based activity."[52] These activities can be usefully mapped by an analytical approach based in cultural studies research, which has sought a fuller accounting of the interrelated "cultural processes"—representation, identity, production, consumption, and regulation—along the "circuit of culture," in which these poles come into relation in ways that shift and transform over time.[53] Not surprisingly, all these analytical categories are represented here. But choices must be made, in light of previous research and the agenda of this book, which focuses on recording companies as institutional cultures, first through their role in genre formations and later in the conflicts and collaborations that emerged as many of those same companies changed in various ways with the rise of radio.

Having said that, audiences, in all their myriad complexities, are central to the shaping and reshaping of generic conventions and ever-shifting listener identifications, in the twists and turns of their cultural circulation and tied to constellations of social and cultural formations.[54] As Thomas Schatz has argued, the study of media industries must not lose sight of the roles of creativity, authorship, and cultural production research based in both film studies and cultural studies, amidst ever-shifting relational concerns about the shape of media industries, concentrations of ownership, and mechanisms of control.[55] While this book focuses on recording companies' roles in genre formations, it is the larger set of participants—sometimes understood as communities—that play no small part in a given genre's emergence, sustainability, and malleability over time. The kaleidoscope of cultural distinctions that make up genre formations are often arbitrary, but it does not then follow that they are of no importance, because our everyday experience suggests otherwise, when for instance we are insulted by a friend's rejection of our own tastes, affinities, and dispositions tied to this or that genre. Even as we critique these distinctions, we continue to live with them. As Simon Frith suggests, "the essence of popular culture practice is making judgments and assessing differences."[56] And as Keith Negus has argued, that process goes on across popular music media culture, including within those recording companies that scout, record, classify, and sell to the larger public. In that sense, I share Negus' intention to move beyond describing a cultural product via "technical and routine processes and institutionalized practices," to "understand[ing] the meanings that are given to both the 'product' and the practices through which the product is made."[57]

In focusing on the expansion of folk and pop scouting in the American South, Karl Hagstrom Miller convincingly argues that it was "scholars and artists, industrialists and consumers" that began to segregate what had not been segregated previously, as "a fluid complex of sounds and styles in practice was reduced to a series of distinct genres associated with particular racial and ethnic identities."[58] While this book greatly benefits from such insights, I will not try to replicate that here, focusing instead on the institutional cultures, discourses, and contexts that exhibited much of what Hagstrom Miller has identified. When possible, I have tried to also provide glimpses of consumers of recordings where possible. Much of the work here focuses on genres' industrial contexts and sociocultural circulations, via the nascent recording companies of the era as well as via recording companies' increasingly close relationships with radio and film. But I do so with the foundational recognition that many of the social and cultural tropes and dispositions date back to colonial America. Genre furthered these existing dynamics as a central organizational (and sales) strategy in media, as an explanatory and narrative model through which a set of proposed distinctions and differences are made "real." Some of those niches emerged into lasting genre categories, lasting through companies' dissolution or absorption into other media industries, while others served more ephemeral importance, at least to the companies involved.

The start of this book focuses on the expansion of U.S. recording companies beyond the Northeast into the Great Lakes Midwest and then maps the aftermath in which they turned to new niches when their initial attempts at mimicking the Big Three's repertoire failed. They turned to the music around them, including that of newly arrived musicians from the American South, which in turn led to recording expeditions there, as well as to points around the United States in years to come. And as Karl Hagstrom Miller has noted, talent scouts and folklorists, invested in the purity of the country's regional vernacular music, were crestfallen to find "Broadway in the backwoods," which had arrived decades before via sheet music, the vaudeville performing circuit, and the easy portability of sound recordings from people and places culturally, if not geographically, far away.[59]

Not surprisingly, some of the musical styles came in with built-in distortions and prejudices inherent in the structures of American racism, which would show up in genre classifications with no small implications for the everyday work of the industry. Nowhere was this more pronounced than in the form that gave birth to both American popular music and the central importance to authenticity in its production, reception, and circulation: minstrelsy and its cruel depictions in working-class theaters in the white urban North.[60] As Hagstrom Miller suggests, the "cultural component of segregation" was "in some ways more successful than its legal counterpart" in that its "narrative of discrete racial and regional cultures remaining ensconced long after the doctrine of separate but equal ceased to be the law of the land."[61]

This cultural segregation has led to differing accounts of how companies negotiated the processes of scouting, recording, manufacturing, distributing, and selling sound recordings. This not only meant "the business side," but also the varying cultural barriers that those involved crossed with care and regularity. The music they found, however divided and subdivided by recording scouts, folklorists, and scholars after the fact, were hybridized when "discovered." "Standard approaches," Hagstrom Miller writes, "assume that the commodification is a problem that must be investigated, that music bought is somehow less true than music made," but I share his notion that music is shaped less by "who they were" but by "what they had the opportunity to hear."[62] This famous dichotomy stubbornly remains common in the present day, though Simon Frith argued decades ago that the music-as-expression vs. music-as-commodity distinction was a deeply flawed way of thinking about popular music. Frith rightfully argued that in the age of sound media, a given media commodity (a sound recording) is often a creative starting point in the hands of others rather than the compromised end of an otherwise pure cycle.[63] At the same time, while the cultural categories that impacted genre formations are arbitrary at their core, their impacts are nonetheless responsible for very real consequences in the lives and careers of those involved.[64]

Technological change is never simple or straightforward, but it is implicated in a constellation of shifting social, economic, regulatory, and cultural contexts. Authors commonly write about technology as an "autonomous" force in society, in no large part due to the difficultly in assigning specific agencies to technological change, as Lisa Gitelman has suggested.[65] Media histories "must be social and cultural, not the stories of how one technology leads to another, or of isolated geniuses working their magic on the world."[66] Nor are technologies as stable as we often suggest, not "the phonograph" nor "the radio" as we sometimes write, taking a "heuristic" approach that cites a "coherence" that falls apart with scrutiny.[67] The definitional instability of technological media is further complicated by the unstable nature of the "social." Derived from widely held practices in sociology, common definitions of "society" treat it as a macro state, which we all inhabit. "Problems arise, however," Bruno Latour writes, "when 'social' begins to mean a type of material," a beginning rather than the result of "actors" (human and nonhuman in Latour's formulation) in which the "social" is the ever-shifting end result of actors and assemblages and networks and so on.[68] Through Latour's development (with Michel Callon) of actor-network theory, he sought to contextualize people on a less hierarchical plane in relationship to technological artifacts and formations. Just as the urge to escape technological determinism is easier said than done, Latour outlines the many difficulties in reformulating the social in the ways he suggests. In this book, I have tried to resist assigning agency to large, amorphous "social" trends and have tried to rely on specific actors, relations, and contexts so as not to make claims so broad as to be meaningless.

Thus the emergence of a range of phonographic technologies was in no way preordained. The paths to adoption were tentative and circuitous for long periods of time, only really recognizable as a narrative after the fact (and even then, we have often got it wrong). To begin: It was in no way preordained that the recording industry would be focused on music. Of all the many inventors who had experimented with sound in different ways (Éduoard-Léon Scott de Martinville, Charles Cros, and Alexander Graham Bell, among others), it was Edison that got credit for the "phonograph." The name for his cylindrical recording device was already in circulation as a system of phonetic shorthand stenography invented by Sir Isaac Pitman by the 1840s.[69] Edison's August 3, 1877, experiment with machinist John Kruesi, almost a year after Alexander Graham Bell's successful telephone experiment, is marked as a beginning of sorts. A patent for Edison's new device was granted on February 19, 1878. The inventor first tried to sell his new technology via lectures in which the inventor or his employees would demonstrate and discuss the phonograph's abilities and possible uses. These spectacle-like demonstrations were meant to garner interest in the technology, but any initial amazement regarding the phonograph's ability to separate human voices from bodies had largely waned, as the Edison ledgers suggest.[70]

"In our own times it has become fashionable to declare that music has become a commodity," David Suisman writes. "Like many truisms, this claim obscures as much as it reveals." Commodification, he adds, is a "social and political process, populated by human actors, and one that includes various dimensions and phases."[71] There is no better proof of this than Edison's much-cited trial balloon, in the form of an article published on May 1, 1878, "The Phonograph and Its Future" in *The North American Review*, in which he suggested various possible market uses for the device, including use in alarm clocks, talking dolls, and in the creation of wills, among other uses.[72] As Leonard DeGraaf writes, "The lack of a clearly defined application for the phonograph affected the company's ability to promote it."[73] And while Edison's 1878 essay is sometimes held up to suggest Edison's failure, most of the uses he predicted came to pass in time. Although recorded music would become phonography's focus, other strands continued on successfully with relatively little attention. The earlier initiative to make phonographs business devices later found success via dictation machines. But after early business applications failed, Edison gave the rights to develop the technology for use in toys and dolls or as music boxes, but such initiatives never amounted to a viable plan for the phonograph writ large, even if both uses for recorded sound would come into use. But after Edison's entrepreneurial efforts failed, the inventor largely abandoned the phonograph during work on the technology and infrastructure surrounding electric light, among other projects.[74]

It was the work of Chichester Bell and Charles Sumner Tainter that prevented the new sound technology from drifting into obscurity. Funded by Alexander Graham Bell (Chichester's distant relative) and encouraged by financier and lawyer (and Graham Bell's father-in-law) Gardiner Hubbard, Bell and Tainter found funding and a place to work at the telephone inventor's new Volta Laboratory.[75] During a period in which Edison was distracted by ongoing projects (primarily electric lighting and iron-ore extraction), Bell and Tainter's patented improvements and the launch of their "graphophone" jolted Edison back into improving his phonograph.[76] After experiments with disc technology, Bell and Tainter decided on an improved cylinder phonograph, with replayable wax cylinders, while also avoiding the wobbling common to Edison's mandrel-style phonographs.[77] This research would lead to formation of the American Graphophone Company (an ancestor of the Columbia Phonograph Co.), both using cylinder technology, after attempts at merging interests with Edison failed.

After weighing the viability of different uses for the phonograph, Edison and others focused on business applications. This makes sense, as Paul Israel suggests, given the biographies of many of those involved: "Edison and the other major figures in the early phonograph industry shared a background in telecommunications" and thus "were ill prepared to imagine and create an entertainment industry."[78] To launch his phonograph business, Edison partnered with Jesse Lippincott, a Pennsylvania industrialist who had made his fortune in glass. The North American Phonograph Company's main charge was to popularize the phonograph as business device. Edison and Lippincott largely built its business model on the telephone industry: regional companies served its areas by leasing cylinder phonographs as a technological replacement for stenographers, who quickly organized against the technology. But the early phonograph was not fit for such business purposes, given the technology's fragility. Repairing machines was a logistical problem, especially for those local companies farthest west that had to send machines back east for service. This in turn made the leasing model borrowed from the telephone industry untenable for the phonograph, as lessors were not renewing after their initial agreements expired.[79]

In a closed-door 1890 meeting of phonograph manufacturers, the nascent industry's assumptions were laid bare. Lippincott brought all the local phonograph companies together at Chicago's Auditorium Building to discuss present difficulties and future prospects. Attendees revisited technological and business assumptions about the new technology and revisited technological and business models in relation to the design and sale of telephones, typewriters, and sewing machines.[80] Raising these questions again was more urgent than fanciful, as the industry's focus on the talking machine as a business device was facing collapse. While thirty-three companies were represented at the 1890 meeting, only nineteen labels returned in 1891.[81] On the second day of the conference, the agenda turned to talking machines for public exhibition, at which point Louis Glass of San Francisco's Pacific Phonograph Company spoke. "Gentlemen, I have very little to say except, that all the money we have made in the phonograph business we have made out of the nickel-in-the-slot machine. . . . I will state that the first one we put out was placed in the Palais Royal saloon on November 23rd, 1889, and we have taken in from that machine, up to May 14th, last, $1,035.25."[82] Only Glass and a few other representatives spoke up in favor of the entertainment phonograph, a move that flew in the face of what Edison and Lippincott wanted. Paul Starr notes that it was distributors like Glass who first recognized "more of a market as an entertainment medium than as a business apparatus," for use in "storefront phonograph parlors as well as in such venues as hotel lobbies and train stations." These parlors became the "penny arcades" in which a variety of amusements and automata were also featured."[83]

Edison would never have the same amount of control over the recording industry that he had in those early years, particularly when compared to his control of later technologies, most notably the motion picture. Compared to the sweeping control Edison had with the Motion Picture Patents Co., Edison never had the kind of control in the phonograph business he was able to exert in motion pictures, especially after his partnership with Lippincott dissolved and the business model built around the phonograph as a business device collapsed. And by 1915, Edison's short-lived trust was ruled in violation of the Sherman Anti-Trust Act, around the same time his phonograph patents were expiring and a new wave of talking machine manufacturers would enter the industry.[84] There are convergences and disjunctures when tracking the development of recorded sound and motion picture technologies in light of Thomas Edison's entrepreneurial attempts to develop both.

"If at first the phonograph record was little more than a novelty, the stuff of entertainment arcades and 'educational' lectures," Michael Chanan writes, "it was also an entirely novel commodity: it turned the performance of music into a material object, something you could hold in your hand, which could be bought and sold."[85] But what that object meant—what it could do—was still being tested in a variety of ways. The growing popularity of recorded sound as entertainment encouraged companies to transform their playback machines from public amusements to domestic devices sold in newly aestheticized cabinets that mimicked furniture styles of the day. This is not to say that other uses were not being pursued, as evident in the abecedary series that appeared in *Phonogram* between May 1900 and June 1901. Each month, another use for Edison's phonographs (ranging from the possible to the improbable) was introduced: an alarm clock; an alert on buoys to warn ships at sea; built-in phonographs on bicycles.[86] Phonographs were increasingly ubiquitous in public spaces. But another article that appeared in the *Phonogram*'s June 1901 issue strongly encouraged consumers to adopt the technology into the home: "The importance—we might say the absolute necessity—of music in the well-ordered home is now generally admitted." The article suggested the talking machine as a replacement for both the piano and music box and as an aid to digestion at dinnertime.[87]

The dissolution of the National American Phonograph Company in 1894 allowed existing companies like Columbia and Edison to focus on the entertainment market, while some of those already in the music business founded labels of their own. Performers like J. W. Myers and Russell Hunting first recorded themselves and associates before formalizing the business into labels.[88] Myers recorded under the Globe and then the Standard moniker, while Hunting (also publisher of the *Phonoscope* from 1896) worked at the Universal Phonograph Company, owned by Joseph W. Stern & Company, a sheet music company that made sure to record many of its artists.[89] The Stern/Universal relationship stands as a tantalizing early example of what might have been if a closer organizational structure between song publishing and sound recording had formed, while also pointing out the largely unmapped relationship between the two fields to date. Stern's activities in the sound recording business ended when Edison stopped supplying cylinders, out of what Dave Laing suggests were growing fears of competition.[90] A symbol of the song as selling point in Tin Pan Alley was reflected in recording companies (most notably Edison) not listing performers on record labels.

But with the increased sale of player pianos and piano rolls on the one hand, and talking machines and sound recordings on the other, tensions regarding composers' copyright claims to music being released in these formats reached the Supreme Court.[91] The Copyright Act of 1831 had extended copyright protection to written musical compositions but had no stipulations regarding the mechanical playback of sounds via machines such as the player piano or phonograph.[92] Concerns were great enough by 1895 that song publishers created the Music Publishers' Association (MPA) to address the changing technological contexts in which they worked. So when Congress began to revise the 1870 Copyright Act, the focus was on making sure that new laws addressed the concerns of both song publishers and composers regarding a lack of compensation for music played on these new technological platforms.[93]

While Congress was already at work, the Supreme Court had received White-Smith Co. vs. Apollo Co., in which the White-Smith Music Publishing Company had sued Apollo Company, maker of pianos, player pianos, and piano rolls. White-Smith had published Adam Geibel's songs "Little Cotton Dolly" and "Kentucky Babe" in 1897, and it was the use of the songs by the Apollo Company, maker of player pianos and piano rolls, that had led to the case.[94] The case was argued on January 16 and 17 and decided in favor of Apollo on February 24, 1908, with the Court deciding that White-Smith had not been infringed upon given the differing technological context.[95] This allowed phonograph and piano roll companies to reproduce songs with impunity, which incensed music publishers, who redoubled their efforts to ensure a positive outcome in Congress's ongoing copyright revisions. Another concern moving forward from the Supreme Court decision was this: Aeolian Company both helped financially support White-Smith's court case and bought rights to mechanical reproduction of music from those publishers involved in the Music Publishers' Association. Thus Aeolian had ensured themselves a victory no matter which side won, which led to congressional concerns that Aeolian would create a monopoly in the player piano and piano roll market.[96] These concerns left congressional representatives pondering a future of collusion and a potential Aeolian monopoly, in which the company might control song publishing rights to such a degree that they could control the phonograph and piano roll markets.[97] Those dissatisfied with the Supreme Court outcome redoubled their efforts to influence Congress's ongoing copyright debates. Songwriters and publishers relied on testimony before Congress from celebrities of the day, including composers Victor Herbert and John Philip Sousa, the latter of whom had been an outspoken opponent of "canned music."[98]

In the 1909 Copyright Act, Congress rejected the Supreme Court decision and came up with a plan that became known as compulsory licensing, designed to give song publishers the right to decide on how or whether their music would be licensed for mechanical reproduction, while not allowing that exclusive right in perpetuity. As Edward Samuels writes, "composers could choose whether or not to allow recordings to be made of their works," while charging a price they deemed fair. "But thereafter, any other record companies would be allowed to make their own recordings of the song by paying a fix rate, set in the statute at 2 cents per copy."[99] While the first provision for mechanical playback (via phonographs and player pianos) garnered the most attention upon the Copyright Act's passing, the additional provision for public performance would be more impactful, particularly with the coming of sound film. As Katherine Spring writes, "In the two decades following the Act's inauguration, the economic potential of the second provision became especially apparent to publishers and songwriters, who discovered that handsome sums could be earned by licensing performing rights to venue proprietors."[100] But in the first years following the Copyright Act of 1909, the issue of performing rights remained a thorny one due to great difficulties in enforcement and the arduous task of monitoring businesses that used music without the rights clearance that music composers and publishers so desired. Attorney Nathan Burkan, with the support of Victor Herbert, John Philip Sousa, and others, had been instrumental in the formation of the American Society of Composers, Authors and Publishers (ASCAP) in 1913. But enforcement remained an issue, so after attempts at negotiating licenses with individual New York restaurants bogged down, Burkan in 1914 filed a lawsuit on behalf of Sousa, which failed.[101] That failure led to another case, this time brought by composer-musician Victor Herbert, who had heard his song "Sweethearts" at Shanley's Restaurant, down the street from the Broadway opera in which Herbert performed the song with each performance.[102] In January 1917, the case was heard by the U.S. Supreme Court, which reversed the decisions of the lower courts. Justice Oliver Wendell Holmes wrote in favor of the plaintiff and on behalf of a unanimous court, saying, "If the rights under the copyright are infringed only by a performance where money is taken at the door, they are very imperfectly protected. Performances not different in kind from those of the defendants could be given that might compete with and even destroy the success of the monopoly that the law intends the plaintiffs to have."[103] The court had decided that music played in restaurants and other commercial spaces was played for profit as part of what consumers were buying. The ruling gave ASCAP a legal basis on which to collect fees on behalf of its membership, which the organization quickly pursued. "From the beginning," Paul Goldstein writes, "the central collecting mechanism was a blanket license that would give the licensee *carte blanche* to perform any composition in the ASCAP repertory as often as it wished for a flat fee."[104] "Performances," as defined in the decision, considered sound recordings and radio to be the same as live performance. And while live performance in early radio reached relatively few listeners and was seen by many as a way to promote performances, sound recordings, and sheet music, the medium's increasing reliance on recordings would lead to increased tensions that would flare up throughout the 1920s, 1930s, and beyond. While the emerging events of the 1920s will focus on the recording industry's expansion and transformation with radio's rise, this book will necessarily return to questions of royalties and compensation as the radio industry buys up the leading recording companies at the end of the 1920s and start of the 1930s.

What the White-Smith case, the congressional debates regarding copyright, and the formation of institutions like the MPA and ASCAP made clear was that music could be "composition, performance, sound, and object" all at once, while the importance of any one of these would shift depending on the specific economic or institutional context.[105] This was further complicated by the advent of radio, which was first understood as a valuable tool for the promotion of musical artists. Music, newly "free" to radio consumers who increasingly adopted the technology through the 1920s, spread in even broader ways, reaching larger and larger groups of the public simultaneously, as opposed to the largely asynchronous listening habits of phonograph listeners.

If sound recordings represented a challenge to the song publishing industry's grasp on copyright, sound phonographs and other playback technologies meant a battle for nascent recording companies in terms of patents. The first decades of the recording industry saw collaboration and conflict between the emerging Big Three: Edison, Columbia, and Victor, as well as various new companies entering the recording market. As early as the 1890s, Columbia and Edison agreed to license each other's patents, which made sense given the close history between the two companies and their sound playback technologies, as well as their shared early focus on the talking machine as a business device. By the 1890s, both Edison and Columbia had debuted entertainment cylinders for the burgeoning home phonograph market.[106] But unlike Edison's phonograph and then American Graphophone Company's (renamed Columbia in 1897) graphophone, Emile Berliner's gramophone debuted in 1895 with a primary focus on the entertainment market. Berliner had understood the difficulties faced by the phonograph and graphophone as a business device and the problems caused by the early gramophone recordings' roughly two-minute playing time. Berliner's gramophone company not only embraced a disc technology rather than a cylinder playback system (such an approach had been a subcurrent in recorded sound research for decades).[107] It also shifted emphasis away from the largely failing business phonograph toward musical (and spoken word) entertainment. The technology and its new focus created both turmoil and opportunity as shifts in playback technologies redefined how recorded sound was produced, circulated, and consumed.

A year after the Berliner disc's debut, his Berliner Gramophone Company entered into an agreement with a Camden, New Jersey, machine shop operator named Eldridge Johnson, who began making entire gramophones for the inventor's company soon afterwards[108] Not long after, Johnson distanced himself from Berliner. In addition to developing his own recording processes, Johnson had good reason to avoid the association. In June 1900, Frank Seaman of the Universal Talking Machine Company sued Berliner for copyright infringement based on discs already being produced by Seaman's Zon-O-Phone (later Zonophone) label, which meant "an end to his American operations."[109] Johnson's gramophone company had worked under various names (Consolidated Talking Machine Company, Improved Gramophone, Improved Record) eventually distancing his own company from Berliner's, buying out the inventor's remaining U.S. holdings and settling on a new name for his company: Victor.[110] The company's new format, tied to an emphasis on musical and spoken-word entertainment recordings, proved successful. While Edison remained dedicated to his cylinder format, other companies were interested in the emerging disc business, as varying inventors and companies fought it out in court.

The skirmishes that would lead to the industry's dramatic expansion wouldn't come from battles between industry leaders, but from upstart companies seeking the advantages afforded by the most popular playback formats. Claiming patent infringement in July 1904, Columbia sued Leeds & Catlin, disc producers who were supplying lateral-cut disc recordings to smaller record labels as well as a few of their own.[111] The company, founded by Loring Leeds, was in court regularly, the first major case being a suit filed by American Graphophone Co. (Columbia) in 1898.[112] Many more suits would follow. In 1909's *Leeds & Catlin v. Victor Talking Machine*, Leeds' company argued that because Victor had not extended one of its patents in the Canadian market (the Berliner-Suess patent of 1899), U.S. Victor patents were null and void.[113] As David Suisman writes, "The Supreme Court ruled in *Leeds and Catlin Company v. Victor Talking Machine Company* in 1909 that Victor's patented technology comprised the machine *and* the records, thus making it an infringement for another company to produce records playable on Victor phonographs."[114] Because Victor controlled both phonographic media's most successful hardware and software (to borrow contemporary terminology), it allowed the company to dominate the U.S. recording industry like no other: "This was the cultural equivalent of the absolute authority initially claimed over both the physical rails and the means of conveyance . . . and it was not until key phonograph patents expired in the second half of the 1910s that independent cultural producers, in this case, record companies . . . could operate in the business with any kind of reliable legal foundation."[115]

Victor and Columbia then entered into a patent pool agreement, while the Edison company was steadfast in its commitment to the patented technologies of its namesake. This was a huge mistake for Edison, as a Victor-Columbia patent pool ensured its control of the markets for disc and machines. As the only company in the Big Three not directly associated with an inventor, Columbia came to be defined by lawyers and businessmen. Columbia's then-parent company, American Graphophone, controlled the Bell and Tainter patents, signed a cross-licensing deal with Edison by 1896, and "handled and licensed" other companies' talking machines as early as 1897 (the year they launched their own machine).[116] In October 1901, Columbia was able to get into the disc business through the sale of the Globe Record Company's Climax series, while former Berliner employee Joseph W. Jones was working with attorney Philip Mauro to get a patent on his wax recording technology, also crucial to Victor's business.[117] Once the patent was granted on December 10, 1901, "The Jones patent put Victor and Columbia on equal legal footing, with each company holding a patent crucial to the other's continued operation."[118] Victor's Eldridge Johnson responded by purchasing the Globe Record Company, which ended Columbia's disc business via Climax, while Columbia readied a patent infringement suit against Victor based on the Jones patent.[119] Negotiations ensued between Johnson and Columbia's Edward Easton, which led to the first co-licensing agreement between two of the Big Three's most formidable companies.[120] As David Suisman suggests, "Through agreements to share their patents, large firms both minimized costly conflict of uncertain outcome and shored up their own position as industry giants."[121]

These legal conflicts and collaborations allowed the recording industry to challenge Tin Pan Alley in ways it could not before, as companies courted new listeners through more affordable phonographs and, however tentatively at first, expanded repertoire. But this did not mean that Tin Pan Alley's structuring influence dissipated quickly. Like vaudeville, Tin Pan Alley's malleability as a form has meant that its influence is with us still, and not just during baseball game sing-alongs of "Take Me Out to the Ball Game" during the seventh-inning stretch. Rather, recording companies entered an entertainment environment largely shaped by practices established by Tin Pan Alley, which had commercialized music decades before.

Tin Pan Alley, Phonographic Media, and Commercial Culture

Tin Pan Alley's commercial function had a dramatic impact on popular music and entertainment media writ large. That impact, David Suisman writes, "lay not in aesthetic innovation but in the relation between aesthetic forms and the industry's modern capitalist structure," which made songwriters into laborers and songs into products, organizing American popular music into a recognizable "variety of genres and idioms."[122] Compared to the music that would follow it, some Tin Pan Alley fare seems calculated towards the maudlin, designed to pull on the heartstrings to encourage sales of sheet music in this newly industrialized formation. But it also brought in decidedly new and cosmopolitan influences to the American songbook, borne in part out of the experiences of Jewish immigrants from Eastern Europe, who added their own musical influences. These publishing companies also brought a "brash sophistication that eschewed the schmaltzy excesses of the past century and most of the Italianate ornamentation, relying for spice on borrowings from brass bands, South American dances, the emerging jazz and European modernism."[123]

The music publishing houses of the 1910s were so powerful by 1920 that the key firms—M. Witmark & Sons, Irving Berlin, Inc.; Leo Feist,

Inc.; T. B. Harms & Francis, Day & Hunter, Inc.; Shapiro, Bernstein & Co.; and Waterson, Berlin & Snyder, Inc.—were named in an antitrust suit for controlling 80 percent of songs that appeared on "phonographs, player piano rolls and other musical reproducing instruments," while fixing prices on recordings and piano rolls sold. The companies formed a joint licensing operation called Consolidated Music Corporation, the entity named in the suit.[124] Consolidated only existed for a few weeks, quickly folding in the aftermath of the federal antitrust suit (quietly dismissed in 1922).[125] The power of the music publishing houses had shaped the U.S. music industry in the first decades of the twentieth century, and they were now jockeying for a place in which the phonograph, the player piano, and other playback technologies were changing some basic assumptions about how music was created, distributed, consumed, and stored. The talking machine entered a commercial environment in which the Tin Pan Alley mode of production, distribution, and performance were steadfast organizing principles.

U.S. journalists often reported on the changes in derisive terms. A 1910 *New York Times* article articulated the how Tin Pan Alley industrialized American popular music, which was as evident in consumers' music habits as it was in songwriting. Music consumption, the article began, "is as constant as their consumption of shoes, and the demand is similarly met by factory output . . . manufactured, advertised, and distributed in much the same manner as ordinary commodities."[126] The talking machine was initially seen by song publishers as a new and novel way to sell sheet music rather than as a threat to Tin Pan Alley's mode of production. In fact, the recording industry adopted many of the same practices already common among Tin Pan Alley song pluggers. They sold sheet music through music shops and department stores, the theatrical and cinematic stages, and restaurants and bars.[127] The business of sound recording would transform the production, consumption, and distribution of popular music completely, even as it adopted key aspects of its business model, including the use of jobbers, who began distributing sound recordings as well as sheet music. In terms of sales, the incorporation of sound recordings in addition to sheet music meant another abstracted means of music (or, more broadly, sound) delivery. As David Suisman rightly notes, "publishers at that time made their money from selling objects (sheet music) that represented something intangible and ephemeral (sounds)."[128] Musicians heard on sound recordings might not know how to write sheet music, as performers played "by ear." In retail music stores, talking machines were added to the showroom floor, alongside pianos, player pianos, and other musical instruments.

But the phonograph was also recognized as a "champion of song promotion," as Katherine Spring notes, citing a *Billboard* article that outlined the extended life of popular songs that had already "earned their share of popularity on the regular sheet music market." It "repopularizes them in a field far more far-reaching and profitable. It also takes songs which would never experience spirited counter sales and popularizes them over the instrument."[129] In the second half of the 1910s, the Peerless Quartet's "I Didn't Raise My Boy to Be a Soldier," the American Quartet's "Over There (released the same year as the U.S. entry into World War I)," Marion Harris's "After You've Gone," John Steel's "A Pretty Girl Is Like a Melody," and Charles Harrison's "I'm Always Chasing Rainbows" were all popular titles. Tin Pan Alley had already developed cross-promotional opportunities with Hollywood for popular music's inclusion in silent film, which added fundamental value to cinema's narrative and affective depth while helping to affirm popular music's uses across media. "Whether disparaged as ephemera or valorized as products of great artists who resuscitated popular music for mass culture, songs of the 1920s became bankable commodities whose musical features were exploited to suit the narrative functions of an established classical cinema."[130] Accompanists became customary as films included a thematic song repeated at key intervals in the film narrative, while sheet music covers promoted connections with the latest films. And with the rise of sound recordings as entertainment, those same songs allowed film fans to purchase an object that harked back to the ephemeral cinematic experience, sometimes including a vocal performance by the star of the film. And as radio developed, that medium would become even more important to the circulation of a much broader repertoire of 1920s and 1930s popular songs. But the radio of the 1910s had yet to enter its meteoric rise of the 1920s or its media dominance of the 1930s.

Back on the shop floors, song pluggers' roles waned over time, since a clerk with no musical abilities could help customers preview a given recording. Recordings were often (but not exclusively) sold in music shops alongside sheet music, pianos, player pianos, and other musical instruments, where customers would come in to speak with clerks about the latest releases. Records were also sold in general stores, pharmacies, appliance and department stores, even in funeral homes for mourners in need of musical accompaniment. The pages of the trade journals offered weekly advice on how to attract customers with eye-catching retail window displays and hypothetical scenarios between clerk and customers, to be studied for successful strategies for future sales. A March 13, 1919, article in

Talking Machine World suggested that dealers should capitalize on testimonial letters from satisfied listeners. One such listener wrote in to say that he hadn't been a phonograph fan heretofore, but after one long day at work, he obliged his daughter and listened to Manuel Romain's version of the Tin Pan Alley standard, "I Will Love You When Silver Threads Are Shining among the Gold." "All at once," the listener reported, "my mind rose upon enchanted wings and swung heavenward to the bosom of a rosy cloud. You have read tales about the effects of opium upon the nervous system," he continued, "Well, Mr. Romain's song must have had the same effect on me." Don't clerks realize, the article continued, that such "ecstasies" coming "streight [*sic*] from the heart of a wonderstruck patron—make corking advertisements?"[131]

The recording industry's reliance on the cylinder was becoming a vestige of the past as companies switched to disc methods by the 1910s.[132] By the close of the 1910s, trade journals discussed concerns that some shops had too few clerks on the floor to keep up with customer demand, particularly in smaller stores not equipped with listening booths. Several articles suggested a solution: self-service talking machines for customers to listen to the latest recordings. A January 1919 article in *Music Trade Review* magazine reported that Victor was recommending self-service stations to their retailers, especially in those shops with a single clerk (sometimes the owner). As the article suggested, "Shortage of clerks first led to its inception, though like many other schemes devised to meet war-time conditions it will likely become a fixture in those departments where it is now being used."[133] A southeast Washington, D.C., shop described "as especially crowded in the evening" served as example in a February 1919 article titled "Let The Customer Play the Records." "The proprietor knows only that he has more customers than he can wait on," the article suggests, "and that too many of them only want to hear records which they ask for and have no intention of buying."[134]

With retail shops like these, the U.S. recording industry's dominant companies sought out consumers in different ways: through the dominant labels' chasing a given trend in audience taste until interest waned and through emerging labels fighting for niches, in a broad array of largely unrecorded styles, after head-on competition with the majors proved fruitless. Changes made by industry giants (Victor, Columbia, Edison) and the tactics of smaller labels (Gennett, Paramount, OKeh, Brunswick, and others) helped develop a new broader mode of production, as many recording companies paradoxically became increasingly national (and international) in reach by focusing on regional niches and contexts. That expansion began to close by the end of the decade as the recording industry moved from one that largely dictated its own course to being part of a larger media industry formation, tied to the rise of radio and the increased importance of film. The connection between the recording industry and radio, which was still in its infancy during the 1920s, would increase dramatically into the mid-1930s.

Through all these changes in the recording industry, its reliance on genre remained constant. This relationship between genre (understood as a configuration of social, cultural, ideological, and aesthetic beliefs) and mode of production (in its most concrete sense, how a given company operates) has continued to be an important one to the record industry, because with each new genre and subgenre the industry has the potential to connect with underserved or unrecognized audiences. My analysis begins in 1920, by which time these new labels had asserted themselves through recording previously unheard music and developing new sales and marketing approaches. The musical foundations of the genres discussed here predate 1920, but the start of the 1920s marks the period in which the foundational genres of twentieth-century popular music emerge as industry categories. Importantly, this was made possible in large part by the emergence of smaller labels in the late 1910s that sought out new genres and niche markets to successfully compete with larger labels. This study closes in 1935 or so, by which time enough had changed in the record industry to make it look like an altogether different proposition than a decade before, as the dual forces of radio and the Great Depression had forced the upstart independents to merge with majors or get out of the record business altogether. The recording industry emerged out of this period of experimentation, consolidation, and economic depression as no longer part of a relatively autonomous phonograph industry but as part of a larger media formation that included film and radio. The story recounted here details the ground-shifting transformations that the recording industry faced as its genre boundaries expanded and its industrial formation underwent great change.

Those companies most likely to serve these audiences first are not likely to be industry leaders. In the 1920s, the two clear industry leaders were Victor and Columbia. Victor, with its ubiquitous "His Master's Voice" ads featuring the omnipresent dog and Victrola in what would become a dominant twentieth-century American brand, had a genealogy that traced back to Emile Berliner and the invention of disc phonography. Victor played no small part in how advertisers "encouraged consumers to understand themselves through their possessions and to fabricate their identities in and through things."[135] Columbia, the oldest of all U.S. record companies, had a history that dated back to the 1890s, when the company sold cylinder phonographs for business communication in Washington, D.C. Having the most to lose as industry leaders, Columbia and Victor therefore had little interest in experimenting with new musical styles. The other major player of the early twentieth century, Edison, was in serious decline by the 1920s due to a reticence to abandon Edison's proprietary phonograph technology, an uneven attention to the quality of music released, and Edison's own peripatetic ambitions. Together, they were the original Big Three.

For a time in the 1920s, major labels like Victor and Columbia found themselves outpaced by smaller companies like Gennett, OKeh, and Paramount, who recorded the new genres of the day. At the start of the 1920s, the recording industry was still largely an East Coast phenomenon, centered between New York, Philadelphia, and Washington, D.C. During the 1920s and 1930s, new industry upstarts had shifted business to what was then known as "the Middle West," with musical production centered in the industrial centers that drew poor blacks and whites north. It's become commonplace for me to inform friends from Wisconsin just how many foundational blues recordings were recorded by the Paramount label of Port Washington. Friends are equally incredulous when I mention that Louis Armstrong's first recording sessions were not in New Orleans but at the Gennett Records studio in Richmond, Indiana. At the same time that many Americans were leaving the rural south for the industrial north, these emerging record companies in the Great Lakes Midwest were there to record them.

This book examines institutional cultures at various recording companies—seen as media-producing, culture-producing industries—and those producers, consumers, and circulators of this sound media who helped shape U.S. popular music beginning in this era. The central questions to which I will continually return are: How did the shaping of genres and shifting recording industry practices and ownership affect one another? How do we account for the practices of the recording industry as reflected in its decision-making and culture-shaping/reflecting mode, both as an industry involved in manufacturing culture and reflecting its own participation as a cultural institution? How did the mediating roles of company go-betweens help to both shape new genres and reflect record companies' attitudes about those genres? How did competing technologies such as radio influence the recording industry? And finally, how did the phonograph industry shift from an industry largely concerned with its own fortunes to one concerned with its role in part of a much larger media formation?

The Recording Industry, Media Historiography, and Cultural Production

An implicit purpose of this book is to suggest that recording industry histories would benefit media industry histories (and, more broadly, media studies) through true inclusion, in recognition of the recording industry's history and its deep and abiding connections to other media industries. As an interdisciplinary endeavor tied to fields each with its own specific emphasis and trajectory, media studies research has been defined in primary terms via its relationships to radio, television, and film (with various digital inclusions in recent years), traditionally "marked more by what it excludes (or what has been excluded) than by grounded inclusions."[136] As Michele Hilmes notes, the media that fell outside of traditional humanities research was traditionally seen as "debased and anti-individual, fodder for sociologists rather than critics or historians."[137] But a focus on media industries—including the recording industries—allows for a focus on "those aspects of cultural production in the twentieth century and beyond that most trouble the humanities-oriented categories of coherence and analysis so central to our understanding of culture itself: the author, the text, the reader." But "the media of radio, popular music, television, and film refuse to conform to comfortable analytical paradigms."[138] "By taking an industries approach to the critical study of media," Hilmes continues, "we are indicating a perspective that is inherently contextual and interrelated."[139] Christopher Anderson and Michael Curtin have also examined the "cultural turn" that has marked the rise of media industry histories, in which media are "not the given facts that precede historical investigation," but the "ongoing social productions, defined and redefined through use, affected not only by events and material conditions, but, crucially, by the language used to describe and discuss them."[140]

Like the study of media writ large, the study of popular music has never had a single disciplinary home. This has been a simultaneous strength and weakness that reflects the multifaceted nature of the cultural form. To date, there has been meaningful popular music research from the study of history, music, sociology, literature, and various interdisciplinary traditions. Each has defined and approached popular music in different ways, but I have found Simon Frith's definition of the term most useful because of his refusal to embrace "popular" in neither its "of the people" mode, common in folklore traditions, nor in its strict commercial sense, too often understood in opposition to music as cultural expression.[141] The contradictory sets of criteria that hover around the notion of "the popular" is fraught, to be sure, especially when popularity is understood to be synonymous with market performance.[142] The U.S. recording industry encompasses a wide variety of industrial practices, given both the traditionally low barrier to entry to those who start recording companies and the variety of practices that define the differences between the smallest and largest businesses. In this study, I focus on the recording industry first and foremost as a media industry as opposed to the larger definitional scope of "the music industry," which includes music publishing, the market in musical instruments, and so on. The recording industry has been unique in its long-term role as its own stand-alone medium, but one that has played a crucial role as raw material to other media. While these dynamics were present early on in the history of recorded sound, the latter sections of this book outline the intensifying nature of these connections from the late 1920s onward.

Within the U.S. context, the relative lack of recording industry histories and precarity of popular music studies has led to a tendency toward omnibus approaches, at least until recently.[143] This book has greatly benefited from those earlier studies, but it also asserts the need for more targeted histories more common in historiographic work in film, radio, and television. Foundational texts opted for broad historical approaches, investigations focused either on either technological histories of sound recording or broader studies that attempted to map as much as possible. These studies incorporated textual analyses, technological and industrial histories, regulatory contexts, and more. Some histories told triumphal "great men" histories, spent scant time on social or cultural contexts, couched technologies in narratives of inexorable progress, or sometimes dismissed present-day popular music contexts as signs of decline. William Howland Kenney referenced this work in arguing the following: "Cultural analysis of the phonograph and recorded music has languished as writers and scholars alike have favored the study of technology in its many changing forms. This has had the effect of diverting attention away from the ways that Americans interacted with recorded sound technologies, both in producing recorded music and in consuming it."[144] But despite such flaws, these studies also laid a groundwork for further research in that they outlined technological and industrial contexts that might have been lost otherwise.

More recently, cultural histories have chosen more targeted and thematic approaches that suggest a turn of sorts, in which the perennially arriving study of popular music appears to get closer to a destination with each study. While largely continuing the division between technological and industrial histories of previous decades, these new studies (largely from scholars in history and American studies) were informed by the "cultural turn" across various disciplines and a thematic approach that no longer tried to do it all. Not surprisingly, much of this work comes from authors in interdisciplinary contexts. Examples of this approach include Evan Eisenberg (1987), Michael Chanan (1995), William Howland Kenney (1999), David Morton (2000), and so on. Two key histories by David Suisman (2009) and Karl Hagstrom Miller (2010) added much to existing scholarship. Suisman focuses on the social, cultural, and regulatory contexts in which music became a commodified object while Miller focuses on the recording industry's role in formulating racialized folk and pop genres in the Jim Crow era. This study has also benefited from collectors' ongoing research and literature. While this work has spent less time on the deeply impactful social contexts in which recording companies work, their dogged attention to detail has added hard-sought evidence, often little known outside this research by those in academia or music criticism.

Within contemporary media studies, popular music has held a precarious place at best, largely relegated to textual analyses of screen-based media. More recently, sound research within media studies has renewed interest in phenomena closely tied to popular music, though the sound studies and popular music research appear to have as many complications as affinities to date. This maps with popular music studies' complicated and curious theoretical and methodological trajectory referenced above, which has followed a decidedly different course from the study of film, radio, television, and other media forms. This may be due in part to the blurring of disciplinary boundaries represented in the subject. While cinema, radio, and television were all able to define themselves as separate from the theatrical tradition, popular music is still associated in no small way with live performance, even though most of us consume popular music via media much more often. Traditional media studies diverge in other ways, via two important historical dynamics in which popular music doesn't easily fit. At its inception, mass communication research commonly focused on mass media's ability to reach large audiences simultaneously, while phonography was most commonly understood as an asynchronous medium and not "mass" in the way that the disciplinary focus on broadcasting had encouraged.[145]

"Common sense tells us that sound recording . . . is a mass medium, just like newspapers, films or television," Pekka Gronow wrote in his 1983 article "The Record Industry: The Growth of a Mass Medium."[146] "The problem is in the message," he continued. "The message of records is usually music, and communications research does not know how to deal with music."[147] Gronow goes on to suggest that the sale of recordings is one primary means of proving the influence of recordings as mass media. This dynamic points to an important distinction between recordings as media and music as content, which can easily be confused through recordings' role as media and music's role as content *across* media platforms (recordings, radio, film, and later media). But it is those traditional connections to other media that were made all the clearer by the first major conglomeration of entertainment media industries in the twentieth century: the 1920s phonograph-radio mergers detailed here, most emblematically Victor's merging with the Radio Corporation of America (RCA).

I use the term "recording industry" here to mean commercial recording companies, encompassing both the largest and smallest recording companies that sought to sell recordings (largely, but not solely, music) to consumers. John Williamson and Martin Cloonan have argued, as has Jonathan Sterne, for the multiplicity of the music industries, within which the recording industry is a constituent part.[148] And if there are multiple music industries there are also multiple recording industries, with various roles across media industries and sociocultural contexts, including small commercial record labels that have benefited from the industry's low barrier to entry, as well as some of the largest and most well-known labels and companies.

Like Keith Negus in *Music Genres and Corporate Cultures*, my understanding of recording companies encourages a view that understands these companies as multifaceted organizations, in which various dynamics can be at play at any given time, via the cultural intermediaries' everyday actions that shape what goes on, rather than any conception of media industries as static, monolithic structures.[149] Pierre Bourdieu's intersecting concepts of *habitus*, *field*, and *capital* have been most important here, as it has been across work in the cultural industries.[150] In an attempt to reconcile the roles of individual actors and larger social formations, Bourdieu sees shifting social contexts (fields) correlating in complicated ways with participants' subjective responses, dispositions, and tendencies, or *habitus*.[151] For Bourdieu, these fields and subfields were not mere production of objects in any simple way, but a process of cultural production that produces "not only the object of its materiality, but also the value of this object," a "recognition of artistic legitimacy."[152] As Keith Negus argues, the recording industry has been in a state of near-constant adaptation in relation to genre, audience, technology, promotion, advertising, and the law. "Since its emergence at the end of the nineteenth century," Negus writes, "the recorded music business (and indeed the sheet-music publishing industry from which many working practices were drawn) has been organized according to small-scale productions and selling changing niche markets alongside the creation of big hits and blockbusters. . . . In addition, since its earliest days the recording industry has employed various legal and illegal, small-scale and team-based, marketing and promotional activities as a way of approaching consumers—practices which might well be labelled 'flexible.'"[153] This is certainly evident in the various strategies employed by recording companies during a period in which American consumer culture was just taking shape. Key among these strategies was a reliance on advertising in the pages of trade journals, newspapers, and magazines. "By 1930," Charles McGovern writes, "advertising was ingrained in everyday life, not only as a thoroughly integrated tool of industrial capitalism, but also as a widely accepted cultural influence."[154] Negus characterizes recording companies as in constant flux, not only in their advertising and promotional strategies, but in adaptations to market conditions, changing competition, technological changes, promotional, and legal precedents (paired sometimes with tactics of dubious legality), through which companies have constantly negotiated.

This historiographical work is deeply indebted to the work of archivists and fans, collectors, and discographers. It is also shaped by their interests and questions, by what they decided to keep or dispose of, what questions they may have asked in oral interviews decades ago, what points of view they brought to the work at hand. In the effort to preserve sound recordings, some archives pay less attention to printed materials—business correspondence, catalogs, contracts—in favor of sound recordings themselves. Those players in the recording industry and other media industries of the 1920s and 1930s may have thought of their labor as ephemeral, even unimportant. What they kept or didn't, what they recalled when asked decades later, shapes this too. Finally, the figures who were understood to be of central importance to the U.S. recording industry were largely white and male, while some researchers have, over time, given voice to others involved. Performers, certainly, but also those worked at companies in varying capacities. I think most clearly here of Harry Pace, Mayo Williams, and others, about whom you will hear more in these pages. Important work remains to be done in identifying women who were involved, especially those doing the work of songwriters, scouts, and recording engineers, without claiming those roles publicly or even identifying with the roles privately. Here I am also thinking of the work of Aletha Dickerson, who worked in varying capacities for the Paramount label in Chicago through much of the 1920s, becoming de facto recording manager for Paramount.[155] We have yet to fully map those significantly involved in recording companies during this era.

A difficult necessity in researching this era is to be aware of the many myths, biases, distortions, and misconceptions that have emerged in layers between the musicians discussed here and the music they made. One such example, of many represented in this book, is Elijah Wald's reminder that we experience the blues (among other genres) through a haze of previous generations' values and categorizations, in which the genre was somehow separated from popular music writ large. "Writers like myself have tended to shy away from the fact that blues was once popular music," Wald writes. "Its evolution as a style," he continued, "and the career paths of most of its significant artists, were driven not by elite, cult tastes, but by the trends of mainstream black record buyers."[156] Decades and decades of genre articulations and reformations, amidst shifting sociocultural and aesthetic tastes, have made this era's commercial music culture increasingly inscrutable. If we take the mythologized blues, for instance, we imbue Robert Johnson with great importance while ignoring Bessie Smith, or Leroy Carr, for that matter.[157] We (mis)understand the blues through the industry's racist segregation of the music, but also through the framing and reframing via the American folk music "revival," or the British Invasion's fetishization of blues performers found via crate digging and various other layers of meaning that have brought the music to us in the first place. Jazz's many iterations have obscured its origins as syncopated dance music, while the cultural connotations of country—and its ancestral genre titles "old time" and "hillbilly"—have similarly permutated at various points and in various ways, lending toward a certain inscrutability to its earliest progenitors. My own interest in this era is mediated in various ways, from the commercial and cultural contexts of the 1920s and 1930s as evident in documents of the time, to the remembrances of those involved, "rediscovered" by collectors, journalists, and researchers who did interviews and published texts created in the 1950s, 1960s, and 1970s. While my agenda of placing the recording industry squarely as a media industry, one that both produces and embodies culture in its products and processes, I also hope to restore at least some sense of what the music meant—and what the work meant—to those involved.

The first chapter combines media history and cultural industries research in examining Gennett Records' institutional cultures of production via archival interviews, business correspondence, and the recording industry trade press. The chapter also seeks to understand that culture through Gennett's early 1920s foray into jazz, in terms of recording, scouting, and retail. I have included as many sources and points of view as possible, from Gennett engineers, scouts, and employees, while also recounting musicians' remembrances of recording for the company, including the King Oliver Gennett sessions, which featured a young Louis Armstrong.

But it was a closely related genre that would encourage record companies to expand their scouting activities dramatically. The second chapter traces the discourse that shaped "race records," from discussions of "race music" in the African American trade press protesting the recording industry's representational inequities to the recording industry's commercial adoption of the term following OKeh Records' success with Mamie Smith's "Crazy Blues." Key to this adoption is the industry's advertising rhetoric, encoded with the era's virulent racism, as well as direct (if misguided) appeals to African American listeners. The latter sections of the chapter map the African American–owned Black Swan label and its attempts to redefine race records, as well as Paramount Records' expansion into race records via scouting, recording, and retailing the music of the Mississippi delta.

The last of the three genres examined in depth came first to be known as "old-time music." Chapter 3 focuses on the emergence of old-time music as a viable industry category, through tracking both trade-press discourse and talent scout mythmaking in recording the rural white music of Appalachia and points beyond. Key to the emergence of old-time music was Ralph Peer's much-mythologized scouting trip to the town of Bristol along the Virginia-Tennessee border, in which both Jimmie Rodgers and the Carter Family were discovered. In addition to tracking the construction of old-time music via recording industry and trade paper discourse, this chapter notes the synergistic relationship between old-time music and radio, a technology that helped popularize the music through performers' radio appearances, as well as a vaudevillian take on rural white music and culture via the "barn dance" shows, such as WLS-Chicago's *National Barn Dance* and WSM-Nashville's *Grand Ole Opry*.

Chapter 4 details the consolidation of the U.S. recording industry that began during the second half of the 1920s. While various entertainment industries had been intertwined since their early histories, the technological conversions to electrical sound in the late 1920s brought together the recording, radio, and film industries like never before. While the biggest companies sought advantages in cross-promotional deals, smaller labels negotiated when and how they might adapt to new circumstances. Labels focused on the most widely popular artists and began to turn away from earlier niche explorations. The chapter offers case studies on Paul Whiteman, Louis Armstrong, and Jimmie Rodgers, all of whom expanded their musical repertoires while seeking cross-platform projects across recordings, radio, and film. Niche artists appeared in film and radio as well, as anyone capable of experimenting across contexts did. The merger between RCA and Victor Records shortly before the stock market crash had tremendous symbolic impact but was just one indication of what was to come.

The final chapter details the Great Depression's cataclysmic impact on the recording industry. What started in the mergers and consolidations intensified in the first years of the 1930s. Despite hard times, the trade press psychologized the Depression by shifting blame from larger societal dynamics to individual retailers' ability to sell in tough times. New realities necessitated the further conglomeration of the recording and radio industries, as well as film studios' ownership of recording companies (e.g., Consolidated Film Industries ownership of the American Recording Company [ARC] and Warner Bros.' brief ownership of Brunswick). The 1929 stock market crash (and the dismal financial reports from the years leading up to decade's end) convinced the Edison company to leave the recording industry in short order. Gennett, Paramount, and other small labels felt the changes most dramatically and sought to persevere by limiting their recording activities and returning to a focus on furniture and appliances, with which both labels' parent companies were most familiar. But it was Columbia's fortunes that would shift most dramatically, as it followed its 1920s successes with financial collapse. The company survived largely on its manufacturing works and then its name, as its ownership passed through various recording and radio concerns (ARC, Grigsby-Grunow, and eventually CBS). In the final pages of the book, I discuss Brunswick's and then Decca's Depression-era success, the emergence of swing, and the advent of the jukebox as some key components of the recording industry's rebirth.

So this research focuses on the recording industry *as* media industry, which media industry histories too often ignore. This also is meant to address the lack of recording industry historiography in popular music studies, which is too often ignored in favor of contemporary texts and audiences. And rather than a sweeping narrative that tries to retell the history of the U.S. recording industry from scratch once again, I have repurposed both media industry historiography and cultural industries research in an attempt to think about the recording industry both as a media industry all its own and one increasingly tied to other media industries through this pivotal period of transition.

CHAPTER ONE

"What Do You Think about Jazz?"

Niche Genres and Recording Culture

Gennett Records

"I am about exhausted trying to recollect things I do not remember," wrote businessman and former record company executive Fred Gennett in a 1962 letter to jazz researcher George Kay. "It is 25 years now since I left the record business and most of the people are dead who worked for the company in executive positions."[1] In the letters and interviews that remain with those involved with Gennett Records, from executives to the company's rank and file, there is a certain reticence and a strong sense that the past is the past. Still others recount how mundane the work seemed. Rarely did anyone have the sense that history was being made, but in later accounts some participants disagreed over individual roles in historic events, after the fact.

Gennett Records, of Richmond, Indiana, was the first of the upstart labels of the 1910s to challenge the dominance of the Big Three (Victor, Columbia, and Edison). Gennett's rise also coincided with the popularization of jazz, which along with the race and hillbilly genres reshaped U.S. popular music. The emergence of recorded jazz can be better understood by tracing the points where everyday business practices met sociocultural attitudes in Gennett's mode of production: its institutional culture, its scouting and recording practices, how it selected releases, and its advertising strategies. Gennett's institutional operations, in other words, offer preliminary answers to these questions: What conditions led Gennett to record a broad array of music, including Chicago jazz? What did Gennett's approach to production culture—as evident in recording protocol, artist selection, sales, and advertising—suggest about the relationship between the label and the emerging jazz genre?

Gennett's lasting reputation is often associated with jazz, primarily because the record company captured what are now considered some of jazz's foundational recordings. Beyond jazz, Gennett's recorded output was stunningly broad. Research suggests that the Starr Piano Company, Gennett's parent company, expanded its involvement to approximately seventy labels internationally during its time in the recording industry.[2] Gene Autry recorded some of his first sides there before his cowboy persona became well known. The great majority of Charley Patton's songs were recorded at Gennett's Richmond studio for Wisconsin's Paramount label. Gennett recorded William Jennings Bryan performing excerpts from his "Cross of Gold" speech to musical accompaniment, a recording that was added to the Library of Congress' National Recording Registry in 2004. Gennett also recorded "personal records"—records not meant for release by Gennett, but one-time pressings—for the Ku Klux Klan.[3] Gennett's mobile truck recorded music across the United States, making sound effects recordings for films and capturing a performance by Hopi Indians documented by anthropologist J. Walter Fewkes at the Grand Canyon.[4] Much of this activity and more has been chronicled by author and former journalist Rick Kennedy, the foremost Gennett scholar, who has been responsible for the increased interest in Gennett.

Judging by the number of jazz recordings released, Gennett must have garnered some economic success. These releases tended to be based on the style and popularity of the all-white Original Dixieland Jazz Band, whose Victor releases first gained popularity in 1917, starting a jazz craze that led to the recording of both black and white jazz groups. Gennett releases include Joe "King" Oliver's "Dippermouth Blues" (recorded on June 23, 1923, with a young Louis Armstrong); Bix Beiderbecke and his Rhythm Jugglers' "Davenport Blues" on January 26, 1925; Hoagy Carmichael's "Star Dust" for Gennett on October 31, 1927, backed by Emil Seidel and Orchestra with the Dorsey Brothers sitting in. Jelly Roll Morton and the New Orleans Rhythm Kings recorded "Mr. Jelly Lord" for Gennett, in what is purported to be the first interracial recording session in the United States on July 17 and 18, 1923.

Despite the label's size and location, Gennett first challenged in court proprietary technological formats shared by Victor and Columbia (after Victor claimed patent infringement), which in turn led to a more cooperative relationship between smaller labels in the face of more powerful companies. Gennett's initial forays into jazz also represented a challenge to the Big Three in one of the first instances of an emergent label attempting to succeed through recording music that the larger labels had previously neglected. Indeed, the Big Three companies experimented with many kinds of music throughout this era, but as the following section demonstrates, they were less adventurous from their comfortable positions of prominence in the recording market, though that dissolved in the face of financial trouble (as in the case of Columbia in the early 1920s) or after new markets had been proven (as in Victor's involvement in hillbilly in the late 1920s).

The Starr Piano Company of Richmond, Indiana, was the result of several precursor companies, the first of which began in 1872. George M. Trayser, who had learned piano building in the Alsace region of current-day France, opened a factory first in Indianapolis, and then Ripley, Ohio, before relocating to Richmond, Indiana, in 1872, with the help of prominent Richmond investors James M. Starr and Richard Jackson.[5] When Trayser retired from the business in 1878 (he arrived in Richmond already in middle age), his former Ohio partner M. J. Chase took over the company, now known as the Chase Piano Company, headquartered on Richmond's Main Street.[6] The Starr Piano Company, which would later found Gennett and its many affiliated labels, was founded after Chase left to establish a piano factory in Grand Rapids, Michigan.[7] With James Starr as president and Benjamin Starr managing factory operations, the Starr Piano Company of Richmond developed an impressive factory along the eastern fork of the Whitewater River.[8] The Gennett name first became associated with Starr Piano in 1893, when Henry Gennett and his father-in-law John Lumsden invested in the company, via Starr's relationship with the Jesse French Piano and Organ Company, whose chain stores were an important distribution outlet for the Starr Company in the Midwest and South.[9] Their new venture inherited a strong manufacturing operation and a built-in retail distribution network that greatly benefited their piano, and later recording, businesses.[10] Starr Piano's growth led the company to be one of the Richmond's biggest employers by the early twentieth century. As Andrew Klein wrote:

> Over the years . . . the Starr Piano, Phonograph and Gennett Records were joined by "Starr Freeze" refrigerators and freezers, synchronizer units designed for use with talking picture projection machines, wooden propellers and a valve for barrage balloons dur-

Fig. 1. The Starr Piano factory and "Home of Gennett Records," Richmond, Indiana, as depicted in a 1928 Gennett Electrobeam catalog. (Berea College Special Collections and Archive.)

> ing World War I, wooden cabinets for radio, airbrakes, and even wooden billy clubs to quiet a local factory strike. No such labour disputes occurred at the Starr, however, the factory frequently employing craftsmen for life. In fact, it was referred to as "the old folks home" and one newspaper article noted that two brothers had 108 years of combined service to the company.[11]

Starr founded its new record label in 1915, during a boom in phonograph and record manufacturing and during a period in which some key phonograph and recording patents had expired.[12] As a company involved in manufacturing domestic furnishings such as pianos, Starr first made cabinets for other companies' phonographs, then made their own phonographs, and later made recordings.[13] The company learned the basics of each new venture as they went along. The new phonograph company was initially named Starr, and then renamed Gennett after the family who had wrested control of the business from Jesse French and James M. Starr, as described in a 1908 deposition that involved Starr Piano Company and one of its competitors.[14] Once they were selling recordings, Starr executives realized that other music shops were not interested in stocking records bearing their competitor's name. The Gennett family who now ran the company; patriarch Henry and sons Fred, Harry, and Clarence Gennett all worked at the company. Recounting daily trips that Henry Gennett would take over to the recording division's purchasing office, Gennett employee Wilson "Tag" Taggart said, "He used to come around in the morning . . . rubbing his hands together, round and round. He'd come by around 10 o'clock and ask, 'How are the checks coming in today?' I'd tell him what we had got in our end of it. He said, 'Well, you're holding up the piano business right now.'"[15]

The Gennett label had a distribution network at its disposal from the start. Starr had music stores in cities around the country, which allowed Gennett records to be sold side-by-side with pianos in Starr stores. Starr often found out about regional tastes and trends in music, as well as popular performers in a given region, through their employees' interactions with customers and the local nightlife. While Starr Piano did not hire professionals from the music industry to make decisions for their Gennett recording division, the company greatly benefited from the social and economic networks already established before its entry into the record business. Gennett proceeded in trial-and-error fashion, recording this or that style and seeing if its releases would find an audience.

But a larger problem faced the company: technological format. After patent skirmishes between Columbia and Victor from 1901 to 1903, spurred on by Columbia's decision to sell discs as well as cylinders, the two companies entered into the first of several licensing agreements in which both shared lateral-cut recording technology. Lateral-cut recordings could only be played on Victor or Columbia phonographs, an aggressive move to increase the two companies' market dominance. Lateral-cut grooves were of "constant depth, modulated from side to side," while "vertical-cut" or "hill-and-dale" grooves were "of constantly varying depth, cut or indented into the playing surface."[16] Vertical-cut grooves began to be used in the United States by several short-lived companies after the Pathé Freres label success with the process in France. The Edison company deemed vertical-cut technology to be superior in its own legendary "tone tests" and then confused the format issue further by introducing its own proprietary vertical-cut Edison Diamond Discs, meant to be played back on its own Edison phonographs. But Victor and Columbia's dominance in the U.S. phonograph market continued, most notably through Victor's popular Victrola series.[17] Edison and Columbia continued to compete in the cylinder business (with a few other companies), but Columbia withdrew its cylinders in 1912, while Edison continued to make Blue Amberol cylinders until the company ceased operations in 1929.[18]

Gennett announced that it would switch to lateral-cut technology in 1919, following their hill-and-dale "Art-Tone" series of 1918.[19] An announcement in *Talking Machine World* (March 15, 1919) alerted Victor of Gennett's decision. Without mentioning Victor or Columbia phonographs by name, Gennett advertisements soon made mention that their records could be played on any phonograph. The subsequent court case that Victor brought against the Starr Piano Company had some unintended consequences, including the support of other small labels, including OKeh, Vocalion, Emerson, and the Canada's Compo label, all with vested interests in the court decision. The case led to greater cooperation among these labels, as they used each other for recording help and distribution deals. The case would be a protracted battle, going all the way to the Supreme Court in 1922. Victor argued that its president Eldridge Johnson had been the victim of patent infringement, but it lost the case.[20]

This legal victory allowed Gennett to expand its market, and it did so through identifying musical niches it could serve. Many labels such as Gennett initially focused on competing by offering their own versions of what the Big Three were producing: dance records in the wake of Vernon and Irene Castle–inspired foxtrot dance tunes, light opera, "military band" records, Hawaiian-themed numbers, Tin Pan Alley standards, and patriotic songs fueled by the U.S. entry into World War I. Gennett came to a relatively early understanding that by creating niche markets—through providing underrecorded musical styles to neglected audiences—the company could make a place for itself in what was and continues to be a volatile market. While it's clear that Gennett never stopped pursuing various musical niches, jazz was an important one for the company in the early 1920s. Gennett, through the Starr Piano Store in Chicago, found itself in an ideal situation to record jazz at an important point in the genre's development, with recording facilities a few hours away in Richmond. During this period in which Gennett recorded jazz, the term functioned as a defining one for a given song style or, perhaps even more importantly, a kind of dance. Jazz in its early period was as a subgenre of dance music, but through the decade it would increasingly become a term for popular music in general. Jazz was the site of controversy surrounding its improvisational style, rhythmic syncopation, and, most importantly, its connection with African American culture.

Jazz as Promise and Moral Panic

While the origins of the word "jazz" are notoriously difficult to trace, its use as a musical term was first popularized between 1913 and 1915.[21] While early uses of the term were found in the popular press from New York to San Francisco, the city most associated with the term was where the music flourished: New Orleans. That jazz is a pure product of American hybridity makes the music's flowering in New Orleans seem appropriate. As Ted Gioia writes in *The History of Jazz*, "Many musicians—mostly black, but also Creole and white—were experimenting with the syncopations of ragtime and the blues tonality and applying these rhythmic and melodic devices to a wide range of compositions."[22] The music drew on a variety of influences that coexisted in that city in the first few decades of the twentieth century.

While jazz was born in New Orleans in the 1890s, it came of age in Chicago in the 1920s. "Chicago was the jazz capital of the nation for most of the 1920s," Lizabeth Cohen notes, citing northward migration for economic opportunity and an escape from the Jim Crow South, the 1917 closure of New Orleans' Storyville district, and a strong Chicago nightlife that paid little attention to the Prohibition laws put in place by 1919 that lasted through 1933.[23] "Jazz in Chicago," William Howland Kenney writes, "gave particularly sharp and memorable musical expression to feelings of giddy excitement and rebellious daring, stimulating powerful emotional experiences which permanently shaped our collective memory of that time and place." In Chicago's burgeoning nightlife Kenney sees a variety of social and cultural dynamics reflected, such as "post-World War I economic prosperity, a heightened interracial awareness, the excitement of city life, urban machine politics, and the ongoing rebellion against Prohibition moralism."[24] The city already had an appetite for dance music that predated World War I, and despite the city's racially complex geography, clubs catering to South Side blacks began to move away from black-only entertainment toward programming both black and white acts for interracial audiences.[25] But it was through recorded media that most listeners experienced the new music. Marshall Stearns traced how recordings from both the Original Dixieland Jazz Band and Bix Beiderbecke's Wolverines greatly influenced Hoagy Carmichael, underlining the importance of recorded media in spreading music: "Phonograph records were probably the most effective single vehicle for the spread of jazz. . . . Around 1917, when the Original Dixieland Jazz Band made their recordings, a high-school boy in Bloomington, Indiana, named Hoagy Carmichael heard them and made an exciting discovery. His mother played ragtime, but this was something different."[26] Beiderbecke had greater opportunities for firsthand exposure to jazz in his hometown of Davenport, Iowa, which was one of several towns in which Mississippi river boats that featured jazz bands would dock (many of the most famous New Orleans musicians worked their way north in this fashion). But Beiderbecke was definitely influenced by the records of

ODJB and others, as he discovered a music that would lead to his own direct role in shaping it. For white youths like Carmichael and Beiderbecke, jazz held a liberatory promise, a life-affirming ticket to the possibilities of the modern world.

But for many others, the music and the people who played it were a threat, a source of conflict. Sometimes these narratives were designed for humor, as in a *Talking Machine World* article, "Don't Try to Burn a Jazz Record," which reported on a marital dispute over an unnamed jazz record, broken upon removing it from the turntable. One of them ended up putting the record into the family oven. "The heat of the jazz and the heat of the coal mixed so perfectly and developed such a flame that the chimney couldn't stand the strain and caught fire."[27] Often jazz was denigrated with greater force: "A strange word has gained wide-spread use in the rank of our producers of popular music," began an article titled "The Appeal of Primitive Jazz" in the August 25, 1917, issue of *Literary Digest*, a few months after the Original Dixieland Jazz Band's recordings had first created a sensation. "It is 'jazz,' used mainly as an adjective descriptive of a band. The group that play for dancing, when colored, seem infected with the virus that they try to instill as a stimulus in others."[28] In the *Literary Digest* article, jazz is equated with the tribal African music of hundreds of years before. "The music of contemporary savages taunts us with a lost art of rhythm," the article continues. "Modern sophistication has inhibited many native instincts."[29] Maureen Anderson argues that these articles framed jazz as modern primitivism. In critical statements about the new music in the white press, Anderson suggests that there is a pervasive uneasiness that at once illustrates a general taste for jazz among white audiences and a dislike or distrust of blacks in general. "Magazines turned to composers, doctors, educators, and even the black populace to explain jazz," Anderson writes. "Stunningly, what remains consistent in the reports on jazz is not the ultimate dislike of the music, but the political and social dislike of the black population."[30]

As stewards of musical taste, the staff of *The Etude* largely avoided such racist takes (with a few egregious exceptions) in favor of class-based, satirical dismissals and socioaesthetic assessments with limited acceptance of jazz, given the authors' biases against American vernacular music in general ("American Folk Music: A Negligible Quality," a 1918 article by W. Francis Gates, set the tone).[31] *The Etude* supported phonography, film, and radio, but largely because each technology afforded ways to further what *The Etude*'s writers saw as art music's edifying influence. From the late 1910s through the mid-1930s, articles discussed jazz's relative dangers to young pianists, jazz's relationship to dance music writ large, and the changing nature of jazz jargon.[32] The magazine was a mouthpiece of sorts for the Music Teachers National Association, placing them as defenders of art music, but often using tongue-in-cheek approaches to make their point. In Ernst Felix Tschudi's "The Immortals Protest against Jazz," composers Brahms, Grieg, Mendelssohn, Rubinstein, and Schubert send "a Marconi telegram" of complaint to a dance music composers' convention, in which they protested "the desecrating use of our divine music for fox-trots, rag-time, and so forth." The telegram's return address: "Milky Way Station, Heaven."[33] But the cornerstone of *Etude*'s writing on jazz was "Where Is Jazz Leading America?," a two-part series of interviews that considered jazz as a *national* concern, with leading arbiters of taste published in *Etude*'s August and September 1924 issues. The cover of the August issue announced a focus on "the Jazz Problem," while emphasizing that their discussion of the topic was not an endorsement.[34] "Prominent Public Men and Musicians" that took up the conversation over the August and September issues included music conductor Leopold Stokowski, composer and pianist H. H. A. (Amy) Beach, conductor/educator Frank Damrosch, novelist Booth Tarkington, satirist George Ade, and bandleaders Paul Whiteman and Isham Jones.[35] Chief among the cultural arbiters may well have been John Philip Sousa, who had famously complained about "canned music" roughly two decades before. The opinions ranged from tolerance to dismissal, but the larger framing of the genre and its social significance was resolutely negative. Despite such opinions from aesthetes and despite derisive and often racist anecdotes from the popular and trade presses, Gennett and other labels made the decision to record jazz, forgoing any social and cultural stigmas tied to the music in favor of identifying a growing (if highly controversial) audience for the music. That decision would require refining existing approaches to scouting, recording, and selecting recordings for release.

Talent Scouting

Gennett's decision to experiment in recording the burgeoning Chicago jazz scene was a concerted effort, and one that paid off, at least for a time. As Herman Gray has written, "The focus on smaller, more specialized audiences, for these companies, means that the financial threshold to break even is much lower than for large major companies."[36] And the label had stumbled upon an under-recorded, domestic, hybrid U.S. music to largely ignored audiences. By the end of the decade, the cultural impact of the new music would be powerful enough to define the 1920s as "the jazz age." Through delving into a new genre, Gennett employees had to learn about the music and its cultural resonances as they went along, while scouting music, recording talent, and deciding what to release. Surviving historical documents suggest that employee roles in Starr Piano's recording division were not particularly regimented or stratified in its division of labor. Starr's relatively flexible, even loose, approach to the recording industry—an industry in which the low barrier to entry has traditionally encouraged experimentation—applied to the industry as a whole. The recording industry was still developing an approach to identifying new talent during the early 1920s, through relying on those who had greater knowledge of a given music: record store owners, show promoters, and, as in Starr's case, company employees. Even the term "A&R" ("artists and repertoire") had not yet been coined, since scouting new talent was still a developing practice within the industry. Historical accounts survive of two noteworthy instances of Gennett's scouting and recording efforts in Chicago as the city's jazz scene took shape: first with the Friars Society Orchestra (later the New Orleans Rhythm Kings) and then with King Oliver & His Creole Jazz Band.[37]

The figure believed to be most responsible for Gennett's jazz recordings was longtime Starr employee Fred Wiggins, de facto talent scout during his time in Chicago in the early 1920s and head of Starr Piano's recording division by mid-decade. Relatively little firsthand information about Wiggins remains. He died in 1948, before a renewed interest in American vernacular music led to the rediscovery of both the musicians and "record men" who, like Wiggins, made those recordings possible.[38] Wiggins had become so obscure by the mid-1950s that authors writing about Gennett couldn't always agree on the spelling of his name.[39] Rick Kennedy writes, "Wiggins, 41 years old in 1922, was a descendant of one of Richmond's founding Quaker families. He had worked up through the Starr Piano ranks, starting in the shipping department in 1906. Before Henry Gennett's death, Wiggins was named manager of Starr Piano's Chicago store."[40] Wiggins was soon acting as Chicago talent scout for the Gennett label, where it is likely he learned about Chicago's music scene through his role as manager of Starr Piano Store. According to jazz researcher George Kay, Wiggins' role as scout expanded over time: "During the early days of Wiggens' [*sic*] career, he scouted for artists but the final decision for closing recording deals was left to Fred Gennett. Later, as his ability became more recognized by the Starr officials, young Wiggens' judgment in signing talent and releasing records was accepted as final."[41]

Wiggins' Chicago initiatives were not only tied to finding niche performers and audiences for Gennett to record and sell their recordings. As Rick Kennedy has argued, Gennett got involved in Chicago jazz because of a song publishing relationship between Gennett and the Melrose Brothers Music Company, Chicago music publishers and agents. Brothers Lester and Walter Melrose (brother Frank was a jazz pianist, while sister Belle was a movie palace organist) began with a South Side music shop on Cottage Grove Avenue and expanded their business via their development of a music publishing business, through which Melrose Brothers Music Company became a powerful force in propagating the city's music in and beyond Chicago. The Melrose Brothers hired Jelly Roll Morton, who was already a known jazz pianist in Chicago and his native New Orleans. Morton was hired to write piano music for Melrose Brothers' sheet music business, and the Melrose connection is the most direct link between Morton and Gennett, where the pianist recorded sessions in 1923 and 1924.[42] As Kennedy writes, "Gennett Records and the Melrose Brothers Music Company, working in tandem in the early 1920s, were catalysts in disseminating the local Chicago jazz scene to the nation."[43] Some of the key performers recorded by Gennett were King Oliver, Jelly Roll Morton, and the New Orleans Rhythm Kings (all of whom are discussed below).[44]

Starr Piano's Chicago location was at 423 South Wabash, where downtown meets the South Side in what is still known as the South Loop, named after Chicago's elevated train tracks' circuit through downtown. The area was close to a number of different recording studios and nightclubs that dotted the avenue from near north to near south, including the famous Marsh Laboratories (308 South Wabash) and evangelist Homer Rodeheaver's Rainbow Records Chicago office (218 South Wabash). It was at one club in particular, the Friars Inn (343 North Wabash) where Wiggins first heard a band then known as the Friars Society Orchestra. With the jazz craze that followed in the years after the 1917 Original Dixieland Jazz Band recordings, opportunistic club owners searched for their own act to capitalize on the trend. Friars Inn club manager Mike Fritzel convinced New Orleans musicians Paul Mares (cornet), who then approached Leon Roppolo (clarinet), and George Brunies (trombone), both of whom had been working the Mississippi River riverboat circuit, to become the house band. Rounding out the lineup was Mares' friends, drummer Frank Snyder, saxophonist Frank Pettis, and pianist Elmer Schoebel, who, as the band's only member who read music, often arranged the tunes. Additional players were recruited: multi-instrumentalist Arnold Loyocano, another New Orleans transplant in Chicago, who set aside his piano playing to play bass, and banjoist Louis Black, who along with a number of the players had been working out of the Mississippi River town of Davenport, Iowa. Over the group's two-year stint at the Friars Inn (they would change their name to the New Orleans Rhythm Kings on their departure), the lineup would change at regular intervals as members would leave for other opportunities and be replaced, but Mares, Roppolo, and Brunies represented the core members.[45]

It is not known how Wiggins originally heard about the Friars Society Orchestra. It is plausible that Wiggins stopped in to the Friars Inn enough to be familiar with the band. It is just as likely that he had heard of the group through a customer's recommendation at the piano store. Regardless of how he initially encountered the group, in the summer of 1922, Wiggins requested that Fred Gennett take the train from Richmond to hear the group play at the Friars Inn, a segregated white club that was purported to serve gangsters like Al Capone as well as bohemian music fans. Wiggins presumably brought him to Chicago so his boss could be sold by the enthusiastic audience for the Friars Society Orchestra as much as by the music. Apparently, Wiggins convinced Gennett, who in comparison with his brothers Harry and Clarence was most interested in the company's recording division and wanted to keep the company involved in musical trends. The result of Gennett's trip to see the Friars Society Orchestra was the first of several sessions for the Gennett label at the Richmond studio on August 24, 1922. By the time of a subsequent Richmond session the following March, the band had changed its name to the New Orleans Rhythm Kings after losing its regular gig at the Friars Inn. The group's recordings quickly garnered attention and influenced a great number of emerging jazz musicians. Suzanne Fischer writes, "The New Orleans Rhythm Kings were one of the hottest jazz bands of the early 1920s, and a strong influence on many later musicians, including Bix Beiderbecke, Muggsy Spanier, Mezz Mezzrow, and Benny Goodman. . . . [Their] smooth, swinging style signaled a departure from the raucous novelty sound of the Original Dixieland Jazz Band and its imitators."[46] In some ways, the New Orleans Rhythm Kings were closer in style to King Oliver and His Creole Jazz Band, through a purposefully nuanced style.[47] The records made by the New Orleans Rhythm Kings were successful for Gennett, and they encouraged Wiggins to continue his search for jazz talent in Chicago. His next major find would be particularly noteworthy to jazz historians, in that the band was led by a trumpet player that provided a link to the early New Orleans jazz scene as well as being a progenitor of the jazz emerging in Chicago. His backup band was known as the Creole Jazz Band and its leader known as Joe "King" Oliver.

As Fred Wiggins continued to investigate the Chicago jazz scene, he again contacted Fred Gennett back in Richmond about another popular jazz act, King Oliver and His Creole Jazz Band. George Kay's 1953 article for the *Record Changer* only mentions King Oliver's Gennett sessions in passing. From Kay's comments, it is not even clear whether Fred Gennett saw King Oliver's band perform at the Lincoln Gardens: "Fred Gennett recalls meeting Oliver in Chicago, through Wiggens [*sic*], and securing a signed contract on the spot."[48] After leaving New Orleans and traveling to California for an extended stay, Oliver and his Creole Jazz Band had taken up musical residence at the Lincoln Gardens, a popular nightspot on the city's south side. Shortly afterward, Oliver sent a letter to Louis Armstrong, whom the bandleader wanted to come to Chicago to play second cornet. With Armstrong in the lineup, Oliver's band was complete.[49]

Once signed to do a Gennett session, these Chicago bands would often travel to Richmond for short stays. Richmond was founded by Quaker families and had a relatively progressive attitude regarding race despite the period's history of racism and violence. Richmond was a segregated town, however, with African Americans and Italian American Catholics relegated to Goose Town (known as the "North End" to those who actually lived there), an area just north of the train station near downtown.[50] While many white performers stayed in local hotels, black performers stayed in Goose Town because Richmond's hotels were whites-only.[51] For many of them, it was their first recording session, done in a few hours, after which they would return to Chicago. Gennett's general approach to recording was practical, frugal, and geared to an imagined audience that company employees had connected anecdotally through the course of their work. By examining what little reliable information that exists about specific sessions, we can better understand the relationship between Gennett's recording artists and those in the business of determining what would be recorded and released.

Recording: "Only a Record Expert Can Tell"

In the first years of Starr Piano's entry into the recording business, their main recording studio was in New York City, on Thirty-Eighth Street between Fifth and Sixth Avenues. In the 1910s, the major recording companies were all still located on the east coast. Columbia began in Washington, D.C., and Edison's operations were based in Orange, New Jersey, near New York City. Victor's base of operation remained in Camden, New Jersey, near Philadelphia. As one of the first companies to attempt to compete with the major companies, it made sense to establish a strong presence in New York City, through both a Starr Piano store and recording studio for Gennett. In Starr's New York studios, the company recorded a great variety of acts and styles, from dance bands to music for German- and Spanish-language audiences. The company even cut a Gennett phonograph demonstration record, recorded by Ernest Hare, named after the Gennett company slogan, "The Difference Is in the Tone," suggesting it doubled as publicity for the new label.[52]

In Austin Lescarboura's article for *Scientific American*, "At the Other End of the Phonograph," the author described an ideal (and unidentified) "phonographic recording laboratory" located in New York's midtown theater district, where musicians would work amidst state-of-the-art acoustic recording facilities, with rehearsal rooms (disguised as sitting rooms) where nervous performers suffering from "phonograph fright" could rehearse before the next session began.[53] But the studio Starr built for Gennett in Richmond afforded no such luxuries, though the shift to a studio in the Midwest marked a new emphasis on recording and manufacturing discs on site. In the mid-1910s, the Starr Piano Company first worked with the Scranton Button Works to produce their recordings but would later build a facility in Richmond to both make phonographs and to press records on the site that had become known as "Starr Valley."[54] The company soon developed a second studio in Richmond, which was practical since so much of their operation (including their manufacturing facilities for pianos, phonographs, and phonograph cabinets) was located there.

There was another practical aspect to the Richmond studio that had larger implications: while the recordings made in New York were primarily designed to compete with similar releases from the Big Three, Gennett's opening of their Richmond studio was also their first move away from direct competition over the most common popular music of the day—New York's "marching bands, orchestras, and stage entertainers"—and a move toward the niche genres that would arise across the decade.[55] I am not making the argument that popular recordings (pop dance numbers, for instance) were not important to labels like Gennett. What I am suggesting is that success for Gennett and other small labels was not success for Victor, and that niches were important sources of revenue for smaller labels, especially in the periods in which the major labels paid such endeavors little attention. The smaller companies' catalogs regularly listed popular titles patterned after offerings from major labels and geared toward majoritarian tastes. But its niche catalogs were full of titles that sold too, including releases from key figures affiliated with each smaller label. But perhaps of even greater importance than the general shift from New York to Midwestern production was this: Gennett's Richmond studio put the studio closer to the music left unrecorded by the major labels of the East and Northeast. Both Gennett and Paramount would eventually shutter their New York studios for an increased focus on recording in Richmond, Indiana, and Grafton, Wisconsin, respectively (both labels would also work extensively in Chicago at various points). For Gennett, the Richmond location meant proximity that allowed the company to record "artists who were appearing professionally in Indianapolis, Cincinnati, Louisville, Chicago, and St. Louis—all locations with a significant black population that could support the theaters and cabarets in which these artists would perform."[56] Moreover, the Richmond studio allowed the Gennett label a chance to record music then emerging in the Midwest, as both black and white southerners moved north for (not necessarily musical) employment, developing and modifying their music in the process. While the New York studios would remain important for Gennett, the Richmond studio was in a perfect position to capture Chicago jazz as it emerged. This didn't mean, however, that the New York studio was happy with the quality of recording in Richmond. Wilson "Tag" Taggart remembered some tension between the two studio operations in a 1970 interview: "There was some friction. You see, their main recording studio was in New York. They had taken an old dry kiln [in Richmond] down in the [river] 'bottom' to make their recording studio out of."[57] Taggart felt that the eventual departure of Richmond recording engineer Ezra Wickemeyer for the Brunswick label had largely to do with New York's dissatisfaction with the recording operation at Richmond.

In 1923, acoustic recording was the industry norm. The process involved a large horn in which the vibrations produced in the studio were funneled through a sound box, and finally through a needle which etched the vibrations onto a metal master disc. Even for the era, Gennett's mechanical process at its Richmond studio was, wrote jazz researcher George Kay, "a throw-back to the dark ages," which used a cable-and-pulley system anchored by a weight, "similar to the grandfather-clock principle."[58] "The cable was wound around the center pin of the turntable and, as the weight lowered itself, the turntable spun."[59] Rena Clark, originally a machinist for Starr, became involved as an engraver, in a process in which he prepared "waxes" for disc recording. MacKenzie wrote, "Wiggins or Butts (Wiggins' assistant) would inform [Rena] Clarke of the number of recordings they were going to make on a certain day. Rena would shave the requisite number of waxes and put them into a wax cabinet to heat up. They had to be warm for recording." Engineer Harold Soule would then take each wax out of the cabinet for recording as needed and would number them accordingly.[60] "This technology," Ted Gioia writes, "while acceptable in capturing the sound of a single instrument or a human voice, was decidedly weak in presenting the delicate balance between different instruments in a jazz band."[61]

In a 1971 letter to a British jazz enthusiast, Robert P. Gennett (one of Fred Gennett's sons), described his remembrances of Gennett studio operations and the musicians who frequented the Richmond studio:

> They came and went. Some were clean and workmen. Others were literally filthy and ruined many cuts by being scared or musically poorly organized. If I remember rightly, we had to aim at 3 min. plus or minus very little. With a good musician we usually ran 2 masters for selection, others we ran three, and then [some masters were] lost as being not fit to issue because of being just poor or with musical or vocal mistakes. The reason for multiple cuts is thus evident. Also sometimes there could be a poor separation in the plating and stripping processes. . . . Things happened all the time—stopping a cut to tell a tuba player to stick a note in a void—a clarinetist hitting a blue note—a singer's voice breaking—sometimes a rack knocked over—speedups—slowdowns—a flaw in the lead-based wax—which also could be too hot or too cold—losing a sapphire in the middle of a cut off the cutter head—a cough—a sneeze.[62]

In order to explain their recording processes to the great numbers of performers who came through the Gennett studio, many of whom had never been in a recording studio, Starr employees developed an instruction sheet that outlined the procedural and technical aspects of their process:

> We start by making a short test which is immediately played back to the performer from the wax and from that we begin to judge positions and tone and arrangement of music and everything that is nec-

essary to make a good record. After the faults have been found we try another test wax and still another, etc.—if necessary will use one hundred test waxes in a date although this would almost be a physical impossibility—however, when we have played back a wax that sounds perfect, we begin to make a master—we make three masters of each number tested out.[63]

The recording studio could still be an intimidating place for musicians of this period. In an article that appeared in the October 20, 1923, issue of the *Pittsburgh Courier*, the city's black newspaper, bandleader Lois Deppe described their trip to the Gennett studios in Richmond. "Judging from their vision of it," the article recounted, "the game of making a phonograph record means hours of painstaking work, eight continuous hours of work, if you please, in a sound proof room to make three records, one or two of which may be ultimately accepted."[64] The article is also unusual in the detail it gives regarding the Gennett studio itself: "The musicians are assembled in a recording room about thirty feet square, with cement floor and padded walls to prevent echoes, flys of materials dropped from above to make better the acoustics and music racks suspended from the ceiling with straps. Through a curtained wall extends two horns each a foot and eight inches in diameter." The musicians too were often spread around the room in order to get the best sound reproduction from the acoustic horn process. This also included the drummer of Deppe's outfit (Harry Williams), as the article notes: "The drummer's chief grief comes from the fact that his prize collection of noise making paraphernalia must be promiscuously strewn about the floor," adding that Williams had to exhibit "the most extraordinary agility and dexterity in skipping around and giving the mystic spell of his artistic touch at just the right time and just the right place."[65] In the final section of the article, an unnamed "director of the Victor Records Making Service" says that "whether an artist's voice or instrumental rendition will record, well, is a gamble until the master test is made," adding that "only a record expert can tell."[66]

But beyond a few sources such as this one, most remaining information about specific recording sessions focus on those that have become the most famous with time, such as the King Oliver and His Creole Jazz Band recordings at Gennett with engineer Ezra Wickemeyer. By the time of King Oliver's first sessions at Gennett's Richmond studio, the band was well prepared. Oliver's regular gigs at the Lincoln Gardens allowed them a regular venue in which to try out new material and hone their arrangements. It's doubtful that the Gennett recording engineers wasted many takes with King Oliver's band, a group known for its musical precision. In their adopted city, Oliver's band was soon the most sought-after, as they further developed the new sound: "One of the supreme ironies of the history of New Orleans jazz," writes Ted Gioia, "is that so much of it took place in Chicago."[67]

King Oliver and His Creole Jazz Band first went to Richmond to record for Gennett on April 5 and 6, 1923. With the success of Gennett's New Orleans Rhythm Kings recordings the year before, the label increased the number of jazz sides it recorded while focusing on the best acts they could find out of the vibrant Chicago scene. In the article Louis Armstrong wrote for the *Record Changer* published in the summer of 1950, he recalled his sessions with King Oliver for Gennett: "When we made those Gennett records Joe wasn't in his prime, like he was before he sent for me. To show you how much stronger I was than Joe: those were acoustical records, with those big horns; Joe would be right in the horn blowing and I would be standing back in the door playing second trumpet."[68] Because there is so little firsthand information about recording sessions with Gennett, these brief comments by Armstrong have been often repeated by jazz enthusiasts for whom the King Oliver sessions represent the Rosetta Stone of the genre. And since sessions were such a routine at the Gennett studios that there seemed little reason to find one any more special than the last, little information has survived about the Oliver sessions—except this anecdote from Armstrong. It led to much hyperbole over the years; soon Louis was playing his horn out the door and down the hall. But here's another firsthand account from Lil Hardin-Armstrong, pianist in King Oliver's band and Armstrong's second wife, in which she discusses the April 5 session: "At the first session we recorded in the great big horn . . . and the band was around the horn. Louis was there, as he always was, next to Joe. Didn't work out; couldn't hear no Joe's playing. So they moved Louis way over in the corner, away from the band. And Louis was standing over there looking so lonesome."[69] Part of the appeal of this story to jazz enthusiasts is the pathos in this Shakespearean order-of-succession narrative. Hardin-Armstrong conjures this in her account, noting that Oliver had earlier told her that Armstrong was better than him and quoting Oliver as saying "as long as I keep him with me he won't be able to get ahead of me; I'll still be the King."[70] Armstrong's superior volume and tone made her feel sorry for Oliver and that it's a "bad sign when you begin to feel sorry for a man."[71] But for our purposes, the anecdote gives us a better sense of the recording process in the acoustic era. The volume balance was struck between the musician (who determined how loud he or she would play, as much as a given instrument would allow for such adjustments); the engineer (in this case, the "they" mentioned in Hardin-Armstrong's accounts were likely Ezra Wickemeyer and his assistants); and proximity to the horn (or horns) crucial to acoustic recording. Like many other musicians' accounts, Baby Dodds was impressed by the Oliver sessions at Gennett: "I've made a lot of recordings, but the biggest kick I got out of any recording session was when the King Oliver band went out to Richmond, Indiana, to record for the Gennett Company."[72]

Musician Ralph Berton remembered King Oliver and His Creole Jazz Band and the New Orleans Rhythm Kings as "the best units" of the Chicago jazz scene. "But believe me," he added, "there were many more orchestras of value whose names are forever effaced from our memories." He rightly suggested that the key to these two legendary jazz bands' long-term success were the recordings they made, the records that were later celebrated by record collectors who made sure this music would not be forgotten. "Imagine that for one reason or another, Jelly Roll Morton had not had the chance to record the numerous records that he made. I am convinced that if he had not had the chance to make those records his name today would totally unknown to the followers of jazz."[73]

Record Selection and Sales

In terms of choosing what material to record, Gennett had a hands-off attitude regarding the repertoire that musicians recorded for the label. From what evidence that survives, Wiggins' role as talent scout in the burgeoning Chicago jazz scene was to seek out music that held the promise of sales, based on what information he had gleaned by attending Chicago-area clubs and the opinions solicited from customers in Starr Piano's Wabash Street showroom. He wasn't particularly concerned with the songs recorded, since like many of those involved early on, Wiggins didn't have a taste for the music being recorded. By today's standards, it is noteworthy that Gennett would allow so much creative control to the artists by not dictating specific material. There is evidence that if a record was recorded initially and didn't perform well, the company was reticent to try it again.[74] Since the overwhelming majority of those working for Gennett could claim their only experience in the music business in their work for the Starr Piano Company, they had little experience in the business of recording. When considering the sheer number of recording sessions logged in the Gennett ledgers and the broad array of musical styles recorded, it would be unlikely that employees would have had familiarity with all of them. With that many sessions, however, the employees surely learned a lot about getting the best technological reproduction they could through the prevailing acoustic recording processes of the era. "Jazz enthusiasts today," George Kay wrote in 1953, "would howl with despair at some of his decisions to destroy cherished jazz masters as 'not suitable for release.'"[75] "The final decision as to what would be issued" at Gennett, John MacKenzie wrote, "was made in Richmond."[76] There was disagreement among Gennett employees about who had the responsibility for making the selection. Harold Soule claimed that Fred Wiggins decided. Joe Geier said that Wiggins was not involved. Clayton Jackson said that although both he and Wiggins had "tin ears" they did have a knack for picking selections that the public would buy.[77] The existing data, when considered comparatively, suggests that while Wiggins wasn't necessarily involved in all decisions in such a fluid and busy work environment, he was one of the central gatekeepers that decided what was recorded.

The kind of decision making Wiggins employed did not always sit well with his staff in the recording division. In an interview with MacKenzie, Harold Soule "complained bitterly about Fred Wiggins' personal method of selection," which reveals that Wiggins' decisions were based as much on technological appropriateness as much as on quality of the performance. "Wiggins had an old wind-up phonograph in his office that he'd use to listen to the test pressings. It was about as lousy a piece of equipment as you could ask for. I couldn't understand why the test pressings had to be played back on this old-style rig. Being a kid in those days, I couldn't see the reason, of course. I can see his viewpoint now."[78] Wiggins understood that their audience was largely working class and, in many cases, rural. Indeed, besides the Starr Piano stores, most Gennett products were sold via local music dealers and others who sold recordings in general stores and the emerging chain store market. Wiggins based his decisions on the likelihood that their audience would not own expensive phonographs, and his decisions were usually final. Florence Gennett said her husband Harry—nominally Wiggins' superior in the company's hierarchy—saw Wiggins as being "a little on the temperamental side and that if he'd made up his mind you couldn't change it."[79]

Recording engineer Harold Soule, who for many years worked in Gennett's New York studio, complained that the best recordings that featured both high and low frequencies were rejected by Wiggins because they sounded bad on the cheap machines used by many of Gennett's customers. As John MacKenzie wrote, "If his cheap machine produced distortion (what Wiggins referred to as 'blasts' in the ledger notations), then the cheap machines owned by the typical buyer of the Gennett family of labels would produce the same effect."[80] Soule remembered, "They sold so many records to hillbillies and farmers and people that just had an ordinary old crank machine." In addition, the records themselves, generally made of cheaper materials, would wear down much faster.[81] Engineer Joe Geier remembered, "Technically speaking, the recording engineers could reject a take—Fred Wiggins wasn't much interested in the technicality or musicality of the recordings. He was really only interested in getting them into the studio to record and making the finished records that would sell and make a profit for the company."[82] If Wiggins' approach was practical in the short term, it would eventually lead to difficulties for Gennett in its ability to compete with leading companies. As John MacKenzie wrote, Victor and Columbia spent "vast sums of money constantly improving their recording process, hiring talented men to evaluate and redesign their system, together with musical directors and advisors to constantly upgrade the quality of their musical output. The Gennett family, with brother Clarence in control of the financial end of things, were reluctant to spend anything on such frivolities—Gennett's failure to keep pace had a telling effect."[83]

After the technological and aesthetic decisions regarding the test pressings were made, the next decision was to choose which recordings would be coupled as sides, a process that was known as "mating." "Strict economics," MacKenzie wrote, "dictate that more profit can be obtained if only one side of a record is a winner."[84] Gennett employee Clayton Jackson claimed he made most of the mating decisions: "I lived there in Richmond. . . . I did most of the mating. I had a room up at 1226 E. Main. I would take boxes full of the test pressings home with me in the evening. . . . I had testing turntables up there and I would take these test records and mate them and make up the catalog."[85] Fred Wiggins also made mating decisions, according to MacKenzie's research.

After 1924, Starr Piano signed an agreement to supply recording to Sears & Roebuck for their Challenge, Conqueror, Silvertone, and Supertone labels, and around the same time they began producing their own discount Champion record line, which sold for fifty cents, as opposed to the seventy-five cents necessary for Gennett releases, and that featured releases targeting the race records and old-time music markets. Starr's recording division would save the best takes of a given session for the Gennett imprint, and those takes considered less than ideal would be made available for these subsidiary labels. While Gennett supplied the master discs—the company routinely took several takes of the same song—Sears employee J. J. Shays was primarily responsible for mating these records. John Mackenzie wrote, "They [Sears] would send a listing down to Richmond from Chicago, and the printing of the labels would be done from that list."[86] Gennett's in-house mating and cataloging focused on the Gennett imprint and Champion, Gennett's popular budget label, which specialized in old-time and race recordings. Many of the same records that appeared on Gennett would also appear on Champion (or various Sears labels). This was a common practice in the prewar recording industry. For these releases, Clayton Jackson would come up with pseudonyms for recordings already released on Gennett, thus allowing the company additional revenue for the same recordings:

> When you mated a couple recordings together for Champion records, you needed a name. So you did it right then and there. Telephone books were handy references, and city directories. Most of the ones I mated, I got out of the city directory and the telephone book. Anything you could find that you didn't already have. What worried you was duplicating, getting too many Howard Williamses, or something like that.[87]

Once the songs were mated, manufactured, and scheduled for release, the sales department would get to work on the latest batch of recordings. The central means by which Starr sold Gennett releases was through the Starr Piano Company stores around the nation and those music stores that carried recordings as well as musical instruments.[88] Gennett's mail-order subsidiary labels were sold through the Sears & Roebuck Company, an important market for the label. Through both these retail points, as well as through direct mail, the distributed catalogs and supplements kept buyers aware of the full line of Gennett releases. By 1920, after a decade in which stardom across media genres had emerged as a potent selling point, recording companies were taking note. In a *Printers' Ink Monthly* article, the author advised "'Bring the star and the public together . . . people who buy records should know the human side of the folks who are actually making the records. The better they are acquainted, the more sales."[89] The company turned to newspaper advertising to both general and niche markets around the nation. Gennett's advertising budget was not those of other labels, even when compared to other small label competitors. In the white press, advertising focused on Starr Pianos and those Gennett recordings that the company deemed most worthy. In trade magazines like *Talking Machine World*, the company took out full-page ads, which usually featured the brand rather than any specific recordings.

The head of Gennett's advertising through much of the 1920s was Roland Ball, who according to purchasing employee Wilson "Tag" Taggart, replaced a man whose last name he knew only as Niles early in the 1920s.[90] During his time in the advertising division, Taggart remembered much of the company's emphasis on promotion through their Starr stores. He recalled working on Gennett window displays for Starr Piano stores nationwide, which were made in Richmond and sent around the United States:

> When I was a kid and started working down there, I always liked art and I began talking about some window displays. They had so many stores. Mrs. Henry Gennett got so interested when I started talking about traveling window displays. She'd get down on her hands and knees on the floor out at her house, with me, and work on these window displays like a common laborer.[91]

Gennett and Starr Piano were also known for their advertisements in *Talking Machine World*, which depicted fantastic scenes that conjured music's otherworldly power, with a Pan-like figure playing flute on a sand dune or gauzy female figures in diaphanous clothing depicted on some faraway beach.[92]

Perhaps the most telling advertising that remains is in the form of record catalogs, which were circulated to retailers, who would then often pass them along to interested customers to keep them informed of the latest releases. From two Gennett catalogs from the 1920s—a June 1922 catalog with bandleader Ray Miller and a 1925 catalog that featured all lateral recordings to date—we can get a strong sense of the way the company cataloged their jazz recordings. Despite the presence of Miller on the cover of their 1922 catalog, the first few pages were given over to patriotic and light opera songs, genres more associated with the 1910s. Ethnic recordings, like Felix O'Day's "Dear Little Shamrock" backed with "The Pretty Girl Milking Her Cow" and Morris Goldstein's "Die Green Kosine," were geared to immigrant communities, while Gennett "Green Label" recordings sought to match the prestige of Victor's Red Seal artists with the Gennett Symphony Orchestra, the Mozart Trio, and others.[93] The next section listed dance favorites and suggested that Gennett's jazz recordings were considered to be a subset of the larger dance category. Gennett promised "the latest, best and snappiest hits of the day" but by "new orchestras": Husk O'Hare's Super Orchestra of Chicago and Lloyd Barber's Green Mill Orchestra.[94] In keeping with the custom of the time, both bands covered popular hits of the day. O'Hare's "Boo Hoo Hoo" was also recorded by Noble Sissle and Eubie Blake (Emerson, recorded January 1922), the Joseph Samuels Jazz Band (OKeh, recorded March 1922), and Aileen Stanley (Edison, recorded April 1922). In Gennett's comprehensive lateral recordings catalog of 1925, "Dippermouth Blues" by King Oliver's Creole Jazz Band is listed as a foxtrot, as were recordings from the Friars Society Orchestra (later the New Orleans Rhythm Kings).[95] But if jazz wasn't yet the genre term of choice, the word came to define the decade. Soon the industry's biggest labels were experimenting with the new music with varying degrees of interest and dedication. Increasingly, the word defining the genre was an invitation to debate.

"What Do You Think about Jazz?"

The jazz music that Gennett had documented as it emerged was transforming as it took its place on the national stage. More than any other jazz proponent, Paul Whiteman sought to spread jazz's presence and respectability. It was a conscious effort in which Whiteman sought to combine hot jazz with symphonic music, an unnecessary hybrid to some, but one that played a key role in jazz's influence on popular music in the United States. The most explicit argument to be made for jazz came via Whiteman and his Palais Royale band's "An Experiment in Modern Music," a February 12, 1924, concert held in midtown Manhattan's Aeolian Hall, on Abraham Lincoln's birthday.[96] This was a homecoming concert for the band after nearly a half year spent in the United Kingdom. Would American audiences remember them? Whiteman and his band's program and this specific locale were both designed to place jazz for serious consideration as art music, though Whiteman would write of his nervousness surrounding the project: "Visions of playing a jazz concert in what a critic has called 'the perfumed purlieus' of Aeolian Hall, used to rouse me up at night in a cold perspiration."[97] George Gershwin was at the piano and debuted *Rhapsody in Blue* for the show. Zez Confrey also performed, and in keeping with his bridging jazz with older art music, Whiteman asked an aging Victor Herbert to contribute *A Suite of Serenades* for the Whiteman band to perform on the bill.[98] Flyers promised "New Typically American Voices."[99] The audience response was effusive, despite previous misgivings from various corners. "But the list of pessimists was a little shorter, I believe, when at half-past five . . . we took our fifth curtain call."[100]

Jazz's reach had surprised everyone. *Voice of the Victor*, the company's in-house magazine for retailers, published "What Do You Think about Jazz?" by Gertrude Swift. "Do you enjoy good jazz?" the article begins. "For it can be very good. It can also be, and frequently is, very bad."[101] While suggesting that many of "the elements which go to make jazz are not new to music," she went on to say that "it is the negro influence which has dragged our popular music out of the rut of silly sentimentality it was in a generation ago."[102] Jazz had arrived. Two years after the Aeolian Hall concert, after Whiteman and others had made jazz into the country's predominant popular music, he was uncomfortable offering up definitions of the genre. "I have been dodging this question for years," he wrote. "Probably it doesn't matter much. Whatever I think, there will be many to disagree with me. Jazz has always made a grand hostess help, for it will throw the most conservative dinner company into paroxysms of discussion and disagreement."[103]

The effect of the widespread availability of jazz recordings, by Gennett and others, had an immense effect on the spread of this new music. Jimmy McPartland was a member of the "Austin High School Gang," a group of white teens who became jazz fans and then musicians, on Chicago's northwest side. Despite being in the same city in which New Orleans jazz was being transformed into "hot jazz," McPartland and his friends first heard this new music on a Victrola at a local malt shop called Spoon and Straw, where the shop's Victrola would play the latest jazz records, usually Paul Whiteman (soon to be known nationally as "the King of Jazz"). "Mechanical reproducing devices," Neil Leonard writes, "hastened the acceptance of jazz not only by familiarizing both listeners and musicians with nontraditional characteristics of the new music but also by helping to bring it into closer alignment with traditional values."[104] The jazz players that had grown up with "the conventions of New Orleans jazz," as Ted Gioia calls them (the syncopated rhythm and "stride" piano, the stylistic move from ensemble-based tunes to soloist-dominated performances) began to change their styles to fit audience tastes. "During the decade of the 1920s," Gioia writes, "popular songs and thirty-two bar forms were increasingly used by Chicago jazz musicians, while blues and multi-theme ragtime structures—so central to the New Orleans tradition—became less common."[105]

And while Chicago remained an important town for jazz, there was a palpable sense that something had passed, as Chicago's small jazz nightclubs were replaced with bigger venues. "Chicago's leading jazz cabarets," William Howard Kenney writes, "provided a powerfully concentrated, highly charged jazz experience. The leading black-and-tan cabarets had presented the best black jazz musicians, singers, and dancers of a talented and ambitious generation to inter-racial audiences that were packed close to the band and to each other on small dance floors and around tiny tables."[106] By contrast, the swing era that emerged later would resegregate jazz audiences in its dance halls, some of which were located at rural resorts far away from the urban venues so central to jazz's emergence. And as the vibrant Chicago jazz scene began to decline, New York jazz was already on the rise. Meanwhile, another musical genre, closely tied to African American culture, was about to engage a number of small labels in music they knew little about.

CHAPTER TWO

"Are These Not Great Artists?"

Race Records and Genre Discourse

Race Music as Cultural Emancipation

The January 1916 issue of the *Chicago Defender* featured a call for more African American opera stars on record: "During the Christmas holidays there were thousands of dollars spent by our people for Victrolas. It was an easy matter to hear one say that they got a $250 Victrola for a Christmas present and that they paid seven or eight dollars for records."[1] The records purchased by African Americans featured white stars of the opera stage, the article continued, including Enrico Caruso, Ernestine Schumann-Heinke, and Geraldine Farrar. But where were the names of black opera stars, such as Anita Patti Brown, Roland Hayes, or Hazel Harrison? "Are these not great artists?" the article continued. "Have they not been trained to the highest degree in their profession?"[2] Brown would finally make her own phonograph recording for Columbia in 1920, as Hayes had done in 1918, though neither would see official release by white-owned labels. Both were available via Columbia's personal recording series, paid for by the individual artists.

By 1919, the first known African American–owned label addressed similar cultural concerns as expressed in the *Defender*: the short-lived Broome Special Phonograph Records, founded by George Broome in Medford, Massachusetts. An advertisement in the NAACP's *The Crisis* newspaper promised "Phonograph Records of our Foremost Negro Artists."[3] The label addressed the dearth of recordings by releasing tracks from the most famous African American concert artists and orators: composer and baritone Harry T. Burleigh and soprano Florence Cole Talbot, as well as a reissue of the "Atlanta Exposition Address" by Booker T. Washington.[4] Broome's labels proclaimed these records "Brown Seal Records," a challenge and a wink to Victor's prestigious Red Seal brand. Like the earlier *Defender* debate, these recordings focused on operatic performers in an appeal for cultural as well as economic capital.[5] Recordings were not understood only as potential revenue for performers, alongside live performances, piano rolls and sheet music, but a concerted African American presence on phonograph records would also represent "an instrument of cultural emancipation."[6]

Unlikely players would facilitate the beginnings of that cultural emancipation without necessarily meaning to do so, through music that neither the *Defender* nor Broome championed. In the absence of sustainable black alternatives in the recording industry, during one of the most volatile periods for race relations in the twentieth century, it was smaller white-owned labels in need of an industry niche that defined and developed race records as a category. This chapter argues that the dramatic rise of race records represented a significant battle over cultural representation, within unique social and institutional contexts. A huge swath of African American vernacular music would be recorded as new labels sought niches of largely unrecorded music for underserved audiences. Label expansion into African American music turned opportunistic amid impressive sales and then expanded dramatically on the heels of a few key successes. The marketing of the new genre transmitted an ongoing ambivalence in the form of contradictory attitudes and representations of black performers (and African Americans in general) as evident in company materials and practices.

Donald Clarke has rightly argued that the country's institutionalized racism was a structuring element in the eventual division of categories into "race" and "hillbilly" genre formations, while Patrick Huber has suggested that the determining factors in this racial separation were practical. Of course, the two have never been mutually exclusive. The tragic fields created and managed by institutional racism in its myriad forms determined to varying degrees the practical decisions and dispositions of those involved.[7] As it emerged, the "race" designation carried more societal than musical connotations, functioning as a catch-all industry term for music recorded by African Americans. A broad series of musical styles were included under this moniker. By the early 1920s, "race records" became the preferred industry term, a phrase that named the commodity, a malleable term first associated with jazz and later with what we now call the blues. The term "race records" would lead to allegiances and disavowals, part of a discursive circuit of companies, musicians, and audiences trying to make sense of musical styles as recording further expanded into uncharted musical territory. This chapter follows the term "race" to designate music through various social contexts, as it moved from debates in the African American press to adoption by white record companies, and as the genre's boundaries changed according to shifting industry notions about their consumers. These changes came with the industry's expanded geographical reach as it began its transformation from a set of regional focuses to a national industry. Through press accounts, industry advertisements and catalogs, and archival interviews with those involved, the cultural dimensions of record industry decisions—and a picture of an industry in flux—begin to take shape.

The emergence of race records as a viable genre category meant it had to be flexible enough to change as the industry discerned shifts in audience tastes. The industry's best guesses were reflected in the companies' advertising campaigns and scouting activities. In an industry that was learning through market experimentation through the firsthand experiences of its scouts, these materials provide concrete evidence of how companies understood their audiences, as well as their own role in the business.

African American Performers and the Early U.S. Recording Industry

In the years following World War I, returning African American soldiers experienced the profound dissonance of having fought for a country that refused to allow them full inclusion. In his 1969 study of jazz and blues in the recording industry, Ronald Foreman suggested that while African Americans were slowly garnering new opportunities in the industry, recording companies were still trafficking in nineteenth-century minstrelsy through "coon songs" and "novelties" by both black and white performers. As Foreman wrote, "Against the 'novelty' appeal of jazz and the sustained progress of 'blue' songs, a firming attitude of racial self-consciousness, through which appeal to pride in achievement re-occurred, spurred Negro citizens to reach new goals or acquire new evidences of their worth as a people."[8]

African American contributions date back to the earliest era of recorded music, although the industry ignored black audiences. After the recording industry's unsuccessful attempt to use the phonograph for office correspondence, companies followed the lead of penny arcade proprietors who had turned to coin-in-a-slot phonographs providing music and speeches as popular entertainment by the 1890s. Tim Brooks argues that African Americans' role in the recording industry begins much earlier than commonly acknowledged. George Johnson's 1890s recordings represent a beginning of sorts, followed by legendary recordings by the Fisk Jubilee Singers, Bert Williams, and others.[9] Despite these performers' roles, the industry remained focused on white listeners, following the industry's nineteenth-century logic that popularized minstrelsy to white audiences.

The recording industry's strategic shift coincided with the rise of a robust African American entertainment industry, as Lynn Abbott and Doug Seroff have usefully recounted.[10] While a relatively small percentage of touring black performers ever made recordings, singers like Johnson and his popular "Laughing Song," the Fisk Jubilee Singers' version of "Swing Low, Sweet Chariot," and above all Bert Williams' famous "Nobody" all helped to make a significant impact in an industry new and open enough to offer African Americans a place in it, albeit on the industry's own racist terms. These recordings were highly mediated events, through which performers often fought to be recorded, struggled over repertoire, and negotiated their place within larger societal discourses.

The *Defender*'s push for inclusion did not stop with opera, though starting there expressed a clear desire to compete with white artists via the conventions of high art. Their rhetorical approach suggested class anxieties both in relation to white cultural production and amidst long-standing conflicts between classes within the black community. The *Defender*'s call was part of an emerging voice of protest, in part echoing early jazz bandleader James Reese Europe's comment that "we have our own racial feeling and if we try to copy whites, we will make bad copies."[11] The music African Americans made, he argued, should reflect the hardships and aspirations of its people. It is paradoxical then that the answer—at least a partial one—came not from black entrepreneurs but from smaller, white-owned record companies looking for niches not yet scouted by larger labels. The first label to make such a breakthrough recording, New York's OKeh Records, did so by happenstance.

"To Hear Is to Buy": OKeh and the Emergence of Race Records

By the early 1920s, the Big Three were showing signs of weakness. Columbia struggled to regain its market share after overproducing phonographs in hopes of selling enough to cover its debts. This led the company into bankruptcy reorganization, which turned control of the company over to

Wall Street investors.[12] The Edison label continued to suffer because of its namesake's refusal to adapt to lateral-cut recordings and the company's relative inattention to the quality of its recorded talent. Only Victor's success in both the recordings and phonograph business continued. Radio's cultural impact was beginning to grow during the early 1920s, though the Big Three remained largely oblivious to the technology's far-reaching implications. Meanwhile, an increasing number of new, smaller labels ambitiously moved into new genres after realizing they could not compete with the Big Three in established genres.

Among these, New York City's OKeh Records benefited first. OKeh Records was founded amidst the phonograph boom in May 1918, as part of the Otto Heineman Phonograph Supply Company, Inc. Heineman had been Berlin director of Carl Lindstrom A.G., a German phonograph manufacturer with business in Europe, Asia, and South America. After supporting Gennett in its legal battle over playback formats, OKeh followed the lead of numerous companies who switched from vertical-cut to lateral-cut recordings by 1920, which allowed their records to be played on most phonographs, including those made by Victor and Columbia. 1920 would be a decisive year for OKeh, as its opera and classical music series showed signs of failure, while a chance taken on "blues" paid off.[13]

"What has been called 'classic blues,'" Leroi Jones (Amiri Baraka) wrote, "was the result of more diverse sociological and musical influences than any other kind of American Negro music called blues. Musically, classic blues showed the Negro singer's appropriation of a great many elements of popular American music, notably the music associated with popular theater of vaudeville."[14] Richard Middleton defines the blues as "a strophic secular song, usually improvised and usually for solo singer, accompanied by an instrument or an instrumental band; the lyric is generally in a three line, aa^{1}b form, and as a rule the musical structure is a harmonic formula, which the singer uses as a basis for improvisation."[15] The blues is most commonly secular as opposed to sacred or gospel music, which also emerged out of "Sanctified" or "Holiness" churches of the South and was often met with resistance from mainline churches. While rock's reframing of the blues made the solo-guitar-playing male blues singer synonymous with the genre, blues' early emergence was through female vocalists backed by small bands: the "classic" or "vaudeville" blues worthy of much more attention.[16] These distinctions suggest not only what kind of music was being produced in terms of its aesthetic habits, but also the social dynamics of who was included or excluded and who the music was designed to serve.

Noting that blues began as a "private and personal music," Leroi Jones (Amiri Baraka) wrote, "It became a music that could be used to entertain *formally*. The artisan, the professional blues singer appeared."[17]

The blues did not emerge with Mamie Smith's 1920 hit "Crazy Blues" any more than Smith—the vaudeville perennial—was an overnight sensation. Smith's chance to record for OKeh came well into her career and was due to efforts by songwriter and performer Perry Bradford, who had first convinced OKeh's music director Fred Hager to record Mamie Smith performing two of Bradford's blues-influenced pop songs, "That Thing Called Love" and "You Can't Keep a Good Man Down" on Valentine's Day, 1920. Hager had hoped popular vaudevillian and OKeh artist Sophie Tucker might sing Bradford's tunes, but Tucker was ill as the recording date approached and Smith was ultimately featured.[18] Both Smith and Tucker were thundering vocalists with stentorian styles useful for both vaudeville stages and acoustic recording studios. Their vocal deliveries were punctuated by earthy humor, counterbalanced with a vulnerability given voice through melancholy lyrics. While Tucker and Smith had marked differences, there were enough similarities that allowed lyricist Bradford to successfully make the pitch. OKeh invited Smith back for an August 1920 session, in which she recorded another Perry Bradford composition, "Crazy Blues," backed with Marion Dickerson and Alex Balledna's "It's Right Here for You."[19]

Smith's "Crazy Blues" OKeh session came during the aforementioned case in which Gennett challenged Victor and Columbia's proprietary lateral-cut recording format agreement, and an emboldened OKeh ensured greater accessibility by releasing the sides in the lateral-cut record format.[20] Bradford's convincing Fred Hager to record with Smith opened up "the potential for both a Black urban and a white southern market for 'blues' records for Black female vaudevillians," as William Howland Kenney notes.[21] Bradford based his hunch on a larger audience for blues on seeing whites give change to black street musicians in the South: "Bradford's testimony suggests that he thought of what came to be called classic blues records, made by Black female entertainers, was designed at least in part for whites."[22]

"I can't sleep at night / I can't eat a bite," she sings over the Jazz Hounds, Smith's backup band, assembled by Bradford for the session. The song is a story of a woman who remains in a love with a man that treats her horribly: "There's been a change in the ocean, a change in the deep blue sea," Smith belts out the song, nearly shouting her lament, soaring up over the band: "But I tell you folks out there, ain't no change in me," followed by "My love for that man will always be." In the last lines of the song, the desperation grows in a few lines from Bradford's original composition that are left out of later cover versions. After a verse in which the song's subject flirts with suicide by laying her head on some train tracks ("I gladly snatched it back"), she seeks other kinds of relief as her confusion and pain shifts focus from internal to external targets: "I'm gonna do like a Chinaman and go and get some hop/Get myself a gun and shoot a cop." Smith's brash lyrics make it all the more intriguing (and perhaps telling) that it was white-owned OKeh that released the recording and not an African American–owned label set on racial uplift in a racist society. Regardless of OKeh's commercial reasoning in the chance they were taking, the societal response was largely positive.

The black press was quick to comment on Smith's recordings. Tony Langston of the *Chicago Defender* wrote, "Well, you've heard the famous stars of the white race chirping their stuff on the different makes of phonograph records. . . . But we have never—up to now—been able to hear one of our own ladies deliver the canned goods. Now we have the pleasure of being able to say that at least they have recognized the fact that we are here for their service."[23] Two months after the release of "Crazy Blues," an advertisement for a Chicago music store owned by pianist Clarence Williams touted Smith as "Our Race Artist."[24] From the first OKeh *Chicago Defender* advertisement for Mamie Smith's "That Thing Called Love," a mail order option offered the recording at its regular price with an additional "17 cents in stamps," which was attractive to both record companies and rural consumers with no easy access to recordings.[25]

Sales of Mamie Smith's "Crazy Blues" were impressive by all accounts. While specific sales figures from this era are largely unreliable, there is no question of the record's general success. The record is said to have sold a million copies within the first year of its release. Ronald Foreman reported that "Crazy Blues" sold seventy-five thousand copies in Harlem during the first few weeks of its release.[26] Public appearances by Smith during the fall and winter of 1920–21 led to further sales. Smith was the first African American woman to appear at evangelist Billy Sunday's Tabernacle, where she played before big audiences, including eleven thousand attendees at a stop in Tidewater, Virginia.[27] Meanwhile, OKeh Records' parent company, General Phonograph, launched an aggressive advertising campaign in the recording industry trade press to make sure that phonograph dealers would continue to stock "Crazy Blues." A sampling of advertisements in *Talking Machine World* late that year illustrates both the importance of the record and the shaping of a race records vocabulary.

The amount of advertising focused on Mamie Smith's "Crazy Blues" in the U.S. recording industry's flagship trade paper suggests the recording's unusual popularity. In the October 15, 1920, issue of *Talking Machine World*, two ads touting Mamie Smith appear. One ad features Smith with the latest OKeh roster of releases, featuring photographs of prolific pop singer Henry Burr and others. The same issue features another advertisement focused solely on Mamie Smith and "Crazy Blues." Ads continued in the November issue, including an advertisement from a record distributor alerting retailers to the availability of Mamie Smith's hit, addressing concerns by some retailers unable to reorder the title elsewhere. The November 15 issue also featured another Mamie Smith advertisement from OKeh Records, underlining the popularity of "Crazy Blues" via "Mr. Public Opinion."

The "Mr. Public Opinion" ad embodies many of the racist notions that guided white-owned companies selling music by African American artists from the earliest days of the recording industry. The ad sings the praise of Smith's accomplishments, while imagining its potential audience with a figure out of the minstrel tradition, an early example of contradictions that riddled the industry as race records emerged.[28] It is likely that OKeh and *Talking Machine World* understood these not as direct appeals to black audiences, but to *Talking Machine World*'s imagined majority white retailer readership. In the December issue of *Talking Machine World*, readers saw yet another full-page advertisement focused on Smith, similar to the ad in the October issue. The ad copy described her as a "Singer of 'blues,' the music of so new a flavor," adding that "to hear is to buy."[29]

OKeh advertisements in the pages of the African American press made direct appeals for listeners. If it was a victory for more black entertainers to record for white companies, it was also noteworthy for more white companies to advertise in black newspapers. In his book of autobiographical sketches, Perry Bradford wrote, "Anyone working for the *Chicago Defender* looking back to 1921 would see the time when OKeh Records was the first company to give an ad to our Negro newspapers." OKeh took out advertisements in other black newspapers as well, including the *Norfolk Journal and Guide*, the *Baltimore Afro-American*, the *Pittsburgh Courier*, and Indianapolis's *Freeman*, known nationally for its coverage of the black entertainment circuit.[30] OKeh's marked presence in the black press was important in an

Fig. 2. OKeh advertisement for Mamie Smith's "Crazy Blues": "To Hear is to Buy," *Talking Machine World*, December, 1920. (Library of Congress.)

industry that had ignored the breadth of African American music for decades (while financially benefiting from the minstrelsy tradition). If industry lore is to be believed, the pride black consumers felt about Mamie Smith's hit record led those *without their own phonographs* to pay the then-exorbitant $1.00 list price, as the recording became a symbol of pride, progress and possibility. "Among the masses buying them [the records]," David Wondrich writes, "were a goodly number of black people; blues collectors used to turn them up as far from Broadway as Vicksburg and the Delta."[31]

In 1920, *Variety* replaced sheet music sales with record sales as the best measure of a song's popularity.[32] Phonographs were increasingly commonplace in U.S. homes, as the industry sold more tabletop models, including Victor's budget-priced Victrola IV and VI. Mail order sales meant that even if you lived nowhere near shops selling records, those records were available to you. If your favorite performer wasn't playing nearby, you could buy a record. As recordings spread through the country, a key dynamic of phonography became evident. In the year after Mamie Smith's first OKeh recordings, she had become aware of the effect of recordings on audience expectations, as evident in this quote from the January 29, 1921, issue of the Norfolk, Virginia, *Journal and Guide*: "[These] thousands of people who come to hear me in my concerts expect much, and I do not intend that they shall be disappointed. They have heard my phonograph records and they want to hear me sing these songs the same as I do in my own studio in New York."[33] Reviews of Mamie Smith's live performances made comparisons to her phonograph recordings, and diffusion of her recordings also began to make Smith's releases synonymous with race recordings. OKeh's success led other labels to follow the trend.

In a context in which the labels knew little about the music they were now recording, the race records market opened up further opportunities for blacks working in music. As Russell Sanjek noted, "Black musicians, songwriters, and publishers played a key role in the production of race records, from signing performers to the final selection of material to be recorded."[34] But because white-owned companies have always controlled the recording industry, black performers, songwriters, and talent scouts always had to negotiate a situation in which they were both essential and marginal. "African Americans had either to gain a measure of control of the process of record production or be colonized in wax," writes William Howland Kenney. "Race records therefore present a dilemma: widely celebrated for preserving and spreading a taste for the blues and gospel music, we can never know how different they might have sounded if Blacks had been able to control their production."[35] Kenney's crucial if unanswerable "what if?" points to ubiquitous racism's impact, but it should not obscure instances in which a measured increase in cultural autonomy was available. But even in those instances, black-owned labels understood their releases in relation to the prejudicial attitudes of the white majority and negotiated their own cultural positions accordingly. The most significant experiment during the 1920s came from New York's Black Swan.

Music from Strivers' Row: The Intervention of Black Swan Records

Alongside OKeh's successes, the desire to represent African American music on its own terms continued with Harry Pace's Black Swan Records, founded in 1921. While Pace was not the only African American entrepreneur active in the early 1920s, his initiative was the most ambitious and well-known venture. Black Swan's short life is useful to consider, not only due to the unique societal difficulties the label faced, which included both the forceful reassertion of racist and xenophobic currents in U.S. society. Black Swan also had to negotiate debates over defining "the new Negro" and how African American cultural production should be represented in larger society. This debate, in turn, highlighted significant class divides within the black community.

Black Swan's abbreviated, meaningful history represented a company that wished to present the breadth of black music. If the industry's modes of production were opaque—or "invisible"—to consumers, as David Suisman argues, then Black Swan's arrival represented a "radical attempt to confront, challenge and disrupt the invisibility of the modern music industry."[36] Black Swan challenged common recording industry assumptions about African American culture through "defetishizing the music and fashioning it instead into an instrument of social and economic justice."[37] It was while Pace was a partner with composer W. C. Handy ("Father of the Blues") in a music publishing business that he became frustrated with the practice of white recording companies buying the rights to black songs solely for white artists. Black Swan's intent was to release recordings that featured black performers only. The label was named after nineteenth-century black vocalist Elizabeth Taylor Greenfield, the "Black Swan." The new company leveraged prestige through its board of directors, which included Pace's Atlanta University mentor, sociologist and civil rights activist W. E. B. DuBois, who by this time was in New York as editor-in-chief of the NAACP's *The Crisis* magazine and a key figure in the Harlem Renaissance.[38] In the March 22 issue of *The Crisis*, a notice on Black Swan's activities and a monthly income of $12,000 a month, while the *New York Age* noted dire annual results for 1921, citing "a profit of slightly over $3,300 on sales of $104,628.74 in 1921").[39] The article in *The Crisis* was coupled with a map, illustrating the "Distribution of Black Swan Phonograph Records," that noted locations of wholesale distributors and jobbers, retail dealers, and "agents" in every corner of the country, though those agents were, as Allan Sutton notes, "independent salespeople who peddled records door-to-door, on street corners, or wherever else they could."[40] Given that Black Swan had to fight to build a business infrastructure within a white-owned industry with only limited support at best, creating even this network was and is impressive.

While we usually think of the Harlem Renaissance in terms of literature and the visual arts, Pace's intervention in the recording industry might be seen alongside Harlem's black cultural producers as part of DuBois's "talented tenth," a phrase that inadvertently underlined class fissures within the black community. Black Swan and Pace Phonograph's office location on 257 West 138th Street planted their operations firmly in Harlem, as did Pace's connections with African American artists and intellectuals. A stretch of townhouses between 138th and 139th Street was known as "Strivers' Row," an area whose residents included musical notables such as Eubie Blake, Fletcher Henderson, and Pace's former business partner, W. C. Handy. Lacking their own manufacturing plant, Black Swan had difficulty finding a white-owned company to press their first records. It was Paramount Records (and its parent company, Wisconsin Chair) that agreed to manufacture Black Swan discs. Paramount had gotten into the recording business after earlier successes making cabinets for Edison's phonographs (a *New York Clipper* notice also mentioned the Fletcher Record Company of Long Island City, New York, was also manufacturing Black Swan discs as of February 1923).[41] Pace's business acumen impressed Paramount's Art Satherley, who still felt it necessary in a 1971 interview to defend his business associate of years ago: "Now then, you might say to me, 'Were there anybody else in this business that you knew that knew something about this business,' and I'll say yes. And he was a black gentleman by the name of Harry C. Pace."[42]

Black Swan's connection to Harlem society may have also influenced its initial recorded output, which veered away from music deemed potentially offensive. Pace's concern about the content and future of race records, specifically blues, was reflected in comments he made in the August 1921 *Talking Machine World*: "While it is true that we will feature to a great extent 'blue' numbers . . . we will also release many numbers of a higher standard."[43] While Pace's goal was to address the slights committed against previous African American performers, Black Swan defined its race records in broad terms, promising to feature "our best singers and high class musicians [who] have had no recognition from the large white companies who furnish all the records that are supplied."[44] The label's first recordings featured "a conservative mix of concert songs, sacred material, and vaudeville-style blues," guided by Pace, Fletcher Henderson as recording director and chief piano accompanist, and classical composer William Grant Still as musical director.[45] An ad in *The Crisis* is reminiscent of an earlier appeal from the Broome label. "The Phonograph Records made by white companies are preserving only the type of Negro voice that sings comic and rag-time," the ad copy read. "If every Colored Phonograph Owner bought as many BLACK SWAN RECORDS as he buys of other kinds he would make possible a reproduction of Negro Music and Voices similar to Caruso's," as the ad asks readers to compare their operatic recordings with the master.[46] Black Swan's self-consciousness led to record selections designed not only to highlight individual performers, or even the company, but an entire race.

The label's initial intervention had an unintended outcome: limited sales of mannered recordings that did not attract black consumers in great numbers. It soon became clear to Pace and others that blues and jazz recordings represented the biggest untapped market. Black Swan's ability to support various styles gave way to a release schedule focused on recordings most likely to sell briskly. Black Swan's first success came with vaudeville veteran Ethel Waters, then known as "Sweet Mama Stringbean" on the Theatre Owners Booking Association, or T.O.B.A., black vaudeville circuit. Waters followed this success with two March 1922 Black Swan recordings, "Oh Daddy" and "Down Home Blues." These recordings, along with those of Alberta Hunter and Trixie Smith, kept those at Black Swan optimistic about the label's future.

But Black Swan's increasing embrace of blues and jazz styles may have also cost it support among middle-class black listeners. A growing black middle class came with urbanization and industrialization, and educational opportunities followed with greater resources (a quarter of all African Americans lived in urban areas in 1910, a third did by 1920, and two-fifths did by decade's end).[47] In an essay in which he outlines the institutional prejudices and economic injustices black artists faced in the recording industry, Norman Kelley is careful to point out the historical reticence of the black middle class to embrace the cultural breadth and importance of black popular music. Kelley cites Amiri Baraka's suggestion that Harry Pace was "castigated by the black middle class for not selling music that was more *racially uplifting*, meaning *not* sounding like music by lower-class Negroes."[48] He also critiques Harlem Renaissance writers (with the notable exception of Langston Hughes) who undervalued new forms of black recorded music, noting that while they understood music as "descriptive of Harlem life" and saw jazz as "the definition of the new Negro," they saw black musical idioms as "folk art—like the spirituals and the dance—the unrefined source for the new art," raw material for a new black bourgeois aesthetic.[49] Kelley argues that the 1920s black intelligentsia "could no more accept the reality of its own folk culture than the white intelligentsia could accept the black basis of American culture."[50] As for white recording companies, it would not be Black Swan's experiments but the success of vaudeville blues performer Mamie Smith's OKeh recordings that would lead them into recording a broader swath of African American culture, the codification of race records as a genre, and a recording industry in transition.

"I Saw This Really Was a Business": Expansion, Diversification, Advertising

After OKeh's initial success, the label followed with race records by blues vocalists Sara Martin, Eva Taylor, and others. The label used Mamie Smith's 1921 follow-up, "Sax-o-phony Blues," as a chance to highlight its race records initiative in both African American newspapers and the trade press. An October 15 *Talking Machine World* article noted the launch of "an unusual publicity campaign featuring the many exclusive OKeh colored artists. As a pioneer in this important field, the General Phonograph Company has achieved phenomenal success and the present campaign is meeting with the hearty co-operation of OKeh jobbers and dealers." A proof of the full-page ad and letter from advertising director J. A. Sieber was sent to OKeh dealers coast to coast.[51]

Other companies quickly released recordings by black female vocalists, many of whom were veterans of the T.O.B.A. circuit. Lucille Hegamin released her first record for the Arto label shortly after Mamie Smith's breakthrough; Lavinia Turner recorded for Pathé; Lillyn Brown recorded for Emerson; Daisy Martin recorded for Gennett; and Ethel Waters recorded for Vocalion and Black Swan. Columbia entered the fray with Mary Stafford, who covered both Mamie Smith's hit and Lucille Hegamin's "Arkansas Blues." Edison was late to respond, perhaps due to Thomas Edison's well-known dislike for jazz and blues. The company did eventually release sides by Helen Baxter, Rosa Henderson, Josie Miles, and others.

Race releases were separated from the rest of a given company's catalog in discrete numerical series of their own, as had been previously done with non-English-language releases for the immigrant market. OKeh recording director Ralph Peer remembered, "I saw that this was really a business, like our foreign record business."[52] While this allowed companies a practical means to highlight race to specific audiences, it inadvertently underlined societal tensions surrounding the music. The urge to separate race series from the rest of the catalog and Peer's comparison to their "foreign record" business suggest how separate white and black culture, especially in the southern United States, was understood to be. As Karl Hagstrom Miller suggests, U.S. recording companies were ready to see rural Americans (particularly those in the South) as potential consumers, in the same way companies had sold discrete lines to international and immigrant consumers.[53] Lizabeth Cohen discounts the notion that the phonograph (and later radio) meant that immigrant listeners were necessarily "sharing more in American, middle-class culture every time they rolled up the rug and danced to the Paul Whiteman Orchestra," instead emphasizing the ways in which ethnic recordings helped them find a group identity via their immigrant experiences.[54] Cohen makes an important distinction between those immigrant communities buying recordings from their home countries and those African Americans consuming race records: "Southerners, black and white. . . . eased the trauma of moving north to cities like Chicago by supporting a record industry of hillbilly and 'race records' geared specifically toward a northern urban market with southern roots."[55] Cohen's last claim of recording companies' focus on a northern market of transplanted southerners in the 1920s is highly debatable; counterevidence exists in scouts' accounts of their labor, the labels' promotional materials, and anecdotal accounts from generations of collectors and where they located extant recordings on their the mid-twentieth-century record searches. But Cohen's larger argument here is important, if initially counterintuitive, that blacks and whites both felt trauma—albeit in markedly different forms, as blacks were fleeing the Jim Crow South—in their migration north. Race records, like old-time music, was the music of homesickness and dislocation as rural blacks and whites made new homes in the industrial Midwest, a place already shaped by institutional racism and classism.

OKeh had started the trend toward separate race catalogs with their

8000 series, while Paramount followed suit with their 12000 series in 1921. Vocalion, under the ownership of Brunswick-Balke-Collender, issued its first race recordings in 1926, promising "better and cleaner" race records, a comment more likely to have addressed concerns over lyrical content than recording quality. Pathé-Perfect started their race series in 1925, while Brunswick launched its 7000 series in 1926. Columbia was the first of the industry leaders to join in with their 14000D series in 1922. They found early success with one of the biggest race artists of the 1920s, Bessie Smith. Smith was discovered by Columbia talent scout Frank Walker and had learned much about singing from her friendship with Ma Rainey. Walker suggested she cover already popular vaudeville blues songs of the period, as was the regular industry custom (she did "Downhome Blues," already a hit for Paramount's Alberta Hunter, and Lucille Hegamin's "Aggravatin' Papa" and "Beale St. Mamma").[56] "With seeming effortlessness," Daphne Harrison and John Cowley write in their essay on vaudeville blues, Smith "sails through any song, blues or not, with a special bluesy quality, leaving a wake that swamps any singer who attempts to imitate her artistry."[57] Having been brought to Columbia by Frank Walker, Smith is said to have saved Columbia Records in industry lore, though the label's rebound was gradual. Smith began recording for the label in February 1923, and the company went into receivership in October of that same year, followed by radio's increasing competitive edge beginning in 1924–25.

The label that approached the new trend with the most trepidation was Victor, who as industry leader could afford the hesitation. The company's "Red Seal" division had a reputation for releasing the best in high-culture recordings, so perhaps the company believed that a foray into blues would tarnish the company's image.[58] The company was slow to enter the field, as Perry Bradford acknowledged, because they "just couldn't afford to lower their prestige."[59] As Allan Sutton notes, they instead chose to feed "blued-inflected tunes to white performers, who often treated them as comic novelties" with exaggerated, racist accents.[60] But Victor could not afford to ignore the growing race records field any longer. Though Victor had previously conducted in-house experiments regarding race records, they began relatively quietly with ads in the African American press, collaborating with manager, songwriter, music publisher, and owner of Triangle Music Publishing Company, Joe Davis—who could supply singers with music and arrangements already prepared—for a number of Victor sides by Rosa Henderson, Lena Wilson, Edna Hicks, and Lizzie Miles.[61] An August 4, 1924, advertisement in the *Chicago Defender* announced "New Victor Records: A

Special List of 'Blues,'" along with a brief announcement that appeared in the September 1923 issue of *Talking Machine World* listing more popular-dance-oriented releases by Arthur Gibbs & His Gang, some Tin Pan Alley numbers from Noble Sissle and Eubie Blake (who also contributed numbers culled from 1921's *Shuffle Along*), and a spoken-word comedic skit from Arthur Moss and Ed Frye.[62] And later in the decade, with the resignation of Victor cofounder Eldridge Johnson in 1927 and the sale of his substantial interests in the company to New York banking interests (Speyer & Company and J. & W. Seligman), the label expanded into previously underexplored genres, including race records, with its V38500 series that year.[63] Victor's formidable publicity apparatus was a benefit to the series, as Victor retail order blanks and advance lists for June and July 1928 announced "Never Better!" "Getting Bigger," "True-Blues," and "Here They Are!" while publicizing new releases.[64] The releases themselves represent a mixture of jazz tunes (Duke Ellington's "Harlem River Quiver" and Jelly Roll Morton's "Jungle Blues"), sacred material ("I Shall Not Be Moved" by the Mount Zion Baptist Quartette), even jug band music ("Bob Lee Junior Blues" by the Memphis Jug Band), as well as what would come to be known as country blues ("Saturday Blues" by Ishmon Bracey, misspelled "Ishman" by Victor).[65]

The race records market grew through the 1920s as companies increasingly released the music in its many variants, including country and vaudeville blues, as well as gospel and hot jazz. By mid-decade, instrumental jazz tunes—once ubiquitous—were now advertised as "instrumental blues" to cash in on the ever-growing market.[66] Jeff Todd Titon estimates that the number of race recordings grew from roughly 50 in 1921 to 250 in 1925, then to 500 in 1927.[67] While Victor catalogs had gone as far as juxtaposing African American musicians with classical composers and musicians such as Leopold Stokowski and Pablo Casals, Victor's race record catalogs commonly featured racial stereotypes based in minstrelsy. But Victor's entry into the race record market was a capitulation of sorts, a recognition of the malleable genre's importance, another sign that the niche category had come into its own.

By mid-decade, the race records category had expanded further to include "sacred," "sanctified," or "gospel" preachers, choirs, and musical quartets performing on record. White and black recordings of sacred music and preaching dated back to the early years of the industry. The expansion of gospel music and spoken-word sermons via phonograph recordings was another sign for white-owned recording companies of the breadth and depth of the African American market. It transformed black Christianity in the process. Lerone Martin has argued that "phonograph religion" commodified black Christianity, laid the foundation for later religious broadcasting via radio and television, and created celebrity preachers, thus changing the nature and tone of "religious authority."[68] Both OKeh and Paramount were active in sacred recordings early on, as was Columbia, a company trying to regain its strength after economic difficulties early in the decade. Paramount had the Norfolk Jubilee Quartet, while Columbia was successful with the Birmingham Jubilee Singers. "The history of gospel music," writes Irene Jackson-Brown, "is a history of the struggle to be accepted as *the* black liturgical music," a struggle to find a place in traditional Christian liturgy that lasted through the 1920s and 1930s until the form became dominant in black churches during the 1940s.[69] Commercial sacred recordings certainly played a role in this transformation. An important early discovery by Columbia was Reverend J. M. Gates, an Atlanta-area preacher discovered by talent scout Frank Walker on a trip to the city. Gates's Columbia recordings included a mixture of sermon and song. Robert M. W. Dixon and John Godrich describe "Death's Black Train Is Coming," an immediate success when Columbia released it in 1926: "The singing had a compelling rhythm; the song concerned a metaphor popular among black Christians, and the Columbia engineers had arranged for a train whistle to sound several times in the background."[70] Gates's subsequent releases were so successful that he recorded new releases a breakneck pace; forty of the sixty Columbia gospel records released between September 1926 and June 1927 were by Reverend Gates.[71] Gospel music, as a subgenre of the larger race market, influenced and was influenced by nonsecular blues and jazz styles (as the lines between these styles regularly blurred). The subgenre stands as an example of the industry's increased depth and diversification during this period of expansion.

The role of advertising was crucial to these emerging labels' sense of their new audiences, how these audiences were identified and reached, and the distribution and diffusion of race records around the nation. Illustrators, copywriters, and photographers were found in a variety of ways. Freelancers, advertising agencies, or in-house publicity departments manufactured advertisements for periodicals and promotional materials for store displays.[72] Newspaper advertising was an important facet of most labels' budgets, and companies would either create their own advertisements for publication or would partner with a local company. In addition to advertising in both African American and trade phonograph newspapers, point-of-sale advertising was an important component for information about specific artists, labels, or genres. At a music shop, Titon writes, a prospective customer might have seen a myriad of approaches to attract attention: "banners, streamers, posters, photographs, catalogs, supplements, handbills, and of course, records." Window displays were designed to catch the eye of a potential consumer, and once a customer was in the store, the salesman was more than happy to play a demonstration copy of the latest record or even sell a customer a new machine. As competition intensified, labels looked for ever more innovative ways to advertise in other social spaces, big and small: "Lantern slides projected in theaters announced the latest discs; record sleeves presented miniature catalogs and sometimes provided order blanks."[73]

White companies based in the urban North had more experience in selling music to immigrant groups in U.S. cities than to blacks. While music critics and historians have rightly emphasized the racial segregation, violence, and class discrimination of the Jim Crow South, Patrick Huber emphasizes the ways in which these conditions were an *everyday* structuring reality that people negotiated given their societal places and positions. Recording company scouts were "motivated primarily by practical and commercial considerations" and reified divisions by "dividing race and hillbilly into special series [that] allowed talking machine corporations to target specialized markets."[74] Huber's account highlights how the routinized structural and institutional racism surrounding the music both dictated and arose out of the realities of the day; scouts routinely worked within the everyday realities of Southern segregation and discrimination in formulating race records as a genre. The two cannot be separated. Businesslike negotiations of racial and class distinctions were a part of what we might call Jim Crow retail, which meant segregated retail opportunities and shopping experiences, as well as different opportunities available to those who were discriminated against in the South's racial and class divide. Lizabeth Cohen has usefully mapped the rise of mass consumption in the 1920s, "the production, distribution and purchase of standardized, brand-name goods aimed at the possible buying public" that transformed everyday lives, as manufacturers targeted an increasingly broad group of U.S. consumers.[75] The talent scouts' work in the Jim Crow South illustrates the deeply uneven development of consumerism during the some of the most violent years of the Jim Crow era, and the work of talent scouts in this era offers another site of such activity.[76] Scouts negotiated Anglo-American reticence to cater to African Americans (and also catered to segregated retail in the form of black-owned businesses within some African American enclaves, though mail order advertisements in African American newspapers were a more common means of reaching black consumers). C Vann Woodward notes that "much ingenuity and effort went into the separation of the races in their amusements, diversions, recreations and sports," which required quite specific brick-and-mortar accommodations in tightly controlled public spaces.[77]

When possible, white recording directors and company owners relied on various cultural intermediaries, including retail record dealers, storeowners, and talent scouts, such as Paramount's Mayo Williams, regarding placement and content of their advertising campaigns. But white companies regularly had little sensitivity, as evident in varying rhetorical strategies employed in company advertisements. It is tempting to see the varying representations of African Americans as a gauge of a given company's understanding of its audience, though it would be an oversimplification. It is no accident that Black Swan's depictions were the least sensational and the most humanizing. Other labels' advertisements ranged from ill-informed and insensitive to vicious. Titon writes, "The advertisements reveal white uncertainties about the increasing visibility of black people in the urban North, and by extension in the life of the nation."[78] OKeh Records, from early- to mid-decade, had one of the most high-profile campaigns in the business, due in no small part to the early successes of its race series.

OKeh was necessarily at the forefront of advertising the race records genre category, as extant OKeh catalogs between 1924 and 1931 suggest. If the advertisements featured derogatory depictions of African American performers and its presumed African American audience, the catalogs were usually worse. The catalogs illustrate some key rhetorical shifts in terms of artist and record presentation, in what was a deeply prejudicial and often violent era for African Americans. OKeh's confusion and ambivalence in reaching out to black audiences is made clear in illustrations often mired in minstrel imagery.

While such images may have appealed to some ignorant or prejudicial white consumers, the label's advertising to blacks was most effective when detailing the talents of specific artists. OKeh's race records catalog supplement from June 1924 announced "Here They Are!" and featured Virginia Liston on the cover, noting the label's long-standing interest and faith in female blues singers in the vaudeville blues style.[79] All three acts highlighted in the catalog—Liston, Sara Martin, and the duo of Effie and

Charles Tryus—were listed as "exclusive OKeh artists," but the term seems to have had little meaning, since artists came and went according to their fortunes, while often recording for other companies under pseudonyms. By 1924, Mamie Smith had been overshadowed by Columbia's powerhouse Bessie Smith (no relation), and whatever exclusivity OKeh had asserted on Mamie Smith was gone. Smith had been picked up by Chicago's short-lived Ajax label ("the Superior Race Record"), where her most famous labelmate was former heavyweight boxing champion Jack Johnson. Her recordings reportedly sold poorly at Ajax, though her name was powerful enough that she later moved on to Victor, then back to OKeh (purchased by Columbia in November 1926).

Given the phenomenal success of jazz as dance music in the early 1920s, it's not surprising that several early OKeh catalog covers feature dancing couples. On the cover of one of OKeh's first "Blue Book of Blues" catalogs in 1925, the figures reference nineteenth-century "golliwog" rag dolls as much as people, with indeterminate racial identities made clearer by the racialized context. In another 1925 OKeh race catalog—the largest listing of any catalogs discussed here—a dancing couple is depicted tripping the light fantastic, well-to-do figures surrounded by jazz-age glitz.[80] The following year's catalog featured a less determinate group of characters; the illustration consists of a woman dancing, flower in her hair, similarly well-heeled and fashionable as in previous catalog images, though more generalized in terms of racial identity. Unlike the couples of previous years, here she appears alone, or, more accurately, in a direct relationship with the source of the music itself, the band. Behind her, an inset image depicts a band, also of indeterminate racial identity. A viewer could just as easily understand the figures to be black or white, which might have benefited the company in its desire to appeal to multiple audiences. But in a series of 1926 OKeh catalogs, references to the minstrel tradition were explicit, in keeping with the practice of using exaggerated racial stereotypes to sell records to white audiences, despite the label's significant forays into courting black audiences by 1926.[81] One "OKeh Truetone Race Records" catalog cover depicted a racist caricature in a top hat, earrings, and bowtie. It embodied Eric Lott's description of minstrelsy as a "spectacle of vulgarity," portraying a figure of white disgust and desire for an abject other, a product of minstrelsy whose "fundamental outcome was to secure the position of white spectators as superior, controlling figures."[82]

That the figure is depicted in the intimate bodily act of eating should not go unnoticed, as food has been an important site in defining racial difference. The figure is depicted eating what appear to be chitterlings, a choice that makes sense in terms of the image's intent. The cultural connotations of chitterlings in the U.S. context make it a deeply racialized food (despite its history as a food of the global poor). It is paradoxically used to attack African American traditions, despite a culinary lineage in which slave owners fed slaves by the cheapest means possible.[83] The following year's catalog featured the same illustrative style, a figure of similarly exaggerated features, but here a banjo being played replaces food as racist signifier.

Inside the OKeh catalogs, the artist depictions had also shifted.[84] In the earlier catalogs, the image-text relationship was much tighter, with text describing the music and the performer's image alongside. But in these catalogs, only a few sentences directed the reader back to the image. Instead of reverent depiction of blues performers common in earlier catalogs, these depictions opted for affect-laden hyperbole and dialect in place of artist testimonials. At times, the artist's image was replaced with an illustration based on the lyrical contents of a given song. Instead of an image of Victoria Spivey to publicize her release "Dirty Woman's Blues," readers instead saw an image of man on his knees, begging his "dirty woman" to come home as she passes by on the arm of another man. The accompanying text reads in part: "G'wan, Spivey, you tell 'em, girl, you're wicked with them shake-down Blues. Be Yourself! We mean you've got low-down patter in a lazy creepy drawl."[85]

While the transition in OKeh materials is explicit, this dynamic was not limited to the label. As Daphne Harrison writes, "The big labels such as OKeh, Columbia, and Paramount shifted the nature of their ads from dignified, attractive photos of the featured singer to caricatures with pickaninnies, big-mouthed 'Sapphires,' men with bulging eyes and oversized lips, and heavy dialect."[86] In some ways, these images seem like a return of the (not so) repressed. This negative shift wasn't the only one that occurred between traditional minstrel songs and the emerging race records genre. Whereas the lyrics of late nineteenth- and early twentieth-century "coon songs" seemed to point back at the blackface performers as either childlike innocents or malcontents, race records allowed performers to point the finger outward, if only to figures in black society. As Tony Russell writes, "the anti-black sentiments of these early songs are generally redirected against sections of the black public—preachers, wife-stealers, fast-life women, and so forth—whereas the originals poked fun, sometimes good natured and sometimes not, at black people in general."[87] There is little improvement in negative depictions shifting from generalized racist lyrics to more targeted complaints at "sections of the black public," even though it allowed more autonomy for performers. While white derision may not have found its way to specific performers as easily, this shift allowed white fans to laud the performer but retain larger racial prejudices. Increasingly, in advertisements and promotional material, the caricatures were not usually of the performer, but increasingly based on the lyrical content of the song.

Not surprisingly, the most respectful representations were courtesy of Black Swan. It stood apart in its edifying, mannered approach, which was "characteristically genteel."[88] Pace and those at Black Swan had an acute understanding of the stakes of their intervention, as they fought against both prevailing societal stereotypes as well as common notions about the blues as an uncultured, even dangerous music. Not surprisingly, controversies focused on blues lyrics, which regularly depicted catastrophic events, hardships, betrayals, murders, abandonment, and addiction, a musical analog to tabloid journalism. In that sense, it was not that different from the wreck songs and murder ballads depicted in the rural white music emerging on record shortly after race records' initial rise.

Competition between labels for female vaudeville blues singers created both an aesthetic glut in the market and an increased interest in recording blues variants. "As early as 1923," Titon suggested, "Black Swan and OKeh thought there might be a market for blues less sophisticated than the vaudeville variety. Black Swan's advertisement for Josie Miles' 'Love Me in Your Old Time Way' sought to expand then accepted assumptions about blues music, noting just "how strongly contrasted are these songs springing from the depths of the laborer's soul to the commonplace dance tunes that we are accustomed to call BLUES. LOVE ME IN YOUR OLD TIME WAY is in the vein of the laborer's songs."[89] It was rural blues, in all its incarnations and regional styles, that would extend the genre's popularity. Indeed, while the vaudeville blues retained its popularity through the mid-1920s, the country blues would serve to redefine the style in the years leading up to the Depression, as well as becoming the paradigmatic blues style decades later, when postwar blues enthusiasts rediscovered the records (and in some cases the living performers) in the 1950s and 1960s. Again it was labels in search of their own market niche that turned to rural blues. Chief among these labels was a small Wisconsin company called Paramount.

The Unusual Thing Occurs: Paramount Records and Redefining Race Records

The great popularity of vaudeville blues expanded race records as both a genre and a marketing strategy, thus transforming the recording industry as record companies delved further into a deeply flawed effort to attract black listeners, however awkward or insulting the means of attraction. Despite strong biases and prejudices, the recording industry's entrepreneurial upstarts continued to experiment with largely unrecorded black musical styles. Of these styles, "country blues" became the most popular, as vaudeville blues was assimilated into the larger industry's pop sensibilities. Country blues was quite literally rural music, most often performed by a lone male singer accompanied by a guitar, or sometimes a piano or banjo. But not all singers accompanying themselves were playing country blues. The name for both blues variants—vaudeville or "classic" blues and country blues—came later. Stephen Calt traces the term "country blues" to jazz author Charles Edward Smith's liner notes to a 1957 Folkways release, *Big Bill Broonzy Sings the Blues*.[90] All genres and subgenres—contemporaneous or otherwise—do categorical violence to the music being made and the musicians making it. This was as true then as it is now. During the era, all blues variants—as well as music we would identify under a plethora of other categories—were most often categorized and understood as race records. Paramount's race catalog grew to represent a vast number of musical approaches, from string bands to jug bands, gospel and secular and variants in between. The aesthetic territory housed under the "race records" banner was vast.

The emergence and expansion of the race records market led to participation that exceeded the intentions of both the African American press and short-lived black-owned labels. The expansion of the race records market led in turn to new opportunities as well as tensions and to unlikely collaborations in unlikely locales. The industry's lack of knowledge about country blues (and rural music in general) is reflected in the industry's increased reliance on outside experts who knew more about the music and the people who made it than most record company executives knew or cared to know. While Paramount's foray into country blues was part of a market-driven industry expansion, it helped to create a vast collection of recorded vernacular culture. And while the racist attitudes of the era were evident throughout the period, the recordings' breadth and depth helped contribute to the centrality of African American culture in U.S. life, even while those who made the music were denied full citizenship in an era known for virulent racial violence.

For scouts at white-owned labels seeking out new market niches, this work meant crossing any number of spoken and unspoken societal boundaries. Expertise was needed to discern who and what to record, which audiences might be interested, what parts of the country might be tested to target. While executives at labels like Paramount had little idea about given audiences, they responded to strong sales and they hired people who had (or could develop) the necessary skills. Those with much-needed outside expertise were company talent scouts, traveling salesmen who convinced retailers to carry their records and spent significant time learning musical tastes, down to regional and even local areas. They acquired performers to record for their labels, they explored and helped manage audience tastes, and they dealt with various social differences in the course of everyday work.[91] In each of these interrelated roles, these scouts were key in the creation of nascent industry genres as effective but pliable ways in which music circulated through society. Through creating a niche via country blues recordings, Wisconsin's Paramount Records emerged from the fringes of the recording industry and took the lead, benefiting from logistical and societal biases that had led to the neglect of a largely unrecorded music (country, or downhome blues) to an underserved audience (rural blacks).

Though it may have had less national impact in the 1920s, Paramount's country blues recordings—sold as "race records" alongside vaudeville blues predecessors—helped define the genre in the second half of the decade as much as OKeh had in the early part of the decade. The label's headquarters in Port Washington, Wisconsin, a small, white enclave north of Milwaukee on Lake Michigan's western coast, made Paramount as unlikely a progenitor of blues as Gennett had been for jazz.[92] Like many other emerging labels founded in the 1910s, its parent company had first entered the record business through manufacturing phonograph cabinets, then phonograph machines, and finally records to play on them. Paramount's parent company was the Wisconsin Chair Company (WCC), founded in Port Washington in 1888. WCC's initial recording division was founded in 1916 as the United Phonographs Company, which made Vista and Colonial phonographs and sold recordings under the Puritan imprint.[93] The advent of United Phonographs is likely connected to Wisconsin Chair's previous dealings with Thomas Edison, who employed WCC (via a Sheboygan, Wisconsin, subsidiary) to make phonograph cabinets after a devastating fire at the Edison Phonograph Works in West Orange, New Jersey, on

December 9, 1914. Wisconsin Chair set up a subsidiary, Wisconsin Chair and Panel Company, to handle Edison's business. By 1919, United Phonographs's operations had narrowed to the "manufacture and sale of phonographs," while yet another division, New York Recording Laboratories, was founded to concentrate on manufacturing and selling recordings.[94] As a company name, "New York Recording Laboratories (NYRL), was misleading at best. While the name went into use in 1917, the company didn't have a license to do business in New York until the following year.[95] The bulk of NYRL's recording activity occurred in Chicago and later at the company's Grafton, Wisconsin, studio (or in some cases Gennett's Richmond studios, which Paramount would occasionally use).

The most authoritative Paramount researcher is Alex van der Tuuk, whose exhaustive *Paramount's Rise and Fall* has uncovered the breadth and depth of the label and its parent company's activities. In discussing WCC's entry into the recording business, van der Tuuk writes that "an early WCC employee, probably [talent scout Art] Satherley, remembered a recording company in Flatbush, Brooklyn, New York, and suggested it as a good place to inaugurate the WCC's own recording activity. This independent recording studio, possibly already named New York Recording Laboratories, might later have leased its name to the WCC."[96] Whether the Brooklyn studio held the name or not, the title gave a marginal player in the record business a connection to the center of U.S. recording industry operations, even though Victor and Columbia's offices were in Manhattan, geographically close but symbolically distant. The Flatbush studio was soon found to be inadequate, however, and NYRL set up a new studio at 1140 Broadway in Manhattan.

Paramount's early recordings, like those of its sister Puritan label, were all over the stylistic map. Like many other new labels, Paramount initially attempted to mimic the output of the Big Three. Some of the label's first recordings were by legendary industry veterans whose recording careers dated back to the turn of the century. Waning interest in these artists from larger labels allowed Paramount to take a chance. The label released recordings by the Peerless Quartet vocal group, one of many groups to have had a hit with the World War I classic "Over There," as well as Arthur Collins and Byron Harlan, whose comedic "coon songs" were performed in blackface and sung in white exaggerations of black dialect. The label also released its share of World War I–era patriotic tunes and instrumental ballroom dance numbers, its own versions of the kind of recordings already offered by the Big Three. Without a niche, Paramount's initial releases did not fare well in the marketplace. WCC executive and Paramount founder Otto Moeser tried to mitigate the recording division's losses with its cabinet business; by 1922, Wisconsin Chair Company had subsidized its recording division by as much as $100,000.[97] NYRL's lack of a clear vision for Paramount, and its inability to compete directly with the Big Three, left the label with a poor economic outlook. In order to find other profit streams for the struggling recording division, the company began to sell master recordings to other discount labels for rerelease. The company sold its masters to National (New York), Grey Gull (Boston), Bluebird (Los Angeles), and others.[98]

But it would be race records that turned Paramount's fortunes around. The label's involvement with race records developed both as a reaction to the genre's popularity and to market opportunities. Paramount was among the labels that jumped into the race market in the early 1920s, following the vaudeville blues craze. By 1923, Paramount had released more than fifty race titles (including blues and gospel titles), with advertisements that promised "A New Hit Record Released Every Week," surpassing even OKeh's output.[99] Then in April 1924, when Black Swan's financial position became untenable, Harry Pace leased the company's masters to Paramount; numbers 12100/99 in the Paramount race series were reserved for reissuing Black Swan material. The announcement released to the press talked of a "merger," but the details made it clear that Black Swan had few options:

> The Black Swan catalogue of several hundred master records is the most valuable of its kind in existence. Instead of the company operating that catalogue, the Paramount company will manufacture and distribute Black Swan records, from which the Black Swan Co. will receive a definite amount each month. After the Black Swan Co. has paid its accounts and obligations, such as every operating company must have, it will be in a position to pay its stockholders a substantial and continuous dividend, or it can retire its capital stock at a substantial premium.[100]

Harry Pace spoke directly to the *New York Age* that April to assuage fears. He assured its African American readers that "by Black Swan's combination with the Paramount company—a white organization devoted to the interests of the Race and specializing in Race records—the continuance of high class Race music is assured." Meanwhile, the *Port Washington Herald* offered the headline "Local Firm Takes Over New York Phonograph Company," reporting, "By this transaction, the unusual things occurs [*sic*] of the merger of a big Negro company and a big white company."[101]

A May 29, 1924, article in the influential advertising paper *Printers' Ink* discussed Paramount's strategy on behalf of the African American market. "Out in Wisconsin," the article explained, "a phonograph record factory is turning out seven hundred thousand records a month, and the business has been in existence only a few years." It explained that the company "believed that if they could get at the fundamentals of negro music and then produce it in the way the colored people liked it, they could soon become a power in this market."[102] Paramount's first concerted entry into the race records market came via Black Swan's masters as well as those leased from the short-lived Arto label of Orange, New Jersey. Both Black Swan and Arto had made their reputations via female vaudeville blues singers. While Arto had garnered significant sales with Lucille Hegamin (whose recordings Paramount would reissue), the Black Swan catalog was of greater interest. As van der Tuuk writes, this series was a "trial balloon to see if Paramount could annex part of this newly discovered, and lucrative market."[103] While the deal with Black Swan Records troubled the most fervent supporters of that company, Paramount's previous agreement to manufacture Black Swan recordings coupled with Paramount's interest in the race genre gave the label some credibility with African Americans. But Paramount's emphasis on country blues was borne out of necessity. As their foray into the blues expanded, the label sought unrecorded blues performers in rural areas.

With the race records moniker easily expanding to include different styles, and industry executives satisfied with categorizing by race, country blues recordings easily fit. Country blues was one of several styles that arose from ragtime around the turn of the century, along with blues' vaudevillian variant. While vaudeville blues caught on via the T.O.B.A. circuit and through Tin Pan Alley, William Barlow notes that the country blues style was "easily the southern black population's most popular form of secular music" well before race records were offered for sale.[104] The popularity of country blues initially led Paramount to market some of its vaudeville blues artists as if they were part of the country blues style, as in the case of Ma Rainey. Rainey is sometimes thought of as the first "real" blues singer, the "Mother of the Blues." Unlike previous singers whose repertoires included blues as well as a variety of other popular styles, she focused on the blues (in its vaudeville incarnation) throughout her prolific career with Paramount. Proud of his work with Rainey, Paramount scout Art Satherley would tell researchers decades later: "Ma Rainey was the greatest blues singer of all time, for my money."[105]

Despite Rainey's close connection with Chicago's urban jazz scene, Paramount found it advantageous to focus on her rural southern roots, and despite being accompanied by a band, Rainey's approach was rooted in her Georgia background. As Titon writes, "The downhome impulse was undoubtedly responsible for Paramount's promotion of Ma Rainey as a southern-styled singer."[106] Rainey's "See See Rider" is representative of her vaudeville blues approach, backed in 1924 by the Fletcher Henderson Orchestra, with band member Louis Armstrong featured prominently. An affecting example of the vaudeville blues style, it fit nicely with the improvisational saunter of the backing bands with a mournful female vocalist soaring over the top. But she also recorded in the country blues style for Paramount, as in the performances with Paramount star Papa Charlie Jackson during a 1928 recording session. In Jackson's duet with Rainey on "Ma and Pa Poorhouse Blues," he backed her up with his hybrid banjo-guitar as he also did on "Big Feeling Blues," in which she sang lines meant to comfort and commiserate with the lovelorn: "I've got the big feeling blues / I mean, I got a man."

Like Rainey, Papa Charlie Jackson was a transitional figure for Paramount. While Titon holds Jackson up as another possible progenitor of the country or "downhome" blues tradition, he suggests that "others have classified him as a vaudeville singer because he accompanied his songster's repertoire (which did include some blues) on a six-string banjo-guitar, which gave it an unmistakable minstrel-show sound."[107] Jackson was born in New Orleans and as a young adult settled in Chicago, where he worked both as a session player for various jazz outfits (Freddie Keppard, Tiny Parham, and Johnny Dodds). Despite working out of Chicago, Jackson's persona was that of a rural figure. His hybrid banjo-guitar (tuned to a standard guitar scale) with a large resonator, affirmed this persona, as did the fact that he accompanied himself on recordings. All this was used in publicity surrounding Paramount's "discovery" of Papa Charlie Jackson—"the only man living who sings, self-accompanied, for Blues record"—who was introduced in a *Defender* advertisement as if he were a relic of a bygone era. Jackson's distinctive solo blues style bridged New Orleans jazz and rural blues.

With the success of Jackson and the ambitious release schedule, Paramount's recording directors had to function both as talent scouts and traveling salesmen to keep up the pace. The company also began developing freelance scouts that had particular expertise in a given region or style. In the following section, the unique roles of these early scouts will be discussed in light of their role as cultural intermediaries called upon to manage relationships in disparate sociocultural settings, as the recording industry expanded further into the race records market.

Recording Directors, Freelance Scouts, and the Search for Talent

The search for new styles within the race genre became increasingly competitive. This meant not only an increased reliance on the companies' own talent scouts but also on regional experts, who were not directly employed by a given label but instead worked for many different companies on a freelance basis. These "record men," whether employed by a company or not, benefited from developing a strong expertise in a given region's music scene. They were valuable to parent company executives, who had little or no understanding of the music's value or its audience and often relied on scouts to discern the best talent and record the best material for them. Paramount executives had to rely on their own scouts as well as those freelance go-betweens to keep them informed of a market of which they had little prior knowledge. The label lacked the network of stores that Gennett's parent company Starr Piano had in its international chain of music stores, which served as both sales outlets and information centers for a given region's market. While Victor and Columbia had branch offices and affiliated retail stores that specialized in their products, Paramount had to rely on independent retailers, not only to stock their merchandise but also to teach them about regional musical styles. Some of these retailers were used along with company staff to help search for new talent.

The role of talent scouts in the race records market was necessarily multifaceted. They were often called upon to be recording engineers and sales people and to fill various other roles should the need arise. They were involved in an area of the record market derided by the black clergy and black middle class as well as by whites in general. In the case of white scouts, their association with black performers and business people made them suspicious to some. For black scouts, they had to deal with the prejudices held by the companies for which they worked as well as the class anxieties connected to blues within black society. It required those scouting the music to gain an understanding of how to acquire the best of this music for white-owned record companies. As David Wondrich writes:

> The music was unfamiliar, seemed archaic, probably was worthless; yet it sold millions of records. The musicians were black, inscrutable, presumably inferior; yet their performances had to be coaxed in the recording studios, for company profits ultimately depended on them. Given these ponderables, the industry dealt with the singers as carefully as it knew how—as the guardian of a trust might treat an underage ward, or as a scientist might examine moon rocks brought back from a mechanical space probe.[108]

While Wondrich's characterization certainly applies to some key figures in the early recording industry, as well as to the prevailing attitudes of the era, it is important to recognize differences between scouts in terms of differing attitudes toward the music and its performers and their roles as cultural intermediaries in the recording industry. A closer survey of talent scouts points to their roles in negotiating new genres as they emerge. In a discussion in which Pierre Bourdieu referenced "new cultural intermediaries" in 1960s France, he rightly noted the necessarily flexible roles adopted by those who work in such fluid circumstances, in which they must manage a variety of social and cultural settings. It is, he writes, in "the newest sectors of cultural and artistic production" where "jobs and careers have not yet acquired the rigidity of the older bureaucratic professions," adding that "entry into these jobs is often via networks of connections and share values and life experiences formed among members of this group rather than the meritocracy of recruitment through formal qualifications."[109]

OKeh already had its own legendary talent scout, Ralph Peer. His role in the successful generic expansion of the recording industry is crucial, but his own attitudes toward the music he recorded were often dismissive. His interest in what he recorded was purely practical, and he said so often. "He had never been driven by any particular desire to contribute something to traditional music," Barry Mazor writes, but by "decades of mounting practical experience about what evoked a positive response, provoked it again and again."[110] Recording industry history has often given Peer credit for the "race records" genre marker, though the first part of this chapter is meant to contextualize the terms in a much longer conversation within the African American press. Interestingly, Peer understood the race records niche in terms of OKeh's records for immigrant or international audiences. Titon writes:

> When the companies saw there would be a continuing market, they set aside special numerical series for black records. A generic name was needed to identify the already segregated offerings. Ralph Peer, head of OKeh's 8000 series, claimed to have coined the name that stuck: "race records." He had a model for separate record series with unique matrixes via OKeh's international releases: "We had records by all foreign groups: German records, Swedish records, Polish records, but we were afraid to advertise Negro records, so I listed them as 'Race' records and they are still known as that."[111]

Peer had been working in Kansas City for Columbia Records after returning from World War I when he left for New York and OKeh Records. It was at OKeh, Peer said in a 1958 interview, where he "first heard of the nigger and hillbilly stuff."[112] Peer and OKeh's discomfort in getting involved with African Americans during this period is telling, though later market successes would minimize such concerns. Peer's understanding of race records (and later hillbilly records) as separate from mainstream American culture as the "foreign" records he references, sets the stage for the separate matrix series and catalogs of their own. Titon suggests that Peer may have avoided using the term "Negro" because the term had not yet been adopted, adding, "several writers in the *Defender* had expressed the opinion that the word *Negro* was meaningless and, worse, could easily be changed into the pejorative."[113] While Peer takes credit for coining the term "race"—which is accurate if limited to OKeh's popularization of the term—he also chose the term with at least some understanding of its broad signification. "It was not a derogatory term; instead, it signified group solidarity. Though the record companies classified Afro-Americans with immigrant groups for whom special series would be provided, they took pains, in the early marketing period, not to insult consumers with an offensive generic name for the records. By 1923 'race records' was the established identification in advertisements."[114] If OKeh popularized the term, other labels quickly followed suit and began encouraging their own recording directors and talent scouts to seek out "race artists" to record.

Race records meant opportunity for small labels willing to enter the market, but labels quickly needed to designate employees to learn about the emerging genre and formulate a successful strategy. As was the case with most other labels working in this speculative corner of the recording industry, accounts vary as to who was the central decision maker on releasing records at the Paramount label. When considering the size of

Wisconsin Chair Company's recording operations and its various locales, it is likely that while the recording director at a given studio had a great amount of influence, decisions could conceivably come from above via the company's owners and upper management, or from below from engineers entrusted with recording prospective material from artists on an active studio schedule.[115]

In WCC cofounder Frederick Dennett's era at the helm, Maurice Supper's role in Wisconsin Chair's recording division was an important one. Supper had designed the Paramount logo and had been entrusted by the Dennetts in developing their new operation. He had designed the New York Recording Laboratories' studio on the seventh floor at 1140 Broadway and held the post of recording manager for the facility, which meant a large amount of travel from Port Washington, where he kept his residence.[116] Supper's supervisory role was quickly expanded, and he became general manager of NYRL. But in terms of the scouting the music Paramount recorded, Art Satherley was crucial. Satherley, who had worked in WCC's phonograph cabinet business and had been asked to design the chemical composition of the company's first discs, was made recording manager at NYRL. Satherley operated out of New York, working up and down the Atlantic seaboard, from Canada to southern Florida. All those charged with scouting and sales spent significant amounts of time on the road. Supper, Satherley, and others took Paramount Records on the road to sell along with supporting promotional materials: "To further promote the product, dealers' lists, advance lists, and monthly supplements were printed to go with the records."[117]

According to van der Tuuk, John Bostwick became WCC president after Frederick Dennett's death, and he "would make the final decision as to what was issued," with additional, regular input from M. A. Supper, WCC executive Alfred Schultz, and Art Satherley.[118] As with Gennett, stylistic quality was not always the deciding factor: "According to Alfred Schultz, three masters were made of a recording and a copy was made of each master. The tests were divided, and they were careful to pick out the right one. . . . When recordings were auditioned as tests, they risked rejection for being too much like something else, for being generally inferior, or for being poorly recorded or dirty."[119]

Meanwhile, due in part to Chicago's burgeoning music scene and the promise of economic prosperity, many musicians and songwriters from around the country who sought work in the city's studios and publishing houses moved to the city. "As the only major recording center outside New

York in the 1920s," William Barlow writes, "Chicago attracted many of the most talented blues artists in the South."[120] Paramount executives began to take notice. As Alex Van der Tuuk writes, many of the Paramount recording personnel worked in Chicago: "By late 1922 or early 1923, as Chicago became more important as a musical center, recording also took place there. Satherley remembered recording artists in Chicago's Marsh Laboratories as early as 1923. Both Supper and Satherley supervised recording activity in Chicago as well as New York."[121] While OKeh and Brunswick/Vocalion were important to Chicago record production, Paramount became Chicago's central label by the mid-1920s.

A key to Paramount's exploiting the Chicago scene was the hiring of a new talent expert. "After a lackluster beginning in the race record market," Barlow writes, "the company hired J. Mayo 'Ink' Williams in 1924 to manage its race artists series. Williams, an aggressive and energetic black promoter, was responsible for recording many of the greatest blues artists of the decade. It was Williams who signed both Ma Rainey and Papa Charlie Jackson for Paramount recording sessions in 1924."[122] Mayo Williams had attended Yale University with Paul Robeson, where they both played football. When he settled in Chicago, he began selling records with another former college friend, Joe Bibb, the brother-in-law of Black Swan's Harry Pace.[123] Though Williams' own tastes veered from the kind of material he would scout and record for the label, he valued the people and the music they created. "In an age when indifference or even antipathy to blues was characteristic of educated blacks," Stephen Calt wrote, "Williams held the heretical belief that blues represented an important aspect of his racial heritage. When his friends sneeringly referred to him and his retinue of blues singers as 'Mayo Williams and his dogs,' he replied: 'My dogs are thoroughbreds.'"[124] Alex van der Tuuk writes, "After visiting Port Washington to solicit a position with the recording company, Williams was offered the task of registering the copyright of the songs recorded by Paramount. The reason M.A. Supper offered Williams the job was almost certainly because as an Afro-American he would have a better 'in' to the newly discovered market."[125] Williams' role in Paramount's operations may also have helped Black Swan take comfort in Paramount's takeover of its operations.[126]

While working with NYRL and Paramount, Williams founded the Chicago Music Company (CMC), which he ran out of the Hygienic Building at 3621 South State Street.[127] In a letter written to record collector John Steiner decades later, Williams described the company's founding: "The Chicago Music Company was established about 1923 as a subsidiary to the

New York Recording Laboratories, producers of Paramount, Broadway, Puritan and other phonograph records as a company to hold and exploit copyrights of Paramount releases as well as those on its subsidiary labels." He described his work with NYRL and the CMC as "talent-scouting, arranging auditions and recording sessions, arranging performance and copyright contracts and purchases, supervising recording sessions, transcribing music and words which were to be copyrighted by Chicago Music Publishing Company and finally assisting in the securing of copyrights."[128] Just a few blocks away was Dickerson's Record Shop, the store that acted as cultural hub for black and white musicians that Williams' secretary Aletha Dickerson (later Robinson) co-owned with her husband Alex Robinson at Thirty-First and State.[129] Dickerson's work in songwriting has been largely ignored, as has as her scouting and recording talent (especially after Williams' departure from Paramount). She had a diverse skillset as accompanist, arranger, songwriter, administrator, and later recording manager as well as a unique knowledge of the musical culture of "the Stroll," the city's premier black entertainment district.

The Chicago Music Company was located in the heart of the Stroll. Founded around the turn of the century, the Stroll was centered on the corner of Thirty-First and State Street on the South Side of Chicago. The Stroll's sociocultural importance is depicted in Jacqueline Najuma Stewart's *Migrating to the Movies: Cinema and Black Urban Modernity*. Stewart's survey of the area shows a rich convergence of black businesses and organizations, including more than fifteen cinemas, several African American motion picture companies, and a variety of black social clubs, restaurants, educational institutions, and banks.[130] The *Chicago Defender*'s offices were at 3435 South Indiana Avenue. Stewart sees The Stroll, and broadly Chicago itself, as illustrating "the high stakes of defining the racial boundaries of urban leisure for migrants seeking new 'modern' freedoms and opportunities, for whites seeking to maintain racial hierarchies, and for Black 'Old Settlers' seeking to police and distinguish themselves from recent arrivals from the South."[131] Williams' location in the midst of Chicago's black entertainment district put him in a position to be an expert in various facets of the city's musical output and gave him the ability to discern new talent. For Paramount executives, Williams certainly represented an "insider" figure in Chicago's black musical community, despite Williams' own musical tastes (he much preferred the high-minded style of his old friend Paul Robeson to most of the music he recorded). Williams was given the role of recording director for the Chicago branch of Paramount's operations.

Fellow Paramount scout Art Laibly's relationship with Williams and others at Paramount was competitive and sometimes acrimonious, which may have been one of the reasons why Williams increasingly became involved in his own ventures (leading to his own short-lived Black Patti record label, in silent partnership with Paramount and Gennett). A Louisiana native, Art Laibly came to work for Paramount first as an assistant to Art Satherley in 1924. Gayle Dean Wardlow has characterized Laibly as "a salesman first and a recording director by default," someone who had little experience with or interest in what he was recording, or the recording process itself. With Williams' departure from Paramount, Laibly took over its Chicago operations.[132]

For recording sessions, Laibly and the other recording directors had a standard way of operating. A given day's sessions would often throw together a group of musicians who had to wait their turn. Sometimes this led to some musical improvisation. Laibly explained to John Steiner that those country blues sides with additional accompaniment—other than the standard guitar or occasional piano—were due to musicians striking up conversations with one another while waiting their turn to record. Musicians did so "as a courtesy to each other or for kicks."[133] Meanwhile, there are various accounts that suggest that scouts would supply liquor to the performers either during or after the sessions, ignoring the Eighteenth Amendment, which had made Prohibition the law at the start of the decade.[134]

Paramount relied on freelance scouts more than other labels to help identify talent. Mississippi's H. C. Speir was perhaps the most famous race record scout not directly employed by Paramount. Speir ran a music store in Jackson on Farish Street, in the town's African American business district. While Speir initially used talent scouting as a way to publicize his store, his later scouting trips made him a reputation far beyond Jackson.[135] He is credited with discovering some of the key blues performers of the late 1920s and early 1930s, including Ishmon Bracey, Tommy Johnson, Skip James, and, most importantly, Charley Patton. Speir enjoyed scouting work and sometimes paid for trips out of his own pocket in the hopes of recouping his money with the talent he found. In an interview with Gayle Dean Wardlow, Speir remembered, "I sent artists everywhere, and tried to work with all the companies but I probably did end up sending the best artists to Grafton, Wisconsin—the Paramount Record Company."[136] Wardlow, who interviewed Speir on numerous occasions, has commented on Speir's role in race records in relation to commonly held attitudes about the music and the people who made it: "You've got to remember . . . this was still 'race music' and good white people just did not associate themselves with that kind of music. And so Speir did not treat blacks bad, but he also did not associate with them socially." Speir knew the race and class connotations associated with the music, but he also understood its fundamental popularity in the South. As Wardlow notes, "He [Speir] said for years that they ought to be recording southern talent, you know black blues—and the record companies weren't. So about 1926, he tried to interest the companies in recording black blues."[137]

Talent scouts had to quickly determine whether a singer was capable of a good recording, and, if so, how to get the best possible performance in a small number of takes. "Though their standards are difficult to reconstruct, it appears they wanted singers with a substantial repertoire of original material and a distinctive performance manner."[138] Speir generally used the same approach with potential artists, telling them, "I heard you're a good singer, how about singing some songs for me?' After hearing a few songs, Speir would say, "You sound pretty good. You know, I make records; if you will keep practicing and get your songs together real good and come to Jackson in a couple of days, I'll see that you get on records." The trip to Jackson was necessary because of Speir's lack of portable equipment. Speir would often help musicians with their travel expenses, particularly if he thought that the artist might represent a potential hit.[139]

While Speir could claim he discovered Charley Patton and others, the other most important blues singer of the mid-to-late 1920s had several scouts all claiming that they had discovered him. Blind Lemon Jefferson made a name for himself in Dallas playing clubs and street corners in the Deep Ellum neighborhood east of downtown. It was Sam Price, a Dallas record retailer, who had sent a scouting letter to Paramount about Jefferson.[140] Price contacted Mayo Williams specifically, perhaps believing him to be the most likely to give a fair hearing to Jefferson's unique sound.[141] Laibly also claimed he discovered Jefferson, suggesting that "he found Jefferson on the streets of Dallas and late in 1925 invited him to Chicago."[142] The aesthetic impact of Blind Lemon Jefferson can be found first in his unusually intricate and fast-paced guitar style, alternating between strumming and finger picking that served to distance his guitar lines from his singing pattern. Through his guitar work, he would bend the song's rhythm to the breaking point before bringing it back again, as in "Black Snake Moan" or "See That My Grave Is Kept Clean." Similarly, Jefferson employed an idiosyncratic vocal phrasing, at times soaring into a high drone, while at other times using his voice as an additional rhythmic device, syncopating the syllables of words to great effect.[143] Jefferson's singular approach to inflection and mood gave him a striking emotional range despite his reliance on only a guitar and his voice.

"While vaudeville blues sounded close to both jazz and tin pan alley, Blind Lemon Jefferson's blues offered a fundamentally different sound," Jeff Todd Titon writes. "The inescapable conclusion is that Jefferson's music confused them."[144] While Paramount executives seemed interested in Blind Lemon, they weren't sure how to present his music to the public. After initially recording Jefferson doing two original gospel tunes under the pseudonym "Deacon L. J. Bates," Paramount recorded four blues songs under Jefferson's own name in 1926. The label knew that Jefferson's approach was uniquely his own. Titon writes, "It is difficult to imagine a better word to describe Paramount's judgement of Jefferson's songs than the one they used continually in their publicity about him: *weird*. They could not understand his words; they had never heard anyone play the guitar as he did. Even his appearance must have been startling—a blind, black street-singer of enormous girth with tiny, steel-rimmed glasses perched on his nose. So striking was his appearance that [Paramount's] advertising illustrators forgot their rule of avoiding cartoon caricatures of the singer's likeness when it came to his advertisements."[145]

While the kind of blues played by Jefferson would have been a recognizable idiom to many listeners, the sound was new to the ears of executives at Paramount in Port Washington. Predictably, Paramount's success with Blind Lemon Jefferson sent other record companies in search of their version of the country blues sound. For a time, Jefferson's success had made Dallas the center of the blues universe. OKeh, Vocalion, Columbia, and others all had scouts there looking for the next big thing.[146] "Sale of more than 5,000 copies of a record made money for the company, and selling the masters to dime-store labels . . . for a flat fee increased their profit. In addition, a royalty of two cents per record sold went to the copyright holder." But despite label credits, few blues singers actually held copyright. "They were paid traveling expenses and a lump sum for each song accepted by the company, in exchange for which they signed over copyright to the company's music-publishing agent."[147]

Scouts paid different amounts for talent, depending on the belief in the artist's potential and what the artist would accept. Some scouts, like Satherley, Williams, and Speir, had a more generous reputation, while others such as Laibly and OKeh scout Ralph Peer were known to pay less for talent. "Apparently, the fees paid to hillbilly singers were higher. . . . The contrast with vaudeville blues was even more striking. Bessie Smith [for Columbia] received $125 per song at first; in 1924, when her records were selling very well, she was getting $200 per song."[148] Speir's [artists] were probably the best paid of all the artists recording for various scouts. His average fee was either $50 a side, or ten percent of the royalties on the sale of issued titles.[149]

Mayo Williams' ongoing tensions with Art Laibly and a planned restructuring of his financial arrangement with Paramount led to Williams' departure in 1928. Paul Swinton writes, "Things finally came to a head when Liably [*sic*] started to use Williams' trusted secretary, Aletha Dickerson, as his own. Williams became disillusioned with the way his position was being undermined" and was also upset that Paramount's Otto Moeser was changing their financial arrangement from one based on commission to a traditional salary structure.[150] After the dispute with Moeser, Williams left Paramount for work in the race records division of Brunswick records and its subsidiary Vocalion label.[151] Meanwhile, Paramount (in an arrangement made with Mayo Williams) made Dickerson, an accomplished songwriter and arranger, its recording manager, apparently without informing her until months later.[152]

The role of recording directors and freelance talent scouts were crucial in seeking new talent as trends turned from vaudeville to country blues. Their activities in bridging cultures defined the changing genre associations with race records, as companies embraced different forms of previously unrecorded music. While black artists were always at the mercy of a white industry, their value to white record companies allowed them a then unthinkable freedom to record their music and to communicate their emotions and ideas to previously untapped audiences.

Race Records and the Recording Industry's Uneven Progress

This chapter has traced the discursive history of race records through debates in the African American press to the great expansion of African American vernacular music during the 1920s via white-owned companies OKeh and Paramount. It began with a call for greater African American representation on phonograph records. That representation came not in the ways imagined, but in a much broader expansion, uneven and contradictory in the way it was scouted, recorded, distributed, and sold. At each point, various players sought to define the genre and its performers in various ways. This is evident in the advertising strategies of white-owned record companies through which musical aesthetics, social prejudice, and market opportunism all came together. That the industry had recorded many other musical forms before turning to the race market suggests misgivings and no small amount of ignorance. Their advertising campaigns reflect tentative and often awkward first steps toward reaching a new audience while experimenting with different modes of address. And the varying roles played by talent scouts reflect a varied approach to their tasks as well as in their investment in the social, cultural, and aesthetic values of their work. Rather than reflecting a monolithic industry uniformly malevolent in its intent, tracing the work of these scouts affords us an interesting view of cultural intermediaries working in a controversial genre in the midst of a particularly volatile moment in U.S. race relations.

The work of white companies allowed disparate communities to hear a much fuller range of black musical production via an unprecedented expansion. This added new voices to the already complicated circulation of recordings in American society. In folklorist John Szwed's 1968 essay "Negro Music: Urban Renewal," he details the phonograph's role in transforming blues from a regional taste to an increasingly national music. For Szwed, the "neutral, depersonalized mechanism of the phonograph record" was "the single most important factor in explaining the direction of Negro music in the twentieth century."[153] While the white industry's limited understanding of the black market led to experimentation in sales styles that sometimes looked back to minstrelsy, the boom in the race records market also allowed black musicians unprecedented access to recording technology and culture, helping to represent African American culture in the United States like never before. "Like no other recording activity before or since, race records were designed to fulfill the musical demands of a given social group. At the same time, the flexibility of the record increased the range of taste of its buyers by reducing geographical and social differences between Negroes of various parts of the country."[154] It also allowed regional black musical cultures to communicate with each other long distance through phonograph recordings, which would in turn irrevocably change musical production as the influence of phonograph culture reached previously underserved groups in U.S. society.

An examination of race records' expansion and diffusion via nascent U.S. record companies also offers a picture of an industry increasingly aware of its own social and economic reach. While some labels experimented and lost, as in the case of Black Swan, such experiments created a larger recognition of the largely untapped African American audience.

Black Swan's lasting impact has surpassed its time in business. It led previously little-known labels that survived, including OKeh and Paramount, to assert their own presence in the larger industry and found a market that for a time made them viable in the face of larger companies. Race records had led the recording industry into rural markets for the first time as competition and innovation led smaller labels into the country blues that regional audiences had longed to hear on record. It also served to make the music released as race records into an increasingly national music, with influence far beyond its regional as well as historical limits.

While white companies had initially willed ignorance of blues as a style, there was another rural music style that had long been ignored. While geography and recording logistics played a part in this decision, there also may have been a class component: a Jazz-age preference for the modern and urban. In terms of commercial viability, the music of rural whites had languished even longer than black rural music had. Industry expansion into this white rural music, first known as "old time music" and then via the "hillbilly" moniker, led to further complications as companies strove to identify and understand rapidly changing constituencies. Both race and old-time music classifications acted as "advertisements and censors," as Karl Hagstrom Miller has argued, broadly defining both "the scope of a musical category" and "a unique segment of the population," the music and the people, with "one defin[ing] the limitations of the other."[155] In both race and old-time music as genre categories, these genres "resided at the edge of the difference between the music African Americans or white southerners loved and what predominately northern corporate record men imagined that they did . . . imagin[ing] buyers who looked a lot like the stereotypes that filled northern stages, record catalogues, and folklore collections."[156]

CHAPTER THREE

"Uninvited and Unannounced"

Old-Time Music and Radio

Early Old-Time Music Recordings

On Friday, June 30, 1922, Alexander "Eck" Robertson and Henry Gilliland arrived at the Victor Records' New York offices in hopes of an audition. Robertson and Gilliland were fiddle players and had just performed at a Civil War veterans' reunion in Richmond, Virginia. "Uninvited and unannounced," Jeffrey Lange writes, "the two must have received a cool reception, but the company's directors permitted the recording of a few tunes, and Victor subsequently released Robertson's solo fiddle version of 'Sallie Gooden.'"[1] Arkansas-born Robertson settled in Amarillo, Texas, and was playing medicine shows by the early 1900s and later toured the vaudeville circuit with his wife Nettie Levy while competing in fiddle contests and tuning pianos for Amarillo's Total-Line Music Company.[2] Henry Gilliland was also a fiddler working out of his home base of Altus, Oklahoma, where he served as city clerk and justice of the peace. Gilliland regularly appeared at fiddle contests and performed at Confederate veteran reunions around the country (Gilliland and all four of his brothers enlisted in the Confederate Army).[3]

Before this time, rural music recordings were made by vaudevillians with a polished stylistic approach more suited to city audiences. But Robertson and Gilliland were rural musicians with direct ties to the tradition and with no small amount of confidence. Robertson and Gilliland chose to approach Victor, the undisputed king of the 1920s record business, rather than labels that might have been more likely to record them, including those who were busy creating the race records market around the same time. Gilliland and Robertson's spur-of-the-moment trip led to their arrival at Victor still wearing their Confederate soldier and cowboy costumes from their performance at that Virginia Civil War reunion.[4] Their presence on the streets of New York must have been wholly anachronistic.

Robertson and Gilliland's brand of nostalgia, from which the musicians had profited in Virginia, had not yet become crucial to the recording industry's definition of their music. But their repertoire was *already* understood as nostalgic, even before they made a commercial recording. Here was music that evoked previous eras, a war nearly seventy years previous and a western expansion that had reached its terminus. As professional musicians, Robertson and Gilliland's repertoire no doubt would have skewed to older material for the Virginia event, but had they given much thought to how they would be received in New York? "We do not know what went on in the minds of the Victor directors," country music historian Bill Malone writes. "[T]hey may very well have been charmed by the music they heard (or by the romantic symbols of the Old South and Old West being displayed before them)—but they did permit the tests, and several selections were subsequently released."[5] The Victor release did not garner much attention, though most historical accounts suggested it sold moderately well. The results were not enough to convince Victor to pursue such unexplored musical territory.

While vaudeville blues and early jazz were initially understood as urban music, the emergence of country—first labeled as "old-time music" and later by the pejorative but long-lasting "hillbilly" moniker (and its "western" variant)—seemed at the outset as an unlikely return to rural tradition. What makes this so striking is the timing of its emergence with the rise of industrialization and urbanism in American society. Old-time music, as a traditional, antiurban, even antimodern genre, was the first that, by its very popularity, critiqued what in the 1920s was largely understood as progress. Its early relationship with radio was even more paradoxical at first glance, as the music of nostalgia found acceptance via the new technological medium of radio in ways that previous genres had not. And those involved, often depicted as hicks and rubes, and those that scouted, recorded, and managed them were adept at using the technology in innovative ways. The rise of this vernacular music at the height of the jazz age seemed oddly out of step with a society transforming from an agrarian, rural society to an industrial, urban one.[6] At the same time, this harking back is much more complicated than is commonly recognized. What is to be avoided is the "nostalgia-tinged, publicity based interpretations of hillbilly music, stressing the traditionalism, authenticity, and southern white rural origins of this music," which risks the reification of U.S. recording companies' arbitrary and ahistorical division of music into black and white categories, while also negating the socioindustrial smarts of the musicians and industry figures that made the music.[7] Musicians from varying backgrounds helped define the music, even as the categorical borders of the industry defined the genre's racial borders narrowly.

Clues to this music's industrial path can be found in both the initial "old-time music" genre and the derogatory "hillbilly" term later assigned to the music associated with rural hicks and rubes. If jazz and blues were construed as base and primitive by some, then hillbilly was old-fashioned, even regressive (a response reminiscent of the black middle-class reaction to the rise of country blues). In this chapter, I suggest that the commercialization of rural white music served a nostalgic function for white listeners amidst the roiling rise of industrialization and modernity, while at the same time those involved in the formation of the genre quickly developed a sophistication of sorts in how to present it to audiences. As had happened with the standardizing yet malleable functions of other commercial genres, a variety of styles and approaches were codified for the market via an understanding gained by the industry through the trial-and-error work of talent scouts, local go-betweens, record company advertising, press accounts, radio programmers and performers, and the musicians themselves. As with black rural music, scouts had to rely on learning local and regional tastes, listening to word-of-mouth, and taking away their own firsthand impressions of live performances.

Old-time music was the first genre to emerge *with* radio's newfound reach and importance, which along with the release of recordings helped define the music via its cultural circulation. In other words, the genre was the first to truly benefit from radio as much as from recordings. While radio's inventors and advocates had high-flung ideals for the new medium, "early radio broadcasting had a distinctly grassroots orientation," as many early stations had a relatively limited wattage and reach.[8] Both recordings and radio spread the genre and its corresponding themes, while transforming the music. I again focus on all-important industry go-betweens—those working in the closely linked radio and recording industries—that translated (and often repackaged) previously unrecorded music into a far-reaching commercial genre. In the case of old-time music, talent scouts again played an important role in defining how the recording industry would produce, market, and sell the music. In addition, those who brought the music to radio helped spread the genre, through early on-air performances and the "barn dance" shows that went from regional to national broadcasts with the rise of networks. As the last of the emerging genres of the 1920s to be recorded, hillbilly's rise was particularly dramatic in its fast growth and breadth of influence.

"In 1922," Gene Wiggins writes, "'hillbilly' seemed a creative term rather than an obvious one, such as 'woolhat music, or 'peckerwood music,' or 'linthead music.' The term," Wiggins argues, "had the advantage of letting people know what you were talking about, something that cannot be said for 'folk music.'"[9] The vernacular music that fell into this category is most usually associated with the American South, though folklorists, musicologists, and scholars have argued that there was also much activity in the rural Great Plains and Great Lakes Midwest, an association at least partially lost early on.[10] Wiggins notes that "'Ancient airs' even had some popularity for a little while," a phrase perhaps related to the "Negro and Plantation Airs" designation that referenced minstrel tunes in earlier decades. Both terms were antiquated before the end of the decade. Finally, Wiggins notes that WSB Atlanta named its first rural southern white music show (which debuted in November 1922) an "old-fashioned concert" in its advertising.[11] But the most popular moniker for the new commercial genre was, paradoxically, "old-time music" or, alternatively, "old-time tunes." While "old-time tunes" initially may have referenced Anglo-Celtic ballads and "hillbilly" suggested instrumental fiddle tunes, the discrete differences between these two terms didn't hold for long. The most famous early recording artist in the genre was not Robertson and Gilliland but Fiddlin' John Carson, who already undid this juxtaposition by being a fiddle player who sang. Perhaps more importantly, the switch from "old-time" to "hillbilly" suggests a longstanding urban appetite for the rural buffoon as well as the socially evocative power of the term. Because of its eventual dominance by the 1930s, I favor the term "hillbilly" in this discussion to reflect the industry's use of it, while recognizing its complicated and often derogatory history, in that some artists bristled at the label while others gladly accepted it and, in some cases, made the humiliating term work for them.

The Hillbilly's Commercial Rise

In his foundational 1968 essay "The Hillbilly Movement," folklorist D. K. Wilgus discusses the difficulties in defining the parameters of the hillbilly genre. This was not because of the commercial genre's liberal use of the term, but because of his own sense of the genre's hybrid nature as well as his desire to honor both its folk and commercial aspects:

> I find it quite difficult to discuss hillbilly music because no one can determine exactly the limitations of the subject. Hillbilly music may be seen as the style and repertory which developed through the interaction of Anglo-American folk music, Afro-American folk music, and American commercial popular music in the uplands and lowlands of the southeastern United States. Hillbilly music may be seen as a commercial enterprise to purvey standardized musical packages to the rural folk of the United States. Both statements reflect reality, and—as usual—truth encompasses both viewpoints.[12]

It is the function of industry genre making to convey a sense that a genre designates a specific social, cultural, and aesthetic formation. But just as importantly, it should not be too limiting of a construction, which would ensure the need to play catch-up to an audience it can never fully anticipate (something that is *de rigueur* for the industry anyway). And in the case of old-time music, the industry had seemed unaware of a potential untapped audience until it began to expand into new genres during the 1920s. But even amidst this expansion, it is surprising that it took the U.S. recording industry so long to embrace the music of southern rural whites, considering that companies had recorded such a diverse array of musical styles across social, racial, and ethnic lines (and, even more broadly, nonmusical speeches and novelty records of various sorts). "In what might even be seen as a colonial economy," William Howland Kenney writes, "southerners readily offered their raw materials—musical talent and entrepreneurship—to northern record producers and engineers in exchange for a major leap forward in the professionalization of their careers and also for what seemed to the musicians to be generous piecework payments and unprecedented regional and national publicity."[13] As Bill Malone writes, "At the opening of the twentieth century, American music was urban-oriented. Both the fine-art and commercially directed groups neglected the rural population (or else tried to 'urbanize' their tastes) and instead catered to the interests of people in cities."[14]

In comparison to race records' emergence—which included a more concerted and organized campaign from musical advocates and the African American press—the emergence of old-time music seems all the more accidental. Old-time music, as a commercial genre, was the last of the three dominant emerging genres of the 1920s to be "discovered." During the first few decades of the twentieth century, even folklorists looked the other way, as they had with race records earlier (choosing to study field hollers and work songs instead of the blues songs all around them). When examining the rural white music of the United States, variants of which could be found from coast to coast, folklorists preferred the British ballads that could be traced from the Appalachian hills back to the other side of the Atlantic. Wilgus adds "A folksong collector usually finds what he is looking for." Here he was referencing the famous Appalachian studies of noted British folklorist Cecil Sharpe, who cataloged the Anglo-Celtic elements found in the folk music of the Appalachian region in his *Folksongs from the Southern Appalachians*.[15] But because Sharpe was looking for a certain kind of music, area residents only brought him the a cappella ballads he had expected to collect, as Wilgus suggested: "This [hillbilly music] was not respectable music, and what respectable person would want to listen to it, let alone call it folk music?"[16] All signs suggested that outside a strong regional interest in the tradition, this music had little value to the rest of the world.

And yet it is easy to see the music's appeal for those rural Americans, many of whom were relocating to industrial cities for work. Old-time music was associated with home, consumed as antidote to—or expression of—homesickness. The music also quickly appealed to urban Americans, some of whom had little or no connection to the country's rural past but found appeal in the range of experiences and emotions represented in the music, when they had the opportunity to hear it performed, often through caricatured vaudeville versions of the white rural musical tradition. The growing national interest in this music during the 1920s and 1930s might also be understood as an expression of dissatisfaction with industrialized, modern, urban society. Still, record companies were not quick to record what was thought of as rural music. As Bill Malone writes in *Country Music, U.S.A.*, during the tremendous growth in the 1910s U.S. recording industry, "record companies in fact seemed to be more aware of the potential of city ethnic subcultures more quickly than they did the native rural cultures of the nation."[17] Throughout the 1920s, most record labels continued recording ethnic music of various kinds for immigrant groups within the United States. "The recognition of an ethnic buying population," Malone continues, "may have made record industry people cognizant that a similar market existed out in the hinterlands."[18]

But the urban record labels of the early to mid-1920s seemed to have little interest in rural music of any kind. "Before World War I," Malone writes, "the music industry—then represented chiefly by vaudeville, the sheet music business, and the phonograph interests—concentrated on urban America because of its population density and easy accessibility."[19] Recording directors at urban record labels may not have thought they were ignoring the rural market, since their releases did feature comedic rural characters. These characters were usually executed by urban vaudevillians such as Cal Stewart (known as Uncle Josh, a humorist who told stories of his mythic small town of "Punkin Center") and others, playing already well-worn hick or rube roles, in a thematic tradition that can be traced back to the beginnings of American literature.[20] "Genuine rural performers," Malone surmises, "were rejected because they sounded strange and primitive to urban music executives, who believed that the urban buyer would be repelled by the raw country sound."[21]

Certainly it made economic sense for record companies, most of which were still located in the eastern corridor between New York and Washington, D.C., to concentrate on local urban subcultures and the general urban market. Companies generally sold to rural customers though small-town stores and mail order business. The Big Three (Victor, Columbia, and a quickly receding Edison) offered rural listeners what they offered to the rest of their consumers, without giving much thought to differing tastes and interests of rural listeners. Meanwhile, the rural economy was in a serious downturn. After World War I, crop prices tumbled and the economic value of farmland dropped dramatically, causing many rural Americans to abandon their homes for the promise of urban prosperity. As Clifford Doerksen notes in his analysis of early rural radio, the 1920 U.S. Census made clear that for the first time in U.S. history, more citizens lived in cities than in the rural countryside.[22]

While this change was reflected in census statistics, there was a decidedly uneven effect on the national geography. While this wasn't the case in all regions of the country (much of the Southwest urbanized in the decades after World War II), the cities of the industrial Midwest and Northeast were being transformed with workers from the rural South. The music that white rural workers brought with them reflected these social changes. Old-time music was the music of displacement and homesickness. Countless songs recounted the short-lived promise of the road and rail, followed by disappointment, wistfulness, and a yearning for home. Rural life, so closely connected with American society in eighteenth and nineteenth centuries, was not only in economic and demographic decline. "There hung in the air a generalized sense," Doerksen writes, "that rural life had somehow been comprehensively superseded, left in the dust by progress," while in the entertainment field, country rube characters were featured in Tin Pan Alley songs and in comic strips, and "vaudeville performers spun infinite comedic variations on themes of rural imbecility."[23] If the agrarian roots of U.S. culture were partially being left behind in favor of the promise of the urban modernity, then aspects of rural culture were being repackaged.

This didn't ease any anxieties on behalf of those focused on the rural economy. In the mid-1920s, *Farm Life* magazine published a promotional booklet for current and prospective advertisers titled *Consider Fifty Million Customers: A Study of the Rural Section of the American Market*. The booklet, like the magazine to which it was connected, strove to highlight the continued importance of the rural economy and, by extension, rural culture. The booklet emphasized increased productivity on American farms through use of technology with fewer workers. It also sought to complicate the view of the rural market by detailing various statistics regarding rural readership of magazines and consumption of various products, while emphasizing *Farm Life*'s role along the way. "When states are listed according to farmers' purchases of automobiles, telephones or farm implements, we get much the same grouping that we did when the states are listed according to their standing on *Farm Life*'s subscription list."[24] In *Farm Life* magazine, articles had previously taken notice of farm use of phonographs, as reported by *Talking Machine World*. In 1920, before companies were recording rural white southern music, the phonograph seller's trade journal reported the results of the *Farm Life* survey. More than six thousand readers replied, detailing their families' "household equipment, farm equipment, personal equipment, live stock, and other possessions of its subscribers." In terms of phonograph ownership—even in relationship to other "musical instruments"—the ownership numbers were impressive. "Of the 6,115 homes from which replies came, 1,965, or nearly one-third, possessed pianos, and 1,345 owned organs, while 2,058 were the proud owners of talking machines [i.e., phonographs]."[25]

While it is easy to lose sense of the phonograph's importance amidst radio's rise, its impact remained important well into the 1920s. And even though the record companies still had little sense of the rural white audience they were missing, a few experiments with recording this music led to a transformed industry. The commercial rise of old-time records in the 1920s helped the record industry weather particularly difficult transitions, amidst a volatile climate with the rise of radio, quickly shifting audience tastes, and increased competition to keep up with innovations in emerging genres. As with race records, OKeh was involved in old-time music early, although it again took an outsider to convince the label to record it. The other label involved was also the leading record company in the United States: Victor. Victor again waited until the new genre was well established before they invested their resources. Both labels were involved in recording early examples of the genre, through sessions shaped by happenstance. In Victor's case, the initial experiment didn't lead to significant sales. In OKeh's case, the company found the central figure of old-time music's first few years.

Early Old-Time Recording Sessions

As with previous genres discussed, the beginnings of rural white music recordings have been heavily mythologized. Depending on how one defines a commercial recording, the debate continues as to which was the first.[26] But two origin stories are repeatedly referenced. The stories remain important due to the role of the artists in making rural white music *commercially viable*, whether they were first commercial recordings or not (which is of little interest to this discussion): (1) Eck Robertson and Henry Gilliland's aforementioned 1922 trip to Victor's Manhattan offices and their subsequent recording session; (2) Fiddlin' John Carson's first Atlanta recordings in 1923 for the OKeh label, which also resulted in a release. Both recording sessions highlight the circumstantial nature of the industry's first forays into southern rural white music. And while the specific details of early events may be permanently lost, the enduring stories and their sociocultural appeal can tell us much about how this music was and is understood. These origin narratives also tell us a great deal about the stakes involved and how the sociocultural aspects of the genre continue to resonate.

A similar quixotic approach to Robertson and Gilliland's visit to Victor's New York offices occurred in March 1923, when Virginia's Henry Whitter approached the General Phonograph Company to cut a record for OKeh in New York. Whitter, who played guitar and harmonica, worked at a mill in his hometown of Fries, Virginia. Unlike Robertson and Gilliland, Whitter's initial sessions for OKeh scout Ralph Peer led to a recording career, though this was not initially evident after he finished recording. Whitter recorded nine tunes at the OKeh session, including "Lonesome Road Blues" and "Wreck of the Southern Old 97," a song he coauthored. The song was popularized by the former light-opera singer Vernon Dalhart and would go on to become a standard of the genre.[27] Peer, in his role as OKeh's recording director, decided against releasing any of the recordings, but he would release the best of them for the OKeh label after the success of another session in June 1923, this time in Atlanta with local musician "Fiddlin'" John Carson. The Carson recordings would prove old-time music as a commercially viable force.

Carson was a professional musician who had been successful on the entertainment circuit throughout the Southeast. "A champion fiddler born in the hill country of northern Georgia," Jeffrey Lange writes, "Carson gained his musical reputation from performances in political campaigns, various social gatherings, and especially at the Old Time Fiddlers' Association Convention in Atlanta, where he won every championship between 1913 and 1922."[28] Ralph Peer had been scouting new race talent for the OKeh label and was scheduled to do a session in a makeshift studio in Atlanta. At the urging of Polk Brockman, a distributor for OKeh (and other labels) who initiated a record department in his family's James K. Polk furniture store in Atlanta, Peer came to Atlanta on a race records trip and agreed to record Carson as well. "The man responsible for OKeh recording sessions" in Atlanta, William Barlow writes, "was Polk Brockman, a white Atlanta native who got into the record business after taking over the phonograph department of his father's furniture store. Brockman had little interest in the blues, but when he realized that money could be made selling black secular and sacred recordings, he made a deal with Ralph Peer at OKeh Records." Barlow suggests that "Brockman became Peer's surrogate in Atlanta" and later in the scouting and recording centers of Birmingham, New Orleans, and Dallas.[29]

In perhaps the most counterintuitive moment, Brockman first decided to record Carson while in New York City while watching movies at Times Square's Palace Theatre. As Gene Wiggins writes, "He [Brockman] was in the Palace Theatre on Times Square when he saw a newsreel on a fiddlers' contest in Virginia. He took out his notepad, and by the light from the projection booth wrote 'Fiddlin' John Carson.'"[30] Brockman later suggested that the movie audience's enthusiastic response was the impetus. The audience couldn't have been reacting to the music itself, since this unnamed film dates from the silent newsreel era, so they were more likely reacting to the look of the event and its players and to the rural subject matter. It is unlikely that Brockman's epiphany in a New York movie theater was the first time he had considered recording the rural white music with which he was familiar. But the validation that the New York audience gave Brockman with its impromptu response convinced him. Here was the possibility of a much larger audience for this music, which suggested that this regional style might even find a national audience.

On June 15, 1923, the *Atlanta Journal* carried a notice of Peer and OKeh's activities in Atlanta:

> "Canned music" recorded by local musicians will be made for the first time in Atlanta by the OKeh company, of New York, it was announced Friday by R. S. Peer, production manager of the company, who is completing arrangements here for the recording of selections by a number of local musical organizations. About thirty recordings will be made at the laboratory of the company on Nassau Street, including selections by the Morehouse college quartet of negro singers, Fiddlin' John Carson, the Seven Aces, and other organizations.[31]

The article also noted that the trip was possible because of newer equipment that made remote recording more feasible and thus allowed OKeh to make its first recording trip to Atlanta. Carson, who was fifty-five years old at the time of his first recording session, already had a strong reputation in the Atlanta area and would have likely been known to Brockman from press accounts or firsthand contact (perhaps via Brockman's music sales through his family's furniture store). Carson's Atlanta reputation was not only based on his live appearances, but also on his radio performances dating back to 1922 on Atlanta's WSB. Brockman would have been familiar with Carson well before the 1923 sessions in a makeshift Atlanta studio. Although Peer was unimpressed and initially did not want to release any of the material, Brockman insisted that a run of five hundred copies of two songs, "The Little Old Log Cabin on the Lane" and "The Old Hen Cackled and the Rooster's Going to Crow," be pressed for him to sell in the Atlanta market. As Lange writes, "When the initial pressings quickly sold out, Peer signed Carson to an exclusive OKeh recording contract, and so began the commercialization of rural white southern music."[32] A *Talking Machine World* article of a few years later suggests what a big surprise OKeh company executives (Peer among them) got when Carson's records were a success: "When Mr. Carson's first OKeh records were released, it was expected that they would be active sellers throughout Southern territory, where this artist is a prime favorite among all music lovers."[33]

Peer later said about the Carson recordings, "It was so bad that we didn't even put a serial number on the records, thinking that when the local dealer got his supply that would be the end of it."[34] The way Peer thought the recording was "bad," however, leaves his response open to interpretation. Following Gene Wiggins, I argue that it is at least as likely that Peer's complaint was focused on the technical quality of the recording.[35] Carson sang and played fiddle simultaneously, which posed a great difficulty in differentiating the two in an acoustic recording, since spatially distancing the two sounds was a near impossible feat. "People like the story of how awful John was thought to be by Ralph Peer," Gene Wiggins writes, "who labeled him 'pluperfect awful.' It is likely that Peer was less than ravished by the live performances . . . but there is much to suggest that the technical results were what mainly bothered him. The engineer had never recorded a person who sang and fiddled into the same horn."[36] And since Peer's interest and knowledge of the kind of music Carson played was scant in 1923, there is at least some possibility that Peer's complaint was technical and not stylistic. At any rate, Brockman was able to quickly sell all five hundred copies between his distribution outlets and Carson's public appearances: "At a fiddler's contest organized by Brockman, Carson played his phonograph record from the stage and vended copies across the footlights. Brockman got in touch with New York, declaring 'This is a riot.'"[37] OKeh quickly pressed more copies and gave the Carson record a proper (re)release. Peer's claim that the record was later rerecorded in New York (Brockman disputed this, as have others) increases the possibility that while indifferent to the music, Peer may have hated the technical qualities of the recording: "When the national sale got up to 150,000 copies," Peer added, "we were so ashamed we had Fiddlin' John Carson come up to New York and do a re-recording of the numbers."[38]

By all accounts Carson's record sold well, sending a number of other labels into the field looking for their own old-time music talent. At OKeh, Peer recalled Whitter's sessions and quickly released several sides and made more recordings. Meanwhile, many of the other labels that had been branching out into country blues and gospel through the race genre further expanded into the old-time music genre.[39] As Bill Malone writes, "In the years leading up to the Depression, such companies as Brunswick, Gennett, Paramount, Victor (hesitantly), and Columbia began emulating the successful tactics used earlier by Ralph Peer and OKeh."[40] Each label had its man or men in the field and they ventured out with the bulky equipment of the era to set up temporary studios in hotel rooms, warehouses, and radio stations. "Essentially, such men as Frank Walker and Dan Hornsby (both with Columbia), James O'Keefe (Brunswick), Eli Oberstein (Victor), Art Satherley ([Paramount and] American Record Company), and of course, Ralph Peer, were businessmen who had little knowledge of music. They were, on the other hand, unwitting folklorists who collected (and therefore preserved) an immense and valuable body of musical, social, and cultural materials dealing with the rural South."[41]

Malone's assertion that the scouts had "little knowledge of the music" is perhaps too broad a generalization. By this point, these record men had been extensively traveling the areas where old-time music was recorded and would have at least a strong familiarity with the music. Talent scout Art Laibly had long argued for Paramount to record rural white music, as had others. While other scouts (notably Peer, though he has perhaps become most closely associated with old-time music) seemed to have initial doubts about the music's commercial viability, Carson's success was enough to convince them. A wave of acts appeared on recordings in the mid-1920s for the first time, often with names that played up the hillbilly persona: Gid Tanner and the Skillet Lickers, the Possum Hunters, the Fruit Jar Drinkers, and so on.

Among those rural performers finding opportunities to record, there were also professional entertainers jumping on to the new trend. Chief among these was Vernon Dalhart, who in 1924 left his previous career as a singer of light opera and popular tunes to embrace the rural music trend. He first recorded "The Prisoner's Song" for the Edison Company, but it wasn't until Victor released that song with a disaster tune, "The Wreck of the Old 97," later the same year that both songs became huge hits. Dalhart's refusal to sign an exclusive contract meant that he could the cut the two songs for as many as thirty other labels in the subsequent years. Dalhart also managed to hit with a third song for Victor, "The Death of Floyd Collins," while the earlier two songs continued their unprecedented sales. In his 1970 essay on Dalhart's early career, Walter Darrell Haden wrote that "The Prisoner's Song" and "The Wreck of the Old 97" remain among the all-time best sellers in the U.S. recording industry: "I personally do not know of any other pairing since then to surpass or even equal Dalhart's 25,000,000-record hit, short of Bing Crosby's *White Christmas*, backed as it has been over the years with a succession of seasonal flip-sides."[42]

These early recordings proved without a doubt that money could be made in the rural white market, perhaps even more than in the race records market that had initially led talent scouts into the field. If the earlier recordings by Carson and others led labels to jump into the old-time music market, Dalhart's groundbreaking hits led everyone into the fray. As the old-time music genre came to be known as an impressive moneymaker, large and small record companies quickly got into the market, combing the South and lower Midwest for their own authentic hillbilly music. Record companies, along with the phonographic trade press, quickly took the lead in capitalizing on the new market. While some of the scouts understood the value of the music intuitively, others (Peer among them) were not as quick to understand the music's appeal. Peer and others initially relied on figures like Polk Brockman and others with an extensive knowledge for guidance and played up the aspect they perhaps understood in terms of consumer appeal: the hillbilly persona.

From "Old-Time Music" to "Hillbilly"

By May 1925, "Old-Time Music" was making waves. *Talking Machine World* featured an article titled "Henry Whitter, OKeh Artist, Real Hill Country Type":

> Mr. Whitter is a real specimen of the Hill country, coming from Galax, Va., and on his first few trips to New York could not be induced to stay overnight, coming in to the city in the morning, making what recordings were necessary and leaving before midnight arrived. Although he has overcome his shyness to some extent, he is still averse to what might be called "seeing the town." He insists that his trips from the railway station to the hotel and then in the recording laboratories are sights enough for him.[43]

Already in the genre's early period, this article (perhaps with no small amount of coaching from the record men at OKeh) exhibits all ingredients central to the hillbilly persona. The "hill country" referred to in the text inexactly refers to the Appalachian territory, though not the full breadth of the region, which begins in New York State and extends down into Alabama and Mississippi. The "hill country" referred to in these texts is more likely Virginia, West Virginia, Kentucky, Tennessee, and North Carolina. In American culture, the "hill country" has long stood for a remote and often backward place, with outmoded and sometimes mysterious customs, a place both denigrated and exoticized. The proximity of these states to the record labels of the Northeast and Midwest meant that these musicians could more easily to travel to company studios in an era during which electrical recording had not yet made portable recording easier to accomplish. In Richard Peterson's authoritative text on the subject, *Creating Country Music*, he differentiates classes between the working poor and the "hillbilly" persona that would become so important to later barn dance radio: "Many who by modern standards might be considered impoverished were then considered relatively well off. Such people fit into the class of 'respectable, God-fearing poor,' and socially they were a clear notch above the 'poor white trash' that provided the model for the hillbilly stereotype."[44]

The expansion of the hillbilly genre was a great opportunity for rural Americans to gain some say in their own representation. As with race records, this was a complicated negotiation between record companies and their advertising departments, their talent scouts, and the musicians themselves. Genre terms were one central site of activity, as were the depictions of performers in promotional material. In the mid-to-late 1920s, both OKeh and Gennett were still using "old-time music" (with its days-gone-by connotations) as much as "hillbilly" (which emphasized class distinction). Meanwhile, Victor had adopted "hillbilly," perhaps on Peer's urging, possibly because of the label's interest in reaching a larger, urban audience already familiar with hick and rube characters from vaudeville. Certainly Gennett's location in the lower Midwest gave it the closest geographical relationship to its target rural audience. Its advertising copywriting also favored a warm and friendly depiction emphasizing audience inclusion as "one of the gang," as opposed to some of Victor and Peer's rhetoric, which seemed to emphasize difference and exoticism. In the "Gennett Records of Old-Time Tunes" catalog of January 1928, a prominent photo of WLS vocalist and banjo player Chubby Parker emphasizes personal communication from him to his audience via the Gennett catalog, which reads, "Chubby has asked us to tell everybody that he has a lot of new songs for the new year."[45] Elsewhere, in keeping with other labels, Gennett emphasizes its performers' connection to the hills. In publicizing Fiddlin' Doc (born Dock) Roberts, the copy suggests that he had won "so many old time fiddlin' contests that he has lost track of them" and that "Doc was raised in the Kentucky hills and has been fiddlin' since a mere boy."[46] In the professional studio photos that appear of both Parker and Roberts, they are wearing suits (whereas some performers were asked to don overalls while posing outdoors in rural scenes). In a publicity photo, both Roberts and guitarist Edgar Boaz are both sporting bow ties. In an OKeh "Old Time Tunes" 1926 catalog, the performers are also all dressed up in more formal clothing, with the male performers in suits and female performers in dresses. While a more in-depth rhetorical comparison would be needed, it may be that a given performer's relative power and success in the field might also allow a greater control over his or her appearance in promotional materials. Certainly, the figures depicted in this OKeh catalog—Vernon Dalhart and Fiddlin' John Carson among them—were powerful figures in the industry. The cover of the catalog is decidedly more rural in flavor: a fiddle player wearing fitted, tidy rural clothing (in a tie and vest, with his pants neatly tucked into his boots) plays a fiddle on a tree stump, while his dog lies at his feet.[47]

"Early country singers exhibited an ambivalence toward their profession," Bill Malone writes, "similar to that which we have observed among instrumentalists. The hillbilly image and label were troublesome to some while repertory and style of performance were problems for others (whether to perform old-time songs or not, and whether to adhere to a 'rural' sound or to blunt it in favor of a smoother and presumably more salable sound)."[48] Despite the clear and inevitable distortions of rural American culture through the hillbilly genre and persona (image, label, repertory, style), I primarily understand the practices and "rules" associated with hillbilly as an industry effort (adopted by performers to varying degrees in different situations) to generate sales based on what company salesmen felt appealed to consumers. "This commercialization," D. K. Wilgus writes, "was not the attempt of music merchants to corrupt the tastes of the American folk. The merchants were merely packaging a commodity already existing. And they were at first hardly aware of what they were doing, except trying to make a profit. They sought out local musicians in the South, recorded and released the results, and hoped for sales. The type of material and performance which sold naturally became the subject of further research, and also imitation."[49]

It would be several more years before "hillbilly" became fully connected with the new genre. Folklorist D. K. Wilgus recounted that when "a group of Southern folk musicians inadvertently christened themselves 'The Hill Billies' at a 1925 New York recording session, they reflected the attitude of sophisticated America towards their music. But they also fostered a term which was applied not only to hill-country folk music but to all white folk music which was to be commercialized by the many facets of the recording industry."[50] At mid-decade, the music of the Hill Billies was understood as largely limited by regional tastes. In the April 15, 1925, issue of *Talking Machine World*, a small notice recognized "Hillbillies' OKeh Records," noting that "a banjo, fiddle, and guitar plus the exceptional ability of the pianist gives the true mountaineer music, and through the South the Hill Billies are welcomed wherever they appear."[51] As the decade progressed, the term "hillbilly" expanded and became an umbrella term that referenced a wide variety of regional styles that far exceeded the music's primary association with the Southeast and allowed record companies, as we shall see, to employ the term in a variety of ways as tastes changed and audiences emerged.

Ralph Peer is credited with popularizing the term "hillbilly." Archie Green, in his foundational "Hillbilly: Source and Symbol," suggests that Peer understated the impact of his work as a scout by understating his role. In a mid-1950s interview in *Variety*, "he identified himself 'as the person responsible for the discovery and development of the hillbilly business.'" Green added that Peer, in a 1957 letter to him, commented further: "It is quite true . . . that I originated the terms 'Hillbilly' and 'Race' as applied to the record business."[52] Green lauds Peer's pivotal role in taking enough interest in hillbilly artists to give the term a meaning, a central term around which a variety of performers and styles coalesced. Although a smattering of rural performers was recorded all the way back to cylinder recordings made around the turn of the century, the term did not attain this meaning until Peer's defining and concerted efforts during the 1920s, from he and Polk Brockman's initial Atlanta recording session with Fiddlin' John Carson to his transformative recording sessions with Jimmie Rodgers and the Carter Family at Bristol, Tennessee, in 1927.

But even more than the phonograph, which had begun being adopted into American homes during the first two decades of the twentieth century, it was radio's potential to transform rural life that was vital in the 1920s and 1930s (even after local stations had to compete with changing federal directives and network radio). And it was the rise of radio that allowed hillbilly music to emerge from a regional taste of the rural South and Midwest to a national phenomenon, and it was radio that helped imprint the hillbilly persona in the national imagination. The rise of hillbilly music as a viable genre for the U.S. recording industry, a genre that could move beyond regional popularity in the South and Midwest, coincided with the rise of hillbilly radio shows during the 1920s and 1930s. These shows operated, as Michele Hilmes has referred to the hillbilly comedy shows of the 1930s, in "the backwoods of modernity."[53] But in adopting hillbilly music for radio, promoters and performers further hybridized entertainment styles in search of greater national appeal.

Hillbilly Radio and the Rise of Regional "Barn Dance" Shows

By the early 1920s, radio was changing from youthful hobby to a powerful new medium. Several record labels initially sought to prevent their big-name stars or "Great Artists" from performing on air. An article in the June 1923 issue of *The Wireless Age* noted that Victor and Brunswick were forbidding their performers from appearing live on radio: "The two exceptions are emphatic in their assertion of their right to prohibit their exclusive stars from performing for the radio. That is why no important Victor artist has been able to broadcast personally in the past months, and why only one or two of the Brunswick artists have been heard by the radio audience."[54] Victor, as industry leader, could afford to risk alienating its fan base in their reticence about radio. Brunswick, which had become the third-bestselling U.S. label (outselling Edison) in 1923, also felt that it could keep its biggest stars off of radio to ensure phonograph sales.[55] "Here we have an entire industry in agreement on the fundamental assumption that radio either does not effect [*sic*] its sales revenue at all, or does so advantageously; yet two important members of that industry have adopted such a policy as might be expected to spring solely from fear of injury."[56]

Early on in the popularization of jazz and blues, radio was still in its infancy. Later, jazz's influence would widen its definitional reach and in its commercial form would become central to American popular music. Race records, and country blues in particular, never got the access afforded jazz for a variety of reasons, including jazz's centrality as dance music (with which country blues could not fully compete in popularity) but also likely due to racial barriers, class distinctions, and a commercial bias against the unpolished rural sound of country blues. Hillbilly's emergence as a genre, however, meant that the music was at the right place at the right time; as local and regional radio stations began to experiment with music that veered from the civilization-by-radio model in which classical music and light opera was broadcast in order to educate the ignorant masses, white rural music began to be heard increasingly on American radio. As Elena Razoglova has argued, "Local stations in the 1920s . . . forged symbiotic relationships with their farmer, immigrant, and middle-class neighbors," even after the rise of network radio.[57]

While many in the recording industry were concerned about what this new media technology would mean for their business, many hillbilly musicians were interested in increasing their careers through radio. Before 1925, most hillbilly performers were not recording artists anyway, and those that were recording for labels were not discouraged from radio appearances, a move that suggested that Victor and Brunswick's best and brightest were not to be found in the hillbilly genre. Meanwhile, hillbilly musicians were anxious to take advantage of radio's emergence; it made sense for rural artists to seek out on-air appearances on local radio, even though they were not usually paid, since their prime revenue came from live performances. "In the early years, the performer used radio time—for which he was seldom paid—to announce his personal appearances and later to sell souvenirs."[58] Radio programmers needed to fill their programming schedules and happily obliged many local performers. One performer to appear on early radio was Fiddlin' John Carson. By September 1922, Carson could regularly be heard on Atlanta's WSB. Various other noteworthy performers from the region followed quickly behind, including "Clayton McMichen, Riley Puckett, and the singing family of the Reverend Andrew Jenkins."[59]

Nashville's WSM wasn't on the air until October 5, 1925. As Charles Wolfe recounts, the first program included the Fisk Jubilee Singers (whose storied career dated back into the nineteenth century, through its ever-revolving lineup) and "speeches by local dignitaries," but no hillbilly tunes. WSM was the second radio station to enter the Nashville market; the first was WDAD, commonly known as "Dad's." As is common in the early years of any new media technology, both radio stations accommodated the other, with the notion that if either station succeeded, then radio as a media technology would succeed as well. Analysis of radio schedules from newspaper sources of the time show that both WSM and WDAD had broadcast "old-time music" or "hillbilly music" before their first month on the air was over. It is interesting that a rivalry between the two stations didn't begin to materialize until both were drawn into hillbilly music due to its unexpected popularity (unexpected because this music was supposed to be outmoded and outdated).

By mid-decade, the most popular hillbilly music style remained fiddle instrumentals, often based on old familiar tunes. Fiddle contests continued to be important social as well as musical events. As Wayne Daniel notes, during the emergence of old-time music on radio, some stations did on-air fiddle contests and relied on audience phone calls to choose a winner.[60] Radio listeners, Lizabeth Cohen notes, were "exposed . . . to new cultural experiences, to different ethnic and religious traditions, or to new kinds of music. But most important, workers discovered that participating in radio, as in mass consumption and the movies, did not require repudiation of established social identities."[61] Chicago's WLS, Nashville's WDAD, Shenandoah, Iowa's KFNF, and Atlanta's WSB had all broadcast fiddle contests (sometimes lasting weeks or months) by 1926.[62]

"Following the lead of the local program, regional shows took shape," D. K. Wilgus wrote. "The earliest was launched in 1924 on WLS, the Chicago station owned by the mail-order house which had already aided in the development of hillbilly music [Sears]. WSM's 'Grand Ole Opry' followed in 1925, and radio became a substantial part of the tradition, which still included all the old devices."[63] Neither show was the first barn dance broadcast. The first show was aired by Fort Worth's WBAP radio in 1923, as Diane Pecknold notes, with a signal that "could be heard as far away as Haiti."[64] But it was the WLS and WSM barn dance shows that would have the lasting cultural reach, on stations with enough wattage at this point to reach much of the country, important in repackaging local musical tastes for a regional (and eventually national) audience. "Reconceptualizing their audience as a regional one," Susan Smulyan writes, "stations in the South and Midwest moved quickly to introduce programs featuring country music performers. . . . The huge popularity of these shows, as evidenced by letters, listener purchases of songbooks, and attendance at special events presented by the stars, pushed station managers to think of programming for a regional audience."[65] There were significant differences in the two shows, as Susan Smulyan notes. "Midwestern barn dance programs differed in content and performers from their southern cousins, with the midwestern shows featuring fewer pure folk songs and singers and more commercial, popular treatments."[66] The negotiations necessary for hillbilly to transform in this way can be discerned in how the *National Barn Dance* and WSM's *Grand Ole Opry* addressed their audiences via the music featured on the show, each show's mode of address to its audience, and each show's ancillary press and promotional materials.

Founded in April 1924, WLS Chicago's *National Barn Dance* became widely popular, first in the Midwest and South, and later nationwide. The show's founder, George D. Hay, was a native midwesterner (born in Attica, Indiana, in 1895) and a Memphis newspaper columnist at the *Memphis Commercial Appeal* in the early 1920s. When the newspaper started radio station WMC in 1923, Hay became a broadcaster. In his short time on Memphis radio, he made a name for himself through his late-night show. Officials at WLS Chicago hired him away in 1924, where he started the *National Barn Dance*. Hay's name is of great importance due to his work in later establishing the only other barn dance show that successfully competed with the *National Barn Dance*: WSM Nashville's *Grand Ole Opry* (another Hay creation launched after he relocated again, to WSM Nashville in late 1925). While the *National Barn Dance* does not live in social memory as firmly as does the *Grand Ole Opry*, it was just as important and influential. I also emphasize Chicago's *National Barn Dance* here because more than any other show, it signaled the intersection of a regional musical style, changing societal attitudes about rural life, and the emergence of country entertainment as a national phenomenon.

WLS's initial owners through the station's first four years of existence (1924–1928), the Sears-Roebuck Company—the call letters stood for "World's Largest Store"—ensured that the *National Barn Dance* could reach listeners in all forty-eight states with its powerful wattage and strong signal. In 1928, Sears sold WLS to *The Prairie Farmer* magazine. Both Sears-Roebuck and *Prairie Farmer* magazine were companies that had important relationships with rural customers. Accordingly, the ownership transition had no effect on the *National Barn Dance*'s popularity. It helped that nearly half of all midwestern farm families had radios by 1930, compared to only 10 percent of white farm families in the South.[67] By 1933, the show was heard nationwide on NBC's Blue Network, years before the *Grand Ole Opry* joined network radio. These radio shows had potential appeal for various constituencies. First, they provided radio for those rural Americans relocating to industrial cities for work, the music listeners associated with a simpler time back home, consumed as an antidote to or expression of homesickness. And for those still living on the farm, these radio shows—and the music itself—offered a nostalgic, idealized view of a rural lifestyle that had already passed, in the unlikely case that it ever existed at all.

A 1926 *Variety* article summed up early attitudes about the emerging hillbilly genre by focusing on what the trade paper saw as its audience: "Theirs is a community all to themselves. Illiterate and ignorant, with the intelligence of morons, the sing-song, nasal-twanging of a Vernon Dalhart or a Carson Robison on the discs, reciting 'Prisoner's Song' or 'The Death of Floyd Collins,' biggest hillbilly song to date, intrigues their interest."[68] The hillbilly hick or rube was not completely without admirable traits in the national imagination, however, as is still evident in the long tradition of these characters and narrative tropes in American society. Richard Peterson offers an appealing description of the traditional hillbilly figure as "a self-reliant (most often male) child of nature, unfettered by the constraints of urban society" that remains familiar even in contemporary culture.[69]

In such a climate, hillbilly radio emerged slowly and against the wishes of most radio programmers, who then preferred what Erik Barnouw (as well as some 1920s industry insiders) called "potted palm" music: light opera and classical tunes. That music was broadcast to rural as well as urban listeners, with the implicit notion that the programming might bring culture to the far reaches of the United States. Such unexpected success with hillbilly led radio programmers to a previously ignored audience. WLS Chicago, owned by Sears-Roebuck, would have certainly appreciated a hillbilly-styled show to attract their rural customer base, many of whom were already loyal to the company through their mail-order catalog business.

It's important to note, however, that as the hillbilly genre gained listeners through barn dance shows on stations around the United States, the programmed music became less and less connected to what it originally referenced. A good example is the *National Barn Dance*'s theme song, which mixed traditional fiddle music with a march. The shows reached a large audience in spite of or because of these changes. Peterson notes that the *National Barn Dance* broadcasts was attractive to listeners "because they evoked in the listener a nostalgic remembrance of community dances, medicine shows, and street-corner singers packaged respectfully and in a fast-paced format appropriate to the 'electronic marvel of the age,' radio."[70]

The *National Barn Dance* included a variety of acts that directly linked the show to the rural vernacular music of the South. But the *National Barn Dance* featured as many vaudeville-style performers and comedians as anything that might be understood as country music. The regular *National Barn Dance* lineup also illustrated the eclectic and hybrid nature of the musical performances. The show featured fiddle bands, square dance callers, folk balladeers, pop vocalists, comedians, and novelty acts. While the Hoosier Hotshots trafficked in the "hick" and "rube" characters often associated with the hillbilly genre, their instrumentation was as close to jazz as to country music and their use of the clarinet recalled the Yiddish klezmer tradition. Closer to the hillbilly prototype was Lulu Belle and Scotty Wiseman, a couple who sang traditional love songs in a traditional Appalachian style. Unlike the earliest of country performers to appear on radio, performers on the *National Barn Dance* were likely to be professionals who toured to promote themselves in a variety of different contexts.

George C. Biggars, a WLS employee involved with the early programming of the *National Barn Dance*, described the show as featuring "traditional country dance tunes, interspersed with heart songs and sweet ballads—popular in the memories of most listeners."[71] Even during the time of its first broadcasts, the show referenced a time and place that was increasingly understood as being part of the past. Perhaps this played some role in listener identification. Rural listeners were invited to sweetly remember bygone days of an idealized rural past, whether they were still on the farm or had moved to the cities for work. The *National Barn Dance* also allowed space for urban listeners through an appeal made easier by the kind of musical repertoire that avoided the more unsavory aspects of the hillbilly music tradition, which was known for murder ballads and tales of catastrophic disaster.

As with other audio technologies, an unexpected phenomenon arose in that radio listeners were deeply curious about the appearance and personal details of those performers to whom they were listening. As in the early film industry's discovery of the "star system," those involved in radio did not anticipate such strong interest in their artists. Entertainers initially saw radio and recordings as a way to publicize their in-person performances but were surprised in the amount of interest their radio appearances garnered. Bradley Kincaid, "The Kentucky Mountain Boy with the Houn' Dog Guitar," became an early success on the *National Barn Dance* through his renditions of songs in the Appalachian folk tradition. As a song collector, Kincaid closely studied the music he performed. He was particularly sensitive to being called a "hillbilly," noting that he was nothing of the sort; he would often cite his education at Berea Academy in Kentucky. Early in his tenure on the show, he arrived at the *National Barn Dance* theater to discover a large group of people lined up to get in. When he asked a fan in line who he was there to see, the fan replied that he was there to see Bradley Kincaid in person.

This interest in the *National Barn Dance* led to several ancillary publications and products related to the show and its performers. Kincaid, like many *National Barn Dance* regulars, used his popularity on the radio show to push his live appearances. Kincaid was also greatly successful in pushing ancillary products. He published a songbook for fans called "Bradley Kincaid's Kentucky Mountain Ballads," which sold ten thousand copies in its first year of publication. Kincaid also partnered with Harmony Guitars to sell a Bradley Kincaid "Houn' Dog" model to the public. Finally, and perhaps most importantly, *Prairie Farmer* magazine would publish a *National Barn Dance* photo album insert each year for fans of the show, with pictures and stories about the performers. These ancillary products not only served to bolster fans' sense of connection with a given performer on a show, it also in some way translated an experience—listening to the *National Barn Dance* every Saturday night—into a souvenir, either directly or indirectly tied to the show, that could be accessed at any time, unlike the show itself.

The emergence of hillbilly radio presents an ambiguous situation in which rural listeners, who had previously been approached only through paternalistic attempts to bring high culture to the hinterlands, now were addressed directly with shows that grew out of rural, vernacular music. But as popular as shows like the *National Barn Dance* were for rural listeners, they limited their depictions of rural life to the joke telling of good-natured hicks and rubes or the nostalgic remembrances of an idealized past. But it was an idealized past both rural and urban listeners were ready to embrace. In *American Babel*, Doerksen suggests that those first barn dance radio programs, meant for rural audiences but produced in urban contexts, were straddling an ever-growing societal split. It was amidst these changes that a new media technology, radio, would link rural cultures with the country's increasingly national identity as embodied with the growth of network radio broadcasting. "This was not a music foisted on the American folk, but a music exploited by a music industry in search of wider markets. And there was certainly a rush of musicians wishing to be exploited. As George Hay would suggest after leaving Chicago's WLS for Nashville's WSM, the listeners wanted not Grand Opera, but "Grand Ole Opry."[72]

The central difficulty in our understanding the early *Grand Ole Opry* (originally titled the *WSM Barn Dance*) is our inability to hear it for ourselves. In the years before magnetic recording on audiotape—a technology that the United States acquired via spoils of World War II—radio transcription discs were the sole means of capturing a given performance. In the early years of the *Opry*, before network broadcasting made transcription disc increasingly common (so one show could be broadcast at the same time in all time zones), the show was broadcast once for listeners and never heard again. Thus information on the early *Opry* broadcasts comes exclusively from secondhand sources. So, Wolfe argues, while we have performers' accounts, press reviews, promotional materials, and fan letters, we lack "this intangible element (the radio Opry), which was apparently a prime force in the popularization and development of country music."[73]

Traditional folkloric research on the *Grand Ole Opry*, Charles Wolfe argued, has fallen short due to its inability to account for the impact and importance of that most ephemeral of 1920s media forms, radio. As Wolfe rightly argued, folkloric approaches to the Grand Ole Opry's success either privileged recordings made by *Grand Ole Opry* stars or sought to track the "song family" in connection with *Opry* sources. Why did these approaches not work? "The reason, of course, is that the Opry utilized a different medium—radio—instead of phonograph records, oral tradition, or print.

It is much more difficult to comprehend the success of a radio show than to understand the success of a singer, a song, or a record."[74]

Radio's ephemeral nature in the years before it was regularly recorded does make it all the more difficult to get a general sense of its power. Radio historian Steve Craig notes that once the barn dance shows began being recorded before live audiences, enthusiasts would make the pilgrimage to *see* the show: "Some shows became so successful that their broadcasts became major tourist attractions—a must-see event for farm families visiting the big city."[75] Fan response also led to the beginnings of country music journalism, often focused on barn dance radio shows. "WLS apparently was a leader in establishing the link between fans and performers. After the *Prairie Farmer* Publishing Company purchased the station from Sears, Roebuck and Company in 1928, they found that so many requests for photographs of the Arkansas Woodchopper and other performers were received that they could not handle all of the requests. As a result, the station put together a portfolio of photographs ('radio pictures') in 1929 in order not to disappoint listeners."[76]

Another example of the show's popularity: George Hay, while still announcing for WLS Chicago, received an award from *Radio Digest* for being the "most popular announcer in the nation" as voted on by radio fans.[77] Hay had learned how to connect with an audience during his newspaper days while writing a column in which he answered readers' questions. By the time he arrived in radio, he had a keen sense of what might interest the public. His mode of address seems to have suggested an ethos of inclusion, whether he was announcing his barn dance show or working on another part of the WLS schedule. While we remember Hay for his involvement in hillbilly radio, this was not the limit of his responsibilities or capabilities at the station. The *Radio Digest* award may well have been a deciding factor in WSM Nashville's desire to hire Hay away from his post in Chicago. Wolfe recounted that Hay was invited to the opening ceremonies of WSM Nashville by WSM's parent company, National Life Insurance, where the parent company tried to impress Hay with its plans and convince him to return to Tennessee (this time to Nashville), where the company would give him greater control over its programming.[78] Just as William Paley got his father's cigar company into the radio business, Edwin Craig (the son of the company's president) convinced the board of directors at National Life (slogan: "We Shield Millions") that radio would be good for business.[79] Craig, whose interest in radio had begun as an amateur hobbyist, now was in the position to shape radio's course.

After a relatively short period of negotiation, Hay left WLS Chicago for WSM Nashville. "Hay," Charles Wolfe wrote, "saw the move as a step up: he was moving from merely being an announcer in Chicago to 'radio director in charge' of the entire station in Nashville. He would be free to craft the new station's actual image and to develop his own line of programming."[80] While Wolfe perhaps underestimates Hay's role in Chicago (while technically "an announcer," he had launched and programmed his own show), the move to Nashville did allow for more creative freedom. From early on, Hay's plan was to take his *National Barn Dance* concept started in Chicago to Nashville. By December 1925, an announcement by WSM had been made: "Because of this recent revival in the popularity of the old familiar tunes, WSM has arranged to have an hour or two every Saturday night, starting Saturday, December 26."[81]

The *Nashville Tennessean*'s description of a "revival" is interesting, however. To many of the progenitors of this music, the standards had never left their repertoire. But this again marks the music—even though heavily rooted in the contemporary culture of the time—as heavily coded in nostalgia for many listeners. In some cases, the music was thought of as a respite from the corrupting influence of jazz, the most popular music of the land by the mid-1920s. Wolfe cites a press release believed to be written by George Hay publicizing an upcoming performance by two figures central to WSM's barn dance show: Uncle Jimmy Thompson and Uncle Dave Macon. Thompson was in his late seventies when he rose to fame on the program, while Uncle Dave Macon was an entertainer already well known through phonograph records. The WSM release positioned hillbilly as a traditional counterpoint to the perceived encroachment of jazz: "Old tunes like old lovers are the best, at least judging from the applause which the new Saturday night feature at Station WSM receives from its listeners in all parts of the country; jazz has not completely turned the tables on such tunes as 'Pop Goes the Weasel' and 'Turkey in the Straw.'"[82] Here hillbilly music is offered up as an antidote to various kinds of jazz. The press release also notes that while not many listeners frequent barn dances themselves (the barn dance motif referencing a passing leisure activity by the 1920s), American audiences "have the habit of clamping on its ear phones and patting its feet as gaily as it ever did when old-time fiddlers got to swing."[83]

By the late 1920s, radio was quickly emerging from local and regional spheres of influence to a national presence. Key to this was the emergence of the NBC radio network, which offered a set menu of programs available to affiliate stations. WSM Nashville was one of NBC's affiliates. "This meant that the station could carry the same network shows heard all over the country, though it could not yet feed shows up to the network" and would not do so until well into the following decade. While the show would regularly garner hundreds of supportive phone calls each week it aired, the *Tennessean* added that the show's attraction "seems a mystery to a number of people."[84] The response to the show and to hillbilly artists, while not uniformly negative, was often similarly lukewarm. If some print sources denigrated the show along class lines, George Hay and his barn dance show landed a few critiques of their own.

Perhaps the most famous example of the defiant class dynamics can be found in the story surrounding the show's name change. The story is well known in country music circles. By the end of 1927, WSM had changed the name of the *WSM Barn Dance* to the *Grand Ole Opry*. "It so happened that on Saturday nights," Hay remembered, "from seven to eight o'clock, WSM carried the 'Music Appreciation Hour' under the direction of the eminent conductor and composer, Dr. Walter Damrosch."[85] Wolfe recounted Hay's impromptu ad lib and the coining of the term "Grand Ole Opry":

> Hay and the Opry cast were waiting for a network show to end, the NBC *Music Appreciation Hour* with noted conductor Walter Damrosch, so they could come on with their locally produced show. As he concluded, Damrosch lamented that there was little or "no place in the classics for realism" and conducted a short classical piece depicting a train ride. Hay, coming on seconds later, proclaimed: ". . . from here on out for the next three hours we will present nothing but realism. It will be down to earth for the 'earthy.'"[86]

Then, as with most programs, Hay introduced Deford Bailey, the black harmonica player that regularly opened the show. Bailey launched into "Pan American Blues," a song for which he was famous, using his harmonica to mimic the sound of a train moving down the tracks and providing the vernacular musical equivalent of realism. After Bailey's performance, according to Hay's own account and *Opry* legend, he referenced the differences between his show and Damrosch's, suggesting, "For the past hour we have been listening to the music taken largely from the Grand Opera, but from now on we will present the Grand Ole Opry."[87] The fact that this story is a central part of the *Grand Ole Opry* historical narrative (it appears in most narratives about the show) reveals how close it remains to the show's ethos, despite all the changes and intervening years.

While in the early years it was performed in WSM's studio, the show began being broadcast before an audience, like WLS's *National Barn Dance* had been doing for years. After experimenting with bringing an audience into WSM's studio in the National Life Building in downtown Nashville, the show began broadcasting from a number of theaters around the city. The addition of an audience must have infused the show with even more energy and excitement, and by this point the show had settled into a fairly rigid formula, which could be upset by special events or the demands of the network. In Richard Peterson and Paul DiMaggio's reading of the show, based on analysis of an in-house WSM show diary, they wrote, "The program usually opened with virtuoso harmonica player, Deford Bailey, or Obed Pickard, who was billed as a 'one man orchestra.' The last segment of the evening was often a 'black-face' comedy routine by George Hay and Ed McConnell."[88] Among musical acts, fiddle players were featured most often, followed by a sizeable number of harmonica players and string bands. Much of this music was instrumental, as William McDaniel explained in his 1952 history of the *Grand Ole Opry*, because of the social traditions of folk dancing that largely called upon instrumental acts.[89] "The only forms of popular music which are conspicuous in their absence are the foxtrot-dance orchestras and jazz groups which used brass, reed instruments, piano and drums. What was apparently acceptable on the Opry, then as now, seems to be determined more by instrumentation than by lyrical, rhythmic, or melodic content."[90]

As the central figure involved in the *Opry*'s first decade, George Hay's role as tastemaker was central to the show, from his often-repeated advice to performers to "keep it down to earth" to his detailed management of the show. The WSM 1928 Diary reveals that Hay was still doing the booking for the show even though, as one of Peterson and DiMaggio's interviewees noted, "To lay it on the line like it was, Judge Hay didn't know music; he couldn't memorize tunes. . . . He liked rapid tunes because he thought the man making the most racket was making the most music."[91] As the authors suggest, "This was probably Hay's greatest talent, to bypass what experts might find the most aesthetically pleasing or technically proficient in favor of that which would hold a radio audience."[92] He also was careful to leverage the vaudeville tradition for the benefit of the show's themes, jokes, and characters. The show employed ethnic jokes and familiar bits easily identifiable to audience members versed in vaudeville humor and mode of address. And the existing vaudeville model of the hillbilly bumpkin act, of which Hay took advantage, funneled a variety of musical and theatrical influences into a neat hillbilly package. Peterson has argued that the hillbilly character (as opposed to variants, such as the "elderly geezer") offered "a much more potent dramatic persona for the popular culture industry and for the audience. Being set off against and often triggered by the intrusion of urban ways, the hillbilly's humor, aggression, and love laments could change with the times."[93] In this sense, the hillbilly persona was defined through his difference from urban characters, which is only possible through some self-awareness of the differences between the hillbilly character ethos and those of the urban culture with which he comes in contact.

The local press in Nashville began to take notice not only of recording artists but also the *Opry*. "By the middle of 1928 the [Nashville] *Tennessean* was running a regular column of record reviews in which some old-time music, including several records by Nashville radio personalities" were reviewed, including Uncle Dave Macon and Humphrey Bate, noting that the hillbilly reviews tended toward the negative and neither Macon or Bate were referenced as local radio celebrities.[94]

While the majority of the *Opry* regulars were not the slick vaudevillians of the previous decade (despite borrowing heavily from the vaudeville tradition for their show), neither were they the country hicks and rubes they often portrayed. "A substantial majority of the Opry regulars were not even rural residents in 1925, a full year before WSM went on the air, but were residents of Nashville. Not drifters either, they worked at a wide variety of urban trades ranging from barber and railroad dispatcher to insurance salesman, watchmaker, and physician. Those who regularly journeyed to Nashville from the surrounding counties were the sons of farm owners or small town tradespeople."[95]

The clothes *Opry* characters wore early on were not the country getups that might be expected. "The earliest pictures of Opry members show them in dark suits, white shirts, and ties. . . . Only when the show began to cater to a studio audience—about 1928—did Hay begin to insist that group members *look* the part of hillbillies."[96] As Peterson has proved in his book-length treatment *Creating Country Music*, he suggested that hillbilly performers would often adopt the hillbilly (or later cowboy) persona for purposes of performing. In the case of Jimmie Rodgers, Peterson details how Rodgers would perform different roles in different venues, appearing in a bowler hat and tie for vaudeville audiences and in his railroad brakeman attire or cowboy costumes according to audience and context.[97] Rodgers was not the exception, but the rule. His case is a bit unusual only in the number of different roles he played, depending on the circuit and the town. Overall, it was the hillbilly persona that sold. "Record jackets and catalogues featured rural or mountain iconography; fiddlers at old-time dances, mules coming down mountain trails, singers sitting on cabin porches. Just as early country music radio program directors like George D. Hay and John Lair would create a hayseed image for the music they promoted, Peer and other early record company publicists sought to authenticate their product by emphasizing its rural origins."[98] This approach would become even more important after a scouting trip Ralph Peer took in 1927, with stops in Bristol, Tennessee; Charlotte, North Carolina; and Savannah, Georgia. While the Savannah and Charlotte legs of the trip have been largely forgotten, what was recorded at Bristol became essential to the genre's future. The way the Bristol sessions were described by Peer and his new record label (Victor) would also add further layers of myth in the service of sales.

The Bristol Sessions and the New Old Sound

Well into hillbilly's increasing national reach through phonograph records and radio, Ralph Peer "discovered" important new artists and transformed the genre. Although Peer was the talent scout most responsible for the emergence of race and hillbilly in terms of early involvement and developing the use of a marketable category, he remains a deeply ambivalent figure who claimed to have little personal interest in the music he recorded. He oscillated between esteem and disdain for the performers with whom he worked, as layers of myth accumulated with each retelling. Although Peer was instrumental in encouraging the recording of race records at the OKeh label, it was Fred Hager who set up the initial sessions. And Peer initially passed on Fiddlin' John Carson, signing him to an exclusive contract only after Polk Brockman convinced Peer to allow a pressing of five hundred copies and the subsequent record sales helped Peer change his mind. Still, it was Peer who recorded the two acts near the end of the decade—Jimmie Rodgers and the Carter Family—that would rejuvenate the recording industry in the late 1920s amidst concerns over radio and a slowing national economy. The decisions Peer made were reflected the selection of music, the repertoire of its performers, and the way in which appeals were made to audiences.

After 1925, the advent of Western Electric's new electrical recording technology fundamentally changed how recordings were produced. Many have written about the changes that came with electronic recording, particularly the replacement of the acoustic horn with the microphone, which functioned as a kind of audio close-up and further fueled the emergence of star personalities (perhaps tied to the feeling of intimacy created by the voice via microphones on recordings and radio, most famously evident in the career of Bing Crosby). However, my interest in the new recording technology lies in its high degree of mobility. Before electrical recording began in 1925, recording in remote locations was highly impractical. As Eric Morritt writes, "Making field recordings at remote locations was only possible with bulky equipment, and the resulting recordings were of limited quality and were not pressed for public enjoyment. Recordings outside of a studio were used primarily for folkloristic study or for experimental purposes."[99] Record companies that could afford the technology—adoption of new technologies always being an issue—found a much greater degree of flexibility during a moment in which the record industry was experiencing a greater degree of competition for new musical acts and expanding genres. This flexibility and mobility led to even greater activity among talent scouts, including Ralph Peer. Peer had been a groundbreaking talent scout and businessman, focusing on the long-term value of publishing rights. He devised one of the most common music contracts of the era, in which all rights went to a given label for a flat $25 fee per side. In addition, the "composer-artist" received $0.005 for each side (not record) sold.[100] But his work at OKeh came to end as the context of his work changed. In October 1926, OKeh's Otto Heinemann sold his company to Columbia, a company interested in OKeh's success in the burgeoning race and old-time music fields and one that could facilitate Heinemann's interest in entering the electrical recording field.[101] Peer left OKeh before the ink was signed, displeased with the proposition of answering to Columbia's leadership, and approached Victor with a unique deal that proved Peer's grasp of the importance of publishing and copyright: he would work for *no salary* if Victor allowed him to retain copyright on the recordings he scouted, recorded, and released.[102] It was within this context that Peer went on a well-planned scouting trip to Bristol, Tennessee, in 1927. Various record labels had been to the area by that year because of its central location along the Virginia-Tennessee border in the Appalachian Mountains—the "hill country" aura played a big role in selling early country records—and the success previous sessions had brought for a number of labels.[103] That summer, Peer and his assistants recorded nineteen different acts, including Ernest "Pop" Stoneman, Henry Whitter, the Johnson Brothers, and Blind Alfred Reed. But the Bristol sessions were most famous for the two acts whose own careers were transformed and whose music redefined the genre permanently. The mythology surrounding the Bristol recordings of Jimmie Rodgers and the Carter Family as the true birth of country music has obscured how these performers must have been received at the time. Rodgers, a Mississippi native, made the trip to Bristol from Asheville, North Carolina, where he had been making regular radio appearances on WWNC. The Carter Family hailed from nearby Maces Spring, Virginia, and had made a regional name for themselves through their public performances. While both Rodgers and the Carter Family were professional musicians, mythologizing by Peer and Victor started almost immediately, emphasizing the remoteness of the town and the people he recorded. In tales told by Victor's publicity department via its record catalogs and Peer's yarn spinning in newspaper interviews, the Rodgers and the Carter Family were both sold to the public as unpolished bumpkins. Here is an example of the talk surrounding Rodgers, already a consummate entertainer by 1927, who had been appearing as part of the Tenneva Ramblers, a group that had sacked him the night before their Bristol session, at which point Rodgers decided to try out on his own, as Charles Wolfe recounted:

> In 1928, barely a year after the 1927 Bristol sessions, Peer gave an interview in which he recalled that Jimmie Rodgers had been "running around in the mountains" before his session in Bristol, and that when he tried out "he was laughed at." In the 1930 Victor Records catalogue, the biographical sketch of Rodgers describes the session as "a Victor recording expedition into the mountains of Tennessee."[104]

And here is a sample of Peer's depictions of the Carter Family:

> Peer later claimed that when the Carter Family first appeared at the temporary Bristol studio, they looked like they had "come through a lot of mud either by horse and buggy or an old car. . . . He [A. P.] was dressed in overalls and the women were country women from way back there—calico clothes on. . . . They looked like hillbillies." Local tales in Bristol depicted the Carters as never having worn shoes and as never having been in town before; such tales even characterized the Carters as having climbed into the studio via the fire escape because they were too embarrassed to be seen in public.[105]

Wolfe noted that Victor used the term "recording expedition" in their 1930 catalog, with all its connotations of the company's emissaries bringing civilization to the wilds of Tennessee. The Carters, he adds, "regularly came into Bristol, and while they may have been modestly dressed, none of the Carters remembered recording barefoot."[106] Similarly, even a cursory knowledge of Rodgers' biography, though not available to most hillbilly fans of the era, suggests that his wandering out of the hills and into the studio was a sensationalized fabrication. But Peer, who had long been of the opinion that much of the hillbilly's genre's appeal was in the exoticism of the remote, sought to capitalize on a rural "hill life" that had already largely sunk into the past by the time of hillbilly's emergence.[107] Despite depictions of the town of Bristol that were sometimes less than flattering, the town embraced its hillbilly image "with an attitude of boosterism."[108] This also suggests that the more appealing aspects of the hillbilly persona (e.g., rugged individualism, stubborn self-reliance, etc.) held a strong appeal.

It wasn't unusual that Ralph Peer's mythmaking would serve to sell the music he recorded, or that his musical tastes did not match the music he promoted and exploited. And that he had little or no affinity for the performers and recording artists with which he worked was not particularly unusual either, although many scouts, H. C. Speir or Art Satherley among them, were known comparatively for their empathy, understanding, and relative generosity toward their recording artists. Peer, however, would pay the least possible amount for a recording, sometimes paying nothing at all. His trick in accomplishing this was a stoic demeanor during scouting and recording sessions. "Peer told his recording artists as little as possible about how much he and the companies he worked for needed them," William Howland Kenney notes. "Peer bet that rural entertainers would conclude that they needed him more than he needed them, that they would prove so eager to record that the majority would, if that were the only alternative, eagerly do it for free."[109]

While Ralph Peer regularly mythologized and occasionally belittled the Carter Family in shaping their public image, his son remembers him expressing his admiration: "My father always said that what amazed him was that they were good, but they didn't seem to know *how* good they were."[110] After the Bristol sessions, word from Peer and Victor came slowly. But soon the Carter Family's recordings started being released on Victor's "Orthophonic Southern Series." Shortly after, the Carters were asked to come to Camden, New Jersey, to record at the Victor studios. Later, Peer and his wife came to the Carters' home in Maces Spring, for which the family made repairs, painted the house, and sent the Peers away with "a country ham and jars of dewberry jelly."[111] Their records for Victor sold very well, as regularly distributed Victor order blanks and advance lists announced their latest releases. A July 20, 1928 advance list announced, "More and More and More of those red hot Southern Specials," which featured "River of Jordan" and "Keep on the Sunny Side," along with other releases including the Tenneva Ramblers' "Seven Long Years in Prison" and "If I Die a Railroadman" and the University of Texas Longhorn Band's "Eyes of Texas" with "Texas Taps."[112] In the late 1920s and early 1930s, the Carter Family's career skyrocketed as they toured the country relentlessly, including a half-year stint doing daily shows at Mexican border radio station XERA, just across from Del Rio. The station, run by celebrated "Doctor" John Brinkley, was founded after the Federal Radio Commission began cracking down on Brinkley's Kansas station, KFKB, over whether it actually served the public interest. For the next several decades, the Carter Family continued to record, tour, and appear on radio around the country. In the process, they became the first family of country music. With varying line-ups, the Carter Family continued performing, even after A. P. Carter's death in 1960.

Jimmie Rodgers' career was perhaps even more impressive although painfully short. Peer also sent for Rodgers to record at the Camden studios, and Rodgers obliged. In May 1928, Victor announced that "another Jimmie Rodgers record" (in this case, "Ben Dewberry's Final Run" with "In the Jailhouse Now") would receive "national listing," which meant that his subsequent recordings would see coast-to-coast release, as opposed to regionally focused distribution.[113] In addition, Victor included a biographical sketch titled "From Singing Brakeman to Popular Exclusive Artist," which described Rodgers' rags-to-riches story: his time working on railroads, his family life in Meridian, Mississippi, and his life as a Victor recording artist. The note from Victor explained, "A knowledge of Rodgers' struggles and final success will be a real help in selling his records."[114] This move by Victor suggests that officials at the company now understood that Rodgers' appeal and that of the hillbilly genre could have a much larger audience than it previously had thought. It also suggests that Victor understood that Rodgers' personal narrative augmented his existing popularity as a talented singer and songwriter. Rodgers' records continued to sell; on July 11, a letter sent from Victor's department of sales and merchandise to its wholesalers announced the early release of Rodgers' "Treasures Untold" and "If Brother Jack Were Here."[115] His recordings were immensely popular and led him to be the focus of a musical short in 1929, *The Singing Brakeman* (he performed "Waiting on a Train," "Daddy and Home," and "Blue Yodel #1," also known as "T for Texas").[116] Rodgers made many recordings, including a hillbilly/jazz hybrid, "Blue Yodel #9" with Louis Armstrong on trumpet and his wife Lil Hardin-Armstrong on piano. For the next few years, Rodgers relentlessly toured, stopping only when he felt his health was in danger. Already suffering from tuberculosis when he recorded with Peer at Bristol in 1927, his condition worsened over the years. In 1929, Rodgers and his wife Carrie bought a home in Kerrville, Texas (and later lived in San Antonio), with the understanding that the dry heat would help his condition. Rodgers' Texas locale would come into play in two promotional Victor recordings made on a Victor scouting and recording trip to Louisville, Kentucky in June of 1931. The recordings were "awkward but much loved," and the "Carter Family and Jimmie Rodgers in Texas" side features Rodgers' only recorded vocal duet, a snippet of "T for Texas" with Sara Carter.[117] The conceit of each record is that we're listening in while Rodgers visits the Carter Family at their Virginia home and the Carter Family visits Rodgers in Texas. The record played up the Carters as mountain folk and Rodgers as a cowboy of sorts (despite his and his wife's arrival in Kerrville in April 1929). In May 1933, Rodgers died at thirty-six years old, while in New York after being summoned by Peer for a recording session.

The songs recorded at Bristol, which had the full support of Peer's new employers at Victor records, showed the maturation toward a genre that would eventually be known as country music. By the end of the 1920s, the hillbilly genre had already steered away from the earlier fiddle bands and balladeers to the updated sounds offered by the likes of Jimmie Rodgers and the Carter Family. The effect that the new artists had on genre originators like Fiddlin' John Carson hastened their descent from the heights of hillbilly celebrity. By the late 1920s, Carson specialized in spoken-word skit songs with his daughter Rosa Lee (or "Moonshine Kate"), and while he remained popular, his style had receded in this new environment. What had started as "old-time music" was now increasingly hybrid, open to influences far beyond its original contexts. Peer's own post-Bristol ambitions seemed to only grow, particularly via his ongoing interest in the international market, as evident in Victor advance lists and order blanks from August and September 1928, which featured titles for languages and nationalities around the globe.[118] Meanwhile, Peer's ability to make money had so improved with his Victor arrangement that he worried it would attract attention.[119] His Victor agreement allotted mechanical royalties to Peer,

Fig. 3. Jimmie Rodgers and the Carter Family in Louisville, KY, for an RCA Victor recording session, 1931. (Country Music Hall of Fame.)

including publishing royalties and any other royalties to songs he owned that were recorded by labels other than Victor.[120] With money coming into his "only marginally operational" United Publishing business, Peer saw these earnings as "a potential embarrassment for him . . . it seemed *too* conspicuous, a potential red flag and point of contention with Victor executives, though they hadn't initially been paying attention."[121] His solution to this was his January 1928 incorporation of his varying business interests: United Publishing (nominally owned by Anita Peer), Ralph S. Peer, Incorporated (his management concern), and his fledgling music publishing company, known as Southern Music.[122] Peer's founding document for Southern Music was a reflection of his sizable ambition, but also Peer's conception of what a music company might be in a nascent age of intermedial entertainment media. As Barry Mazor writes, Southern Music's mission was described in its incorporation papers as being involved in some or all of the following: "publish, print, and sell sheet music books; buy and sell copyrights; manufacture, buy, and sell recording equipment, records, record players, radios, and cabinets," along with more speculative language about involvement in nascent television experiments, as well as the operation of "commercial broadcasting stations, theaters, or a motion picture distribution company, produce shows, manage performers or bands, act as a booking agent, and more."[123] Eventually Victor would own Southern Music and receive performance and sheet music royalties, while Peer would keep mechanical royalties and incentives to keep Peer on the search for new talent.[124] But whatever the recording industry and its old-time music had been, it was clear that something new was emerging, created as much by the recording and radio industries than by vernacular cultures large and small.

Synthetic Hillbillies

"The mystery is that not only rural people, but city residents," Susan Smulyan writes, "not only southerners and westerners, but northerners and midwesterners, sought out this music."[125] Ethnomusicologists and folklorists might suggest that Smulyan overstates the musical differences between the rural South, Midwest, and Northeast; still, Smulyan's surprise at the urban taste for this rural music remains relevant. Smulyan marks the important class components involved in translating the genre across a variety of sociospatial and cultural borders. As I have argued, she suggests "country music fused the conflicting responses to industrialization with the contradictory, somewhat romanticized feelings many urbanites had about rural life in the 1920s."[126]

While for a number of years there was a moment in which relative unknowns and amateurs did have unprecedented access to the recording industry, by the late 1920s and early 1930s that window had begun to close. By 1935, the music first recorded by amateurs at record labels like OKeh, Gennett, and others was ensconced as a formula. Those record labels that had survived the first years of the Depression turned to cheaper and cheaper record prices—often twenty-five cents or less. In such a climate, scouts looked for material with the greatest chance for mass popularity. And on radio, the hillbilly sound was also shifting. By the end of the following decade, a *Billboard* magazine interview with Meyer Horowitz, owner of the Village Barn—"a Hillbilly club in Greenwich Village"—could suggest that the synthetic hillbilly was much better at entertaining than those "real" hill country folks sought by record companies earlier. According to Horowitz, the article notes, "Real hillbillies rarely have good night club acts. . . . Jewish and Italian hillbillies usually outshine all others on showmanship, he says." "Native-born hillbillies from the sticks," the article continues, "usually provide entertaining music, but in most cases they're short on comedy and showmanship. As a result, synthetic hillbillies are as a rule more desirable in a night club than the real ones."[127]

Hillbilly had clearly joined the ranks of popular music by the 1930s. While jazz increasingly had become the U.S. commercial music of the previous decade, the new hillbilly genre allowed the industry to negotiate the beginnings of economic downturn and the rising power of radio, while turning Jimmie Rodgers and the Carter Family into household names. Behind the scenes, however, executives at the biggest two labels, Victor and Columbia, were facing systemic changes to stem a complicated tide of events. The focus of the next chapter is the ability of these two labels—indeed the entire recording industry—to weather the myriad difficulties that swept over the industry and the country with the coming of the Great Depression.

From its very beginnings, the hillbilly genre was understood as music imbued with the values of a quickly vanishing past, even though that music was already carrying the influence of music it sometimes was juxtaposed against (as in the case of jazz). One of the primary ways these values were cultivated and circulated was through record company practices and narratives, from scouting trips to publicity materials. In her analysis of country music in mid-twentieth century Nashville, Joli Jensen suggests that "we ratify the conflicts of modernity by ascribing to country music the authenticity we see evaporating in the present," following Raymond Williams' argument in *The Country and the City* that suggests we tend to project on the country all those things we feel are being lost in the city.[128]

Hillbilly music and radio had already been domesticated for broader audiences, often heard in cities far removed from the mythologized contexts the music referenced. In Los Angeles, singer, actor, and radio host Stuart Hamblen had made a name as one of the genre's first "singing cowboys," selling the romance of rural, western adventure to listeners to his various radio shows on which he hosted or appeared, including *KFI Jamboree*, *Covered Wagon Jubilee*, *Family Album*, and, after his sobriety, *Cowboy Church*. Unthinkable and incongruous just a few years before, Hamblen rubbed elbows with Hollywood stars like Roy Rogers and John Wayne, even hunting mountain lions with Clark Cable and Gary Cooper in the hills surrounding Los Angeles.[129] If country and city were ever truly separate spheres, the two had been drawn together via media—recordings, radio, and film—in unprecedented and altogether new ways. These connections between media forms would become increasingly commonplace at the end of the 1920s as the recording, radio, and film industries realized the importance of media cross-promotion and the possibilities inherent in cross-ownership. But the anxieties and contradictions tied to rural identities from the inception of old-time music and barn dance radio would only be furthered in increasingly hybrid, mediated forms, while propelling both the music and radio that sprung up around it into the national consciousness. Of the new genres that had been embraced by the recording industry during the course of the 1920s, hillbilly was discovered last. It also was the most industrially viable genre form heading into a period of technological change, media consolidation, and great economic turmoil, in which the shape of the entire industry would change.

CHAPTER FOUR

"On with the Dance"

Media Industries' Jazz-Age Convergence

Music across Media

Radio's promotional appeal to musicians has long been understood by the music industry. A local radio appearance might help sell show tickets or an artist's latest recordings. With the rise of network radio in the late 1920s, additional promotional opportunities emerged—along with much higher stakes—for those artists afforded the chance. Network radio did not emerge all at once. As Alexander Russo writes, "Radio stations drew upon not only national network feeds but also a wide range of programming sources including regional networks, sound-on-disc transcription recordings, and national produced scripts performed locally."[1] Local and regional radio continued to be an important force in communities small and large. Musicians made regular stops into radio stations to publicize upcoming shows and newly released recordings. Some of the most entrepreneurial artists extended their live performance and recording careers into hosting radio shows that extended their reputations in new ways. For recording companies, the reach of network radio meant that artists could publicize national tours, and Hollywood's interest in sound recording was a central engine by which the recording industry extended its financial power and cultural reach. But the collaborations between the recording and film industries were not limited to electric recording technologies, as important as they were. Hollywood's perennial interest in cross-promotion via commercial sound recordings grew in sophistication and frequency. The movie studios also invested in music publishing companies to fuel their voracious appetite for songs as musicals became the key genre of the early talkies era. Following Andre Gaudreault's call to reposition cinematic history in the intermedial contexts out of which cinema emerged, I argue here that the changes in the recording, radio, and film industries in the late 1920s and early 1930s cannot be understood discretely but must be considered together to better understand the scope and depth and breadth of media industries in this (or any) era.[2]

These changes were also reflected in the evolving strategies of performers, who sought out opportunities not in recordings, radio, or films, but all three at once. Some performers found great success with this, while others flirted with such possibilities to varying degrees. In an era when the recording industry's influence was no longer as singular as it once had been, extending one's reach into emerging media platforms became crucial not only in promoting shows and selling records, but in establishing a larger celebrity identity in an increasingly competitive context. In what follows, I will argue that the dual dynamics of generic consolidation as well as industrial and technological convergence played interlocking roles as recording companies sought to remain profitable and relevant. These changes are best illustrated through the cross-media careers of the era's musicians. Moving from the larger structural changes to the careers of individual celebrities such as Paul Whiteman, Louis Armstrong, and others can help exemplify how material changes led to shifting tactics among the era's most popular performers.

Media industry histories are most commonly told within the confines of a given industry. There are understandable reasons for this. It is a formidable task to get a given industry right, let alone hold two or more industries in relationship simultaneously. But a comparative approach is crucial to understanding the changes occurring within the recording, radio, and film industries in the late 1920s. Because media industry histories tend to be discrete rather than comparative, the scope of the industrial, technological, and generic changes is often missed. It meant no less than a recalibration of the recording, radio, and film industries, well before the economic collapse that would mark the beginnings of the Great Depression. This chapter details these changes in their specific impacts and broad implications, changes that have often been obscured by the cataclysmic changes to come at decade's end. But first, I will briefly map the popular music of the late 1920s to illustrate the ways in which the field had changed amidst the mainstreaming of the insurgent styles of the era.

Mainstreaming Genres

By the late 1920s, jazz was synonymous with popular music. Up from its hot roots, bands performed in bigger and bigger configurations, as the music became symphonic and sweet. In 1928, *The Voice of the Victor*'s cover commanded "On With the Dance," which pointed to a lavish advertising spread inside designed to alert retailers of jazz's money-making possibilities: "the need for it, the demand for it, the vast quantities of it that are consumed, the money that is spent on it, the hundreds of musicians who live by and for it—you will realize that jazz is a highly *profitable product with possibilities!*"[3] The ad ensured Victor's supremacy in its signing up all the most popular jazz bands on Broadway (George Olsen, Fred Waring), in the cabarets (Ted Weems, Jacques Renard) and on radio (Jean Goldkette, Nathaniel Shilkret).[4]

Starting in the latter years of World War I, upstart U.S. recording labels had experimented with unprecedented breadth and depth, recording a vast array of sonic styles, much more than I can recount here. This limitation is felt all the more when we account for the various spoken-word recordings, supplemental recordings meant for synchronized playback with motion pictures, and sound-effects recordings of the era. The cross-media activity was not limited to musical activity. But the exponential growth *was* musical, and with each new genre constructed and assimilated via the cultural circuit, the recording industry expanded its aesthetic range of production as well as its own audience. In turn, they learned from consumers via trial and error as those listeners educated them on how to proceed. In the hands of a wide variety of listeners, the impact and influence of recordings and increasingly hybrid commercial music styles began to accelerate as the industry's recordings passed between all strata of American society and permeated regions from coast to coast.

Histories of U.S. popular music too often discuss the insurgent genres of jazz, race records, and old-time music as distinct from what we understand as the "popular music" of the era, understood as songs and styles descending from the Tin Pan Alley hits of the 1900s and 1910s. But it is untenable to separate these for several reasons. In addition to the era's various media convergences, the emerging genres, artists, and songs of the late 1910s and early 1920s were fully part of recording industry logics and practices by the middle of the decade. This stylistic hybridity, long a part of the pop ethos, is evident in a variety of ways: (1) increased connections between the recording, radio, and film industries in ownership, technology, and cross-promotion; (2) the major labels catching up with the emergent genres. While the nascent labels of the late 1910s had been central in discovering and cultivating the emerging genres of the 1920s, the major labels had caught up and had hired away the smaller labels' scouts and launched their own niche-genre series—as I have discussed in relation to race records and old time music—even as those niche markets inched ever closer to the center of popular music as the decade turned; (3) performers' increased diversification into radio and film, which became key to the popular music of the 1920s and 1930s and was a strategy pursued by both the biggest celebrities of the era as well as figures once tied to niche genres alone. All these dynamics played a part in the generic expansion and hybridity evident by decade's end.

Part of the expansion's power came from song publishing as well as the recording industry. Though we tend to associate Tin Pan Alley's golden age somewhere just before recording culture's dominance, song publishing played an important role in the Jazz Age through adapting to new media outlets. As Tin Pan Alley evolved, music meant for sheet music and the family parlor piano was increasingly meant for the phonograph and radio, the Broadway stage, and finally film. Tin Pan Alley built an impressive song distribution network to capitalize on as many contexts as possible, as Katherine Spring suggests: "Predicated on an aggressive commodification of the song as a product, the network in turn influenced the kind of music likeliest to succeed in the changing entertainment marketplace."[5] Publishers associated songs with stars of phonograph, radio, and screen. Even in the silent era, sheet music featured movie stars' photographs on the cover. Movie theme songs appeared on phonograph records and shared titles with silent motion pictures. "By the time the major motion picture companies prepared to convert to sound," Spring argues, "both the music and the film industries perceived motion picture theme songs as commercial products."[6]

In fact, the 1920s could be understood to represent a high-water mark for Tin Pan Alley's influence based on the lasting importance of the era's songs, whether or not the hits actually emanated from New York's Twenty-Eighth Street and Fifth Avenue. George Gershwin and Irving Caesar's "Swanee," popularized by Al Jolson (1919–20) and Van and Schenck's "Ain't We Got Fun?" (1921) reverberated from coast to coast. Crooner Gene Austin popularized both "Bye Bye Blackbird" (1926, lyrics: Mort Dixon, music: Ray Henderson) and "My Blue Heaven" (1927, lyrics: George Whiting, music: Walter Donaldson). Gus Kahn wrote lyrics for Isham Jones' "I'll See You in My Dreams" and Eddie Cantor's "Yes Sir, That's My Baby," as well as Cantor's version of "Makin' Whoopee," his popular version of the Walter Donaldson tune (1928). In 1929, both "Ain't Misbehavin'" and "Stardust" were hits. The former was written for the musical that would become *Hot Chocolates* (lyrics: Andy Razaf, music: Thomas "Fats" Waller and Harry Brooks), while the latter began as an instrumental by Hoagy Carmichael, which he recorded with Emil Seidel and the Dorsey Brothers for Gennett in 1927. Mitchell Parish wrote lyrics for the tune and it became a national hit in 1929.

The Recording Industry in the Age of Radio

The simultaneous benefits and threats represented in radio's ascendance made recording companies both curious and reticent about the medium's promotional possibilities. Radio promised a new kind of simultaneity (roughly speaking) that marked a key difference when compared to common consumption of phonograph records.[7] "Insofar as popular music conflated entertainment and indirect product promotion, it was an ideal vehicle for broadcast advertising," in that, on radio, "songs were endowed with audience appeal, and their capacity to seize a listener's attention and capitalize on the power of association made them effective agents of indirect selling—not only of affiliated products but also of themselves." In a 1924 *Radio News* interview, then secretary of commerce Herbert Hoover ("chief of police of the ether") pronounced radio programming (in its local and what he saw as inevitable network forms) as an edifying social force and bolster for family life: "For in the comfort of our own firesides, father can smoke, mother can knit and the family can make remarks upon the performance with informality not permitted in public places."[8] By 1928, popular songs (recorded as well as live) had come to play "indispensable roles in the acceptance of radio as a medium for mass advertising."[9] By the mid-1920s, novelty tunes like "I Wish There Was a Wireless to Heaven (Then Mama Would Not Seem So Far Away," "Mister Radio Man (Tell My Mammy To Come Back Home)," "Tune In L-O-V-E," and "Love Her By Radio" were in circulation.[10] But amid these adjustments and assimilations, Thomas Edison emerged as one of the recording companies' loudest voices in denouncing radio as a medium. Edison, eighty years old by 1927, had largely steered his company to focus on increasing audio quality and playing time length, while dismissing radio's growing influence. "It's awful—I don't see how they can listen to it," Edison said. "There isn't ten percent of the interest in radio that there was last year. . . . It's a highly complicated machine in the hands of people who know nothing about it," adding that Edison dealers were "rapidly abandoning" the technology.[11] Radio audiences were fiercely loyal and often wrote in to discuss their preferences on what they wanted to listen to and how they wanted to do so. With the end of local radio's dominance in the late 1920s, listeners "articulated an anticorporate moral economy implicit in the reciprocal relationships independent stations forged with their audiences."[12] Meanwhile, phonography's oldest champion continued to speak out against radio in general. In a 1925 item that could not have surprised anyone, a *Talking Machine World* article announced, "Thos. A. Edison, Inc., to Not Enter Radio Field."[13] As radio continued to gain popularity, Edison increasingly became a target for satire in cartoons that largely appeared in radio magazines, as in an October 15, 1926, *Radio Digest* cartoon titled "Tommy's Only Foolin'," in which the inventor announces, "Radio Is the Bunk!" to which a radio fan replies "Aw—So's Your Old Phonograph!"[14]

Recording companies continued to follow both long-term and short-term generic niches, though they were not always clear on which was which. After incorporating the earlier insurgent musical styles of the 1920s into marketable genre categories, Victor and Columbia had by the second half of the decade increasingly sought a broader generic hybridity that sought the national audiences that radio had started to identify and the talking picture had reinforced in certain ways. By the mid-1920s, Brunswick-Balke-Collender replaced Edison at second position (ahead of Columbia, behind Victor) as the U.S. recording industry's Big Three. In 1924, Brunswick signed Al Jolson away after many years at Columbia. The announcement was meant for March 1, but Jolson let his enthusiasm get the best of him in an informal announcement at a mid-February Chicago stage performance.[15] Jolson was paired with Isham Jones' orchestra, one of the label's (and nation's) most popular bands, which helped ensure the greatest impact for Jolson's first Brunswick releases.[16]

The Big Three were predictably concerned by the emergence of radio and eventually sought ways to make radio's reliance on music work for them. As radio entered American life, fierce battles were waged between advocates for commercial vs. educational uses (which are ongoing today).[17] For much of the 1920s, music on radio meant live performances and an emphasis on "quality music," "potted palm" tunes of classical and light opera that were such a part of the medium in those days.[18] At least initially, not a lot of the music released on recordings were heard on radio. But as early as February 1924, an article in *Voice of the Victor*, the company's in-house trade publication geared towards retail dealers, addressed the radio boom. "When radio first began to fill the ether it was thought to be the chief bogey of the instrument and record business. Yet, little by little, the contrary is proving true, and radio is helping to increase business, for the Dealer who knows how to use the advertising power which a radio broadcast brings into the radius of his business."[19] The article emphasizes the "evanescent" nature of the radio song, heard once and perhaps not again, which was an area in which the phonograph record—which could be played anytime a listener desired—had a distinct advantage: "At this point the record seller steps in with a reproduction of the selection which can be called forth whenever wanted," by the finest bands and vocalists that Victor could offer.[20] As David Ewen writes, "Hearing a good song or favorite singer was one thing," but "to renew that pleasure at will through a record was quite another."[21] As we will discuss later in this chapter, radio sought to distinguish itself from phonography, both in terms of radio's superior sonic qualities compared to the acoustic phonograph and in the dual discourses inscribed into the medium's regulations: the centrality of live transmissions and coast-to-coast networks' ability to unify the nation.

While cross-promotional relationships between music publishing companies, recording companies, and film companies had some substantive historical precedents by the mid-1920s, radio's emergence forced recording companies to decide upon their relationship to the technology. Would it be seen by the recording industry as a new and unique promotion opportunity or as a hostile interloper that would cut into companies' profits or even topple its dominance in entertainment media? Recording companies' initial takes were closer to the latter than the former, though by the mid-1920s, companies were especially eager to offer combination phonograph-and-radio cabinets for their consumers as well as to convert and adapt to "electrical recording," which radio had made a new gold standard of sorts in discourses surrounding audio quality.

The Wireless Age polled phonograph dealers around the country, reporting on possible phonograph-radio synergies in August 1923 with the headline, "Yes, the Radiofan Buys Records!" "The broadcasting of music is selling records of that music, and radio is thus making profits for the phonograph industry. Radio has even led to the purchase of talking machines for some homes that were without music until radio brought it into them."[22] Signs began to emerge that the recording industry might actually be able to partner with radio in certain ways (through publicizing their artists as well as each new record release) while continuing to compete with the technology in terms of hardware. While radio's emergence had disrupted the phonograph's role as the primary home entertainment medium, the recording industry began to understand that radio could work for them in certain ways, and it slowly figured out how to make that so. Meanwhile, radio adopted an increasingly commercial model, which appealed to sponsors who imagined "the family gathered around the set, expectant, impressionable, and above all receptive to the ad messages along with the entertainment."[23] By the end of the decade, radio executives were largely in agreement that musical programming above all built the "largest and surest audiences."[24]

To some of the oldest players in the recording industry, these changes didn't instill confidence, and some of them were convinced that the recording industry's peak had passed. Even Victor was impacted. Victor's Eldridge Johnson had long been considering buyout opportunities and finally decided to sell in 1926. Though it was a sound financial decision, it was something that Johnson regretted the rest of his life. Rather than selling to radio or film interests or a rival phonograph company, Johnson accepted an offer from banking interests in New York: Speyer and Company and J. & W. Seligman & Co. in December 1926.[25] Roland Gelatt reported that in 1927 (the first year under Speyer and Seligman), Victor made $7,269,523, down a bit from the 1926 figure but not so much to cause any alarm. He added that "It looked as though the optimists had been right when they predicted, during the early days of broadcasting, that radio and the phonograph could prosper side by side."[26] The question increasingly became *which* radio or phonograph companies would prosper. These banking companies did not retain ownership of Victor long. "To offset losses," Russell Sanjek wrote, "the company was obliged to enter into serious negotiations with Sarnoff [whose offer of a more sweeping radio partnership had previously been rejected by Victor] for the manufacture, in its modern cabinet factory in Camden, of combination RCA radio-phonographs. Victor's engineers may have been unable to create a satisfactory radio of their own or Victor's top executives may have felt a cooperative deal with a radio manufacturer was cheaper, but they finally did enter in to a partnership. Agreement was reached in 1925, and in an additional twist, RCA barred Victor from its patents in the same way that Victor had (endlessly) lorded its patents over competitors before.[27]

The effect of the sale of Victor to the Radio Corporation of America

(RCA) was disturbingly symbolic to many longtime proponents of the recording industry, amidst cheery forecasts for the future in this royal wedding between radio's most powerful network and the recording industry's premier company. "The young RCA of the late 1920s was a glamour stock," Barry Mazor writes, "running up in price and hoopla in proportions equivalent to those of the Internet or mobile communication explosions (and bubbles) of our time." He noted that RCA's David Sarnoff had ambitious plans across media platforms, including a Sarnoff-driven merger with the Keith-Albee-Orpheum theater chain and Joseph Kennedy's Film Booking Offices of America to form Radio-Keith-Orpheum, or RKO Radio Pictures, later shortened to RKO Pictures.[28] Sarnoff was also busy with ongoing work to establish national radio networks through NBC Red and NBC Blue.[29] The new RCA Victor was clearly the most powerful and prestigious organization in the recording industry. Those loyal to the old Victor, however, had reason to believe the new company's interest in that facet of the recording industry was marginal. In fact, RCA was more interested in Victor's Camden complex and the company's manufacturing, distribution, and retail capabilities than in the record business. As Roland Gelatt wrote, "No sooner had the ink dried on the agreement giving RCA control of Victor than plans were set in motion to begin the manufacture of radio sets in the Camden factory. For despite the assurances about radio and phonograph joining together on equal terms, the management at RCA had no real interest in the talking machine."[30] "It took months of corporate fighting and intrigue for Sarnoff to get control," Russell Sanjek wrote, followed by his effort to "reorganize and consolidate the combined RCA Victor Company, of which he became president."[31] The focus on radio should not have been a surprise: RCA's institutional culture was built around the radio, which continued to perform well in the late 1920s. As Russell Sanjek suggested, "With total retail sales of $69.9 million in 1927, $72.6 million the following year, and $75.1 million in 1929, the record business was returning to the prosperity of the postwar period. Victor was, as usual, in the forefront, with profits stimulated by the improving quality of both players and discs, to which advances in the continued development of radio receivers contributed significantly."[32] But the recording industry, while having bounced back from mid-1920s sales lows, was hardly the vital economic force it once was.

If Victor's sale was symbolic, the recording industry did have some remaining reasons for optimism. Despite radio's rise and the industry's early-to-mid-decade low point, record and phonograph sales were on a slight upswing, imbuing industry insiders with a sense of guarded optimism. A January 1928 *Talking Machine World* editorial recognized a few of the problems in the phonograph business but were careful to couch any concerns in an unflagging rhetoric of boosterism: "So far as the talking machine is concerned, there are those, and they are in the majority, who believe that the peak of the revival is yet to come."[33] The editorial went on to talk about the maturation of the radio business, which suggests that phonograph and radio sales were increasingly thought of as part of the same industry, considering the number of radio-phonograph combination cabinets that were being made for the U.S. market. The same issue featured an article titled "Workers Should Profit in 1928," which featured optimistic statements from executives at Victor, RCA, Brunswick-Balke-Collender, and Edison, among others. Their forecasting suggested that the stabilization of the radio market and the improving signs regarding record sales would lead to further profits. But reports in *Billboard* voiced concerns over both sheet music and phonograph records sales, though they saw radio as one possible culprit (joining crossword puzzle books and other domestic diversions).[34]

Unlike Victor, Columbia had no suitors in the wings in the 1920s. At mid-decade, the company was still recovering from its early 1920s financial problems. Wall Street investors had overpriced Columbia stock, which along with overproduction and the 1921 recession, had led the company into receivership by 1922.[35] The rising popularity of radio did not make Columbia's recovery easy, though race records were a bright spot, most clearly through Bessie Smith, who by 1924 was paid $200 a side as opposed to the customary $25 to $50 a side (a flat fee in lieu of royalties).[36] In 1925, the Columbia Graphophone Company, Columbia's British subsidiary ("British Columbia"), purchased its U.S. parent company, with the support of J.P. Morgan & Co. In the following year, a retooled Columbia Records purchased OKeh, while the label continued to record jazz and blues until the end of the decade. As Allan Sutton explains, "Columbia led the field from a technological standpoint, while OKeh was a leader in marketing to the burgeoning race, country, and ethnic markets.[37] Paramount also sought a variety of niches. But the label's strength remained in its race records recordings of what is now called the country blues, adding Blind Blake and Big Bill Broonzy (and a lesser-known Charley Patton) to their roster. Meanwhile the Edison Company finally entered into an agreement with Splitdorf Radio Company of New Jersey to manufacture combination radio-phonographs. But it was too late, given both the early entry of other recording companies into radio agreements and an overabundance of radio manufacturers.[38] Meanwhile, Gennett's early jazz niche had taught the company how niches could help sustain the company (along with producing some of the foundational recordings of the genre). But as jazz had become increasingly attractive to more powerful record labels, Gennett continued its pursuit of jazz while looking for value in a broad variety of niches (most successfully, hillbilly music).[39] As hillbilly recordings began to sell more than the race record niche it had pursued, Paramount joined Gennett's Champion imprint and other labels in pursuit of old-time artists as well.[40] Behind these strategies, in one way or another, recording companies responded to shifting conditions brought on by radio's newfound power.

Radio's emergence mirrored the course of sound recordings in no small way, in what was becoming a recognizable pattern of media adoption and assimilation: inventors positioning the medium and its possible social, cultural, and economic applications; technological experiments and battles; businesses replacing inventors as the medium emerges; and various design changes as the technologies meet consumers in burgeoning retail environments. Most radios ran on direct current (DC) and were therefore battery-based and "portable," though what counted as portable in the 1920s would not commonly be understood as such now. It was Westinghouse that made alternating current (AC) a dominant platform for retail radios, placed in ornate cabinets already familiar via decades of phonograph cabinets that convinced consumers the technology was a benefit to the home.[41] "The coming of the radio craze opened up vast opportunities for the most American of institutions, the entrepreneur. Radio shops appeared in every city, and some—like New York—had hundreds, stocked with countless new brands."[42] Like the talking machine upstarts of the 1910s, many new radio manufacturers debuted during the 1920s, though few would survive the pitched competition of the era, while even fewer would survive the Great Depression to follow.[43] Liveness was privileged, both in nascent radio regulations and in everyday practices of radio, given that most recording companies did not welcome sound recordings to be played on air out of fear that consumers would not pay for sound recordings they could hear on radio without a visit to the music shop. Radio technologies—"interfaces," to use contemporary parlance—were redesigned for simple usage and included a cord that could be plugged into electrical sockets largely used for lamps and a single knob for frequency tuning.[44]

With the increasing influence of network radio, the listener's relationship with the medium began to change. "If in the 1920s," Elena Razoglova writes, "fans addressed their listeners to the artist care of the local station, now the growing network broadcasting system and its production process had become much more complex." "[N]etworks had to restore public trust, which had been damaged by the destruction of many local stations" with the Federal Radio Commission (FRC)'s reallocation of frequencies in favor of network radio.[45] Radio's stabilization into its lasting "American System" (private companies, sponsored shows, public airwaves, government regulation) and the emergence of coast-to-coast networks had real and lasting impacts, including the ability of musicians to benefit from the medium in different ways.[46] The American public increasingly preferred the one-time cost of radio as opposed to buying recordings to hear each new hit, while the radio industry looked for a way to make money besides the sale of receiver sets. With NBC now truly functioning as a national broadcasting system, radio's power continued to increase with networks' embrace of the emerging sponsorship model. "Leery of direct selling in the home," Timothy Taylor writes, "the industry first settled on a 'goodwill' strategy, trying to provide music that listeners would like, which, it was hoped would generate goodwill for the advertiser's product."[47]

While Russell Sanjek described the era's radio fare as "previews, presidential speeches, and championship prize fights," an early NBC national program, *The Lucky Strike Orchestra Dance Hour*, was launched in September 1928.[48] The show featured bandleader B. A. Rolfe, who had successfully returned to the music business after running Rolfe Photoplays, Inc., a Fort Lee, New Jersey-based motion picture production company in the 1910s. But unlike previous orchestras on radio, Rolfe's outfit specialized in sweet jazz for dancing, which represented a departure from both 1920s hot jazz and the light symphonic concert music common on radio: "Rolfe was instructed to play only dance music, in contemporary jazz or waltz tempo; no concert interludes were warranted," which was innovative for a time in which "jazz and dance music was principally relegated to the after-11:00 p.m. remote broadcasts from hotels, dance halls, and night clubs."[49] The advent of shows like these led to a symbiotic and powerful relationship between jazz dance bands and radio, as radio embraced jazz in its commercial form. The radio broadcasts from hotels in New York, Chicago, and Los Angeles conjured a world of glamour and romance to which people listened and danced. Suave masters of ceremonies, beguiling vocalists, bandleaders, and dance orchestras all created a sonic world served to propel the performers, the music, and the medium into great success. By the end of the decade, radio had become the dominant medium in U.S. culture. While not yet at the height of its powers, radio's increasing dominance in

U.S. culture by the end of the 1920s was a prime mover in changes facing the recording industry.

Charismatic singers fronted the dance bands. Their vocal styles were deeply impacted by microphones, on stages, on the radio, and in recording studios. Brunswick had the best of the old guard (Al Jolson) and the rising star of the new guard (Bing Crosby), Victor had the reigning king of the moment in Rudy Vallee. As Allison McCracken writes, "Although historically Bing Crosby has been awarded the title of 'America's Crooner,' from the years 1928 to 1931 radio crooner Rudy Vallee was the first national star created by radio and the first performer to become a star in all three forms of mass media: radio, records, and films."[50] But Vallee's popularity was divisive. "His singing's intense romanticism, its continual and invasive presence within the home circle via radio and its strong emotional effect on listeners were new phenomena," she writes. "By the early 1930s, crooners had become the objects of widespread condemnation by critics, who referred to them most frequently as 'emasculated, effeminate whiners.'"[51]

Columbia's jazz band and crooner roster allowed it to compete successfully with Victor and Brunswick. Despite ongoing economic and managerial issues, Columbia also had the most respected and celebrated musician among other musicians: Louis Armstrong. After moving from Chicago to New York, Armstrong greatly expanded his repertoire and had moved from the instrumental jazz through which he made his reputation as a premier jazz trumpeter to popular song. The influence of Columbia's management as well as Armstrong's own willingness to experiment across musical styles led to his renditions of the most popular tunes of the day, and he became known as a vocalist. In Armstrong's embrace of singing standards and moving away from his hot jazz roots, he still managed to make the material his own much of the time, and while song selection occasionally served him poorly, his record sales proved his ability to work across genres skillfully. As Russell Sanjek wrote, "Louis Armstrong's highly individual interpretations of Tin Pan Alley's major plug tunes, which made near classics of the music, sold more for Columbia than the combined output of his closest rivals on the label, Guy Lombardo and Ted Lewis."[52] Though lagging behind Armstrong, Lombardo's and Lewis's importance to Columbia was illustrated by contracts to keep them on the label. The other labels were also anxious to lock in their talent as best they could. As Russell Sanjek wrote, "Hoping to increase sales and stem their losses, the new Big Three, Victor, Brunswick, and Columbia, rushed to make large offers to dance-band leaders and vocalists who might move records in large quantities."[53]

The recorded dance music of the 1920s came labeled as "foxtrots" or other popular dances of the day, whether the tune earned the association or not. Jazz was an even broader category. Some of the central hits of the 1920s were dance numbers of varying tempos—"Dardanella" by Ben Selvin's Novelty Orchestra (released in 1919, but a 1920 hit), "Always" by Vincent Lopez (1926), and a decade's worth of hits from Paul Whiteman and his Orchestra, whose 1925 version of James P. Johnson and Cecil Mack's standard "Charleston," was a mid-decade number synonymous with the dance craze. Race records and hillbilly songs had percolated up into national popularity during the 1920s (e.g., Mamie Smith's "Crazy Blues," Bessie Smith's "Downhearted Blues," Vernon Dalhart's "Wreck of the Old 97," Jimmie Rodgers' "Blue Yodel"). Whereas jazz was once a niche genre, a specific kind of regional dance music, it was now the predominant music of the era. The jazz of the late 1920s and early 1930s bore less and less relationship to the music associated with the term's emergence. Its spread was in no small part due to new cross-media platforms.

Recording and radio companies' tense but increasingly close technological relationships through the mid-to-late 1920s did not mean that concerns surrounding intellectual property, royalties, and licensing had been settled. Earlier schisms within music publishing and licensing had set the stage for conflict. "Throughout the 1910s and 1920s," Katherine Spring writes, "the two main targets of ASCAP were radio broadcasters and motion picture theater owners."[54] While ASCAP was less successful in regulating the use of music in motion picture contexts (though the organization would ultimately prevail), radio became a key focus when ASCAP's earlier policy of granting licenses for little or no money gave way to infringement suits.[55] In February 1922, both the Music Publishers Protection Association (MPPA), founded to combat vaudeville theaters that asked publishers to pay them for music use, and ASCAP were concerned about radio's rise, fearing that its use of "canned music would lessen the sale of phonograph recordings in a short time."[56] Throughout the 1920s, ASCAP had tried to assert its members' rights to be paid for radio's broadcast of commercial sound recordings. As early as 1922, ASCAP's Nathan Burkan wrote an open letter stating that "a radio broadcast was a public performance for profit," to which broadcasters responded (some reportedly with surprise) that "the broadcasting of records and songs helped to popularize them."[57] After some negotiations and temporary agreements, ASCAP issued a November 1922 ultimatum that all radio stations should get a license to play ASCAP artists' music or face copyright infringement suits in court.[58] "The majority of stations rejected the proposed contracts and banded together to form the National Association of Broadcasters" (NAB), with the intent to make "free music available to all stations" through amending the Copyright Act of 1909 to exempt radio from fees from performances (including sound recordings), while ASCAP threatened, but didn't follow through, with creating radio stations all their own.[59] Meanwhile, ASCAP won a court victory in proving that Newark, New Jersey's WOR radio, housed in that city's Bamberger Department Store, meant to make profits in piping its broadcasts in for customers to hear through the sale of radio receivers on the premises.[60] By November 1925, Russell Sanjek wrote, many of those in radio "began to concede that those who owned a commodity that occupied between 60 and 70 percent of all air time should be compensated," though exactly how to do so was still up to debate.[61] Throughout the 1920s, even after Victor's merger with RCA, tensions between radio interests and music publishers—crucially connected, but only in part, to the nation's recording companies—continued unabated. The NAB's judicial efforts were largely thwarted amidst ASCAP's broadening sphere of influence, as were legislative initiatives through much of the 1920s. The Radio Act of 1927 did not address any of these tensions, neither did it address network broadcasting nor the emergent model of paid radio advertising.[62] Within these contexts, the tightening relationship between recording and radio companies would not mean a cessation of concerns regarding royalties from radio performances, even as radio companies bought recording companies beginning in earnest in the second half of the 1920s and into the next decade.

Meanwhile, tensions between music publishers and music composers continued, based on ASCAP's earlier decision that publishers would garner half the profits for mechanical reproduction rights, while composers and lyricists "split the remainder."[63] Vitaphone's release of *Don Juan* (Crosland, 1926) and *The Better 'Ole* (Reisner, 1926) brought existing tensions to a boiling point amidst Hollywood's talkies conversion. A series of agreements were reached, most notably the 1926 Mills Agreement, which dictated that "Any studio that took out a license from Electrical Research Products, Inc. (E. R. P. I.) to use Vitaphone equipment would obtain synchronization licenses to an enormous collection of copyrighted music," in exchange for the MPPA receiving "guaranteed minimum payments based on the number of seats in theaters where ERPI products (i.e., Vitaphone films) were projected."[64] But shortly after the agreement, and in the midst of roiling change across industries, both the MPPA and motion picture interests were unhappy with the result. MPPA's E. C. Mills once again protected publishers over composers and lyricists, while motion picture executives disliked that the Mills Agreement applied to the United States and Canada alone, nor did it compensate film producers whose films spurred the sale of related phonograph records and sheet music.[65] As so often has happened regarding royalty disputes, the fix was temporary at best, but underlying tensions between a growing constellation of constituents meant that something had to be done. The answer for the movie studios' rapidly increasing need for music meant investing in music publishing.[66]

Tie-Ins and Tie-Ups: The Recordings-Radio-Film Nexus

Cross-promotional practices had long been part of the entertainment industries, but by the second half of the 1920s they had gained new importance. As Katherine Spring notes, "Certainly tie-ups among motion picture exhibitors, music publishers, retail outlets, and other agents of entertainment existed before the coming of sound, but those earlier arrangements were erratic more often than not and tended to satisfy the immediate interests of the relevant parties rather than any long-term forecasts."[67] Tie-up as a term most commonly used in cinema, but it was a practice also employed across the recording and radio industries. These strategies (the more novel the better) were regularly shared in the pages of entertainment trade papers.[68] *Exhibitor's Trade Review* featured a "national tie-up section" with stories about promotional possibilities for retailers tied to new releases. "Exploit-O-Grams" was a regularly occurring feature in *Film Daily*, in which exhibitors shared ideas. A sampling of a few 1928 tie-up promotions: To publicize the release of the Mervyn LeRoy–directed *Harold Teen*, a Davenport, Iowa, soda fountain featured the "gedunk," said to be the favorite sundae of film and comic strip character Teen himself.[69] For Fox's *Chicken A La King* (1928), a Spartanburg, South Carolina, restaurant featured the dish when the film played the local Rex Theater, suggesting diners "eat it here and see it there."[70]

Beyond the plethora of tie-ups, promotional giveaways, and various hullabaloo-generating stunts, there was a much larger dynamic born out of growing industrial, technological, and cultural ties between media companies. While cross-promotional activities between music and film predate the recording industry (film songs in sheet music is one example), recording artists increasingly worked across media platforms for broader and deeper impact by the late 1920s and early 1930s. Dissatisfaction from the Mills Agreement led to attempts by film studios to acquire song catalogs. Studios that could afford to do so entered agreements through 1928 and 1929. Fox bought exclusive rights to Sound Music Co., via its Red Star Music subsidiary; Loew's/MGM made a (nonexclusive) deal with Robbins Music; and Warner Bros. (the studio most closely associated with talkies) made two bold forays into music publishing, founding the Music Publishers Holding Corporation, or MPHC, in a deal with T. B. Harms and subsidiaries, while negotiating exclusive rights to the biggest name in music publishing: M. Witmark & Sons.[71] Each "affiliated with music publishing firms to create or designate subsidiary companies expressly for the purpose of publishing motion picture songs.[72] As radio's dominant entity, RCA was no stranger to conglomeration by 1929, with broad investments in communications (radio and television research), motion picture technologies and exhibition via Photophone and RKO, radio broadcasting, radio sets, and phonographs and sound recordings via NBC and RCA Victor.[73] RCA "undertook more extensive deals by acquiring the majority or entire stock of massive publishing firms."[74] The overall impact, Spring writes, was an early example of "a hallmark characteristic of media convergence: the increasing concentration of corporate power."[75]

Amidst this increased media concentration, many of the biggest recording, film, and radio stars were crossing platforms and extending their star power in various ways. Some of them would use this tactic to extend their careers far beyond what might have happened otherwise. Singers and bandleaders such as Bing Crosby, Ruth Etting, Ted Lewis, Guy Lombardo, and Vincent Lopez would all appear in film features and shorts in the coming decade. Hoagy Carmichael and Fats Waller created an additional career of sorts on the motion picture screen. The advent of sound-on-film talkies, which rejuvenated Tin Pan Alley's fortunes amidst new musical competition, quickly led to musicals, a dominant film genre of the late 1920s and early 1930s. As Russell Sanjek wrote, "In the course of a single year, Hollywood changed its attitude toward the value of popular music in promoting its products, and some segments of the music business were made aware of the awesome power of the talking picture and its weekly audience of 70 million to catapult songs into success."[76] Studios created music departments. "With sound movies and radio," James P. Kraft noted, "business firms reproduced and marketed musical performances on a far wider scale than was previously possible. The mass dissemination of music eliminated thousands of jobs in silent-movie theaters and radio stations across the nation, while simultaneously creating a much smaller number of new opportunities in a few expanding media centers, especially Los Angeles," where musicians competed to find (often piecemeal) work in orchestras working for film studios in various capacities.[77] Tin Pan Alley songwriters flocked to movie studios and signed contracts, though the most high-profile talent (Katherine Spring notes Irving Berlin, Richard Rodgers, and the Gershwin brothers in particular) got the best deals.[78] Other songwriters managed to work by stringing together short-term contracts or arranging deals between music publishers and film studios.[79] Hollywood's newfound exuberance for song seemed to provide opportunities for recording companies big enough to do business with movie studios, but the motion picture industry's embrace of music appeared excessive at times to those in the recording business. While ASCAP now claimed 750 songwriters in its membership ranks and songwriters drawn to work on the West Coast, studios soon found that they had hired too much talent. As Spring suggests, "Hollywood created a glut of as many as six or eight songs in a single production," which "led to a superabundance of recorded movie music, which sold pictures but not recordings."[80]

But generally speaking, if musicians were on film then movie stars were increasingly appearing on records and radio. A key example of this was the sophisticated campaign for United Artists' *Ramona* (Edwin Carewe, 1928), which hoped fans would buy the record and see the film. United Artists' promotional campaign gave as much attention to the song as to the film, amidst a "picture song vogue" that proved that the silent film theme song garnered public interest as Hollywood's sound experiments came to fruition.[81] The film starred Dolores Del Rio, who also recorded a song of the same name for RCA Victor. Before the film's opening, Del Rio sang the song on a coast-to-coast radio transmission. Victor heavily promoted "Ramona" and its B-side ("Ya Va Cayendo," which translates "And It Falls"). The song became a standard of the era. Close relationships between the recording, film, and radio industries allowed musicians to extend their own reach as performers and celebrities, while managing their own shifting personas across media platforms.

Years before *Ramona*, recording industry tie-ups with the film industry were already commonplace. Silent-film actors and singers recorded songs related to specific films and sheet music featuring their images on the front cover. And while Victor and Columbia's tie-ups with radio had become commonplace, concerns about tie-ups' long-term effects on the radio industry were very real. Of course, all musical appearances on radio were tie-ups in one way or another. Individual musicians benefited greatly by appearing on radio before appearances in a given city and were also able to boost record sales in the same way. But radio did not yet offer royalties or performance fees, and retailers worried that consumers wouldn't bother to replace their old phonograph and invest in a radio instead. By the end of the 1920s, articles in *Talking Machine World* and *The Voice of the Victor* assured retailers that all was well, despite concerns over radio's implicit threat and unprecedented shifts in ownership. While emerging recording companies sought out the niches that would become jazz, blues, and country (among other genres and subgenres), larger economic, technological, and cultural realities took shape. Increasingly, the era's music was designed not for regional or aesthetic niches, but a variety of media platforms and purposes, coast to coast. And the recording and publishing wings of the U.S. commercial music industries would be increasingly interdependent with the radio and film industries with which they had already collaborated and sometimes struggled.

The shifts were tied all at once to specific technological, economic, regulatory, and cultural contexts. In addressing cinema fans' favorable response to sound-on-film talkies when they were so used to silent pictures, Steven Wurtzler suggests that a comparative context is crucial to understanding change not only in film, but across media industries: "Hollywood's conversion to sound and cinema audiences' enthusiasm are merely components of the larger pervasiveness and reaction to a new technological mediation of sound. This new technology applied electrical principles to sound recording, transmission, and reproduction in a number of different media forms (predominantly radio, the phonograph, and the cinema)."[82]

Listening to a phonograph record vs. a radio broadcast were different experiences in the late 1920s, though research in and adoption of "electrical" (later, "electronic") sound recording would eventually bring the industries together. While experiments in electrical recording have been traced back to 1903, research in wireless telephony during World War I produced the first viable microphones and amplifiers.[83] In postwar Britain, His Master's Voice (HMV) and British Columbia experimented with electrical recording, while U.S. Victor and Columbia did not. In the United States, it was AT&T Laboratories ("Bell Telephone Laboratories," after a 1925 merger with Western Electric's research wing) that would conduct its own secret experiments during the early 1920s, but its prototypes remained inefficient in converting mechanical to electrical energy to amplify sound.[84] In 1924, AT&T invited Victor's leadership for a demonstration, but Eldridge Johnson's mental state (he suffered a nervous breakdown the same year) and Victor leadership's anxieties toward radio meant that little came of the meeting.[85] Meanwhile, 1924 was a disastrous year for both

Victor and U.S. recording companies. While Victor had been strong between 1920 and 1923, despite Columbia and other labels' troubles during a period of economic deflation and downturn, the company brass assumed going into 1924 that the demand would continue. But Victor executives paid insufficient attention to RCA's improved radio lines, including new versions of their Radiola and Super models, and had an abysmal sales year amidst increasing radio sales.[86] Radio ads had become ubiquitous in the trade and popular press, while radio industry insiders worried that jobbers were carrying so many different companies' products that they couldn't possibly serve them all well.[87] An anonymous talking machine wholesaler was quoted in the November 1925 issue of *Sales Management*: "Six months ago, we were in a dead industry with no hope ahead. It looked as though we were up against a stone wall."[88]

But a change in recording companies' fortunes came with a concerted effort from Victor, beginning in early 1925, to sell its overstock from the previous year. To sell the stock on dealers' shelves, as well as those talking machines still sitting at the Camden headquarters, Victor executives devised a plan to use the company's power and prestige as one of the country's flagship corporations to lean on jobbers and retailers to no small degree, an effort that lasted through much of 1925. "It was a combination campaign," journalist A. W. Hilton wrote, "involving newspaper advertising on a large scale; preliminary research work; actual canvassing among Victrola prospects and the use of pennants, window signs, special front of the store signs, window displays and interior display units."[89] They partnered with their local distributors in the effort, such as St. Louis's Koerbner-Brenner Company, which allowed Victor to benefit from their local expertise on both local consumers and local retailers while also solidifying their relationships with distributors in an increasingly volatile market.[90] The company took out advertisements in cities around the United States, promising "Opportunity for Ambitious Women" from a given locale, assuming that women willing to canvass door to door (based on dealer leads) would speak to female stewards of a given home, taking a "musical census" with no products to sell, at least directly.[91] If the home had no Victrola, or lacked Victrola's "Musical Arts Library" collection, a (commonly male) dealer would follow up in search of sales. A *Printer's Ink* article reported that in St. Louis and surrounding communities, more than twenty-five thousand homes were canvassed.[92] U.S. Victor distributors invited their local dealers in to hear new technology and "assist in developing plans for its introduction."[93] The article also mentioned in passing the radio-phonograph relationship in salient terms: "Radio may, in time, stand shoulder to shoulder with the phonograph, as a sales ally. Today, however, all talk and press agentry to the contrary, the two are competitors."[94]

The following year, Victor was ready with a transitional technology known as the "Orthophonic" method, the first mechanical talking machines designed to play back electronically recorded discs. The machines, along with Victor's aggressive advertising campaign and tight control on its dealers, marked a turnaround. "The leader in this revival of the industry is the Victor Talking Machine Company. With their new machine, the Orthophonic Victrola, the company hopes to revive the business and push its volume of sales to new high records."[95] Victor was not the first to experiment with electrical recording in the laboratory or on the market. A rudimentary process had been devised by Chicago's Orlando Marsh of Marsh Laboratories, though he was unable to develop the technology.[96] More serious competition came from the insurgent Brunswick-Balke-Collender Company, which had entered into an agreement with RCA to develop their Panatrope, beating Victor to market with new electronically recorded process. Brunswick was included in a "Big Four" of the phonograph and radio industries, with Columbia, Victor, and RCA.[97] *The Magazine of Wall Street*'s report on the phonograph and radio industries announced increased collaboration among the warring phonograph and radio factions and the phonograph business's rejuvenated energy: "Funeral orations," the article noted, "are sometimes premature."[98] "Radio Corporation," as RCA was then commonly known, was noted as having "previous harmonious and profitable relations with Brunswick-Balke-Collender Co." via its Panatrope, but that its new relationship with Victor's Orthophonic "is the determining factor in the future of radio in the United States."[99]

The Victrola Orthophonic and other competing machines were meant to shore up the difference in sound quality between older phonographs meant to play back acoustic recordings and what was generally understood as the superior quality of electrically recorded radio broadcasts. Advertising for the Orthophonic began just ahead of a special September issue of Victor's in-house publication for its dealers, *Voice of the Victor*, dedicated to the company's new line. Built around the tag line "Have You a Victrola?" the Orthophonic advertising campaign was extensive and revealed much about how twentieth-century media businesses used massive marketing campaigns while aggressively dictating to dealers to privilege their products and initiatives. Victor Records held a "Victor Day" promotion on November 2, 1925, to publicize the process. Bold advertisements claimed a dramatic difference in quality, a "*character of tone*" altogether new in phonography, "as radically different as the modern motor-car in comparison to the "horseless carriage."[100] The October 15, 1925, *Talking Machine World* carried reports on both the Victor Orthophonic and Brunswick Panatrope lines, while also offering retail dealers a how-to guide to canvassing.[101] Though the realization came late, recording companies had finally grasped radio's formidable powers, as both a promise and a threat.

It was innovations in radio that spurred changes in both the phonograph and film industries. Early Western Electric condenser microphones required enclosures to amplify sound via phonograph "pickups" for recording purposes or transmitters for radio broadcast.[102] These microphones were also used in the Vitaphone process developed for film. Radio's approach was electrical and not mechanical, which meant a whole series of technical and socioaesthetic changes. Electronic recording in both radio and phonography, as Shawn VanCour argues, meant adherence to a "fidelity model aimed not at objective transcription but instead at effective filtering and representation of a performance" constructed for its respective media contexts.[103] VanCour argues that radio's sound production approaches to musical performances were "a necessary foundation, rather than foil to, production and performance styles in conversion-era film and phonograph industries," via its control of reverberation, scaled down performances (musical and otherwise), and a "close-up" sound approach that compressed and articulated all component parts to the fore.[104] Such changes to the technological circulation of electronic sound, as Wurtzler has argued, had huge direct and indirect impacts across media. These included "the film industry's convergence to synchronous sound, the rise of radio networks and advertiser-supported broadcasting, the establishment of a federal regulatory framework on which U.S. communications policy continues to be based, the development of several powerful media conglomerates with substantial interests in the film, radio, and recording industries, and the innovation of a new acoustic commodity . . . in which a single story, character, song, or other product is made available to consumers in multiple media forms and formats."[105] So while the increase in cross-media strategies was industrial, the introduction of electronic sound recording and playback technologies connected what had been disparate industrial and cultural practices in altogether new material practices.

Transcription discs filled the liminal space between debates about the sound quality inherent in radio transmissions vs. the improved phonographic sound of Victor's Orthophonic and similar technologies as the phonograph industry slowly transitioned from acoustic to electrical approaches. If there was an institution that provided a public face for radio transcription discs as technology and business, the World Broadcasting System served that role after its founding in 1929. Transcription discs held a central role in early network radio despite holding a marginal status in industry discourse then and in media studies discourse now.[106] The differences between phonographic and radiophonic cultures are made clear by the hard-fought definitions and distinctions surrounding transcription discs, a term meant to differentiate 16-inch, 33 1/3 rpm discs from 10-inch, 78 rpm phonograph discs designed specifically for radio use.[107] Unlike most phonograph discs in the mid-to-late 1920s, transcription discs were both recorded within electrical (as opposed to acoustic) studio contexts and were meant for electronic playback via radio. While these transcription discs undercut arguments then and now that championed network radio as a national unifying force, my interest in transcription discs is to emphasize the malleability of transcription discs as a medium, "with roots in phonography, cinema, telephony, sound amplification, and radio," culturally defined by "uses, practices, effects, and identities," linked to other technologies via "economic, technical, and social contexts."[108] Transcription discs were developed for electrical reproduction and playback by Western Electric's audio laboratory researchers Joseph Maxfield and Henry Harrison. Their work had implications for the joint Western Electric-Warner Bros. Vitaphone project.[109]

If the phonograph's principal strength was its repeatability, access to a given recording on demand and at will, radio's defining characteristic was its "liveness," a feeling of near simultaneity and of listening not as an individual or in a small group, but with a much broader community of listeners. After some early experiments that included radio on discs for home consumption, broadcasters solidified into a stance opposed to excessive reliance on prerecorded content in favor of live performances and transmissions.[110] Attitudes opposing discs were solidified into Federal Radio Commission (FRC) policy and rhetoric, including General Order No. 16 (1927), which dictated that the "nature of" transcription discs "must be announced," and General Order No. 52 (1928), which ruled that failure to announce the "character" of a transcription disc was tantamount to deception.[111] The FRC's attitude was hardly alone in the push against transcription discs in U.S. radio broadcasting. As Michael Biel writes, opposition to transcription discs "emanated from different sources," including "the federal government, the 'quality' broadcasters, the networks, the musicians union, the record companies, phonograph recording artists, music performance licensing organizations, and some of the listeners."[112] Despite these protestations and more years of negotiations and assimilations, transcription discs found a place in U.S. radio culture, which is evident in the reversal of General Order 52 (1928) by the FRC and the eventual decision of NBC network radio to launch its own transcription service, after complicated internal battles between RCA and Victor once the former bought the latter in 1926.[113] Emboldened by these changes, the World Broadcasting System increasingly positioned transcription discs into the early 1930s as another form of network broadcasting, touting affiliated stations from coast to coast (the company would be purchased by Decca Records in 1943).[114]

But the larger narrative of radio's newfound power was heralded with the advent of coast-to-coast radio—hardly an instant transition—alongside the ongoing importance of local and regional radio. NBC's national broadcast debuted on New Year's Day 1927 (the network had debuted in November 1926) with the broadcast of the Rose Bowl. Radio's newfound dominance was now impossible to ignore and the recording industry treated radio's rise with a mixture of optimism and irritation, confidence and dread. What had been an upstart competitor in the first half of the decade had become much more, evident not only in the new medium's commercial and cultural importance, but also in a regulatory apparatus put in place around it. "By the end of 1927," Michele Hilmes writes, "radio broadcasting had taken on most of the structural, economic, and regulatory features that were to characterize the radio, and later television, industry for the next fifty years."[115] The start of radio's golden age coincided with technological, business, and cultural changes that allowed for coast-to-coast network radio to emerge.

Changes brought on by innovations in electronic recording and playback in radio led to significant changes in phonograph culture's approach to studio recording and later, in terms of phonograph technologies more broadly, as machines went from mechanical to electrical playback protocols. As Susan Schmidt Horning writes, "The greatest benefit of electrical recording from the artist's perspective was its easing of the cramped conditions of playing into a horn ," while also "introducing the possibility of recording larger ensembles with better results than had been achieved in acoustic recording."[116] The role of the studio engineer, or "recordist," shifted from "cut-and-dry empiricism to mathematical and scientific control and measurability," which meant new skills to acquire for all involved, while providing a space for amateurs excited by new possibilities to get involved. It also allowed stand-alone studios—apart from recording companies—to focus on recording, as opposed to the manufacturing, "marketing, sale, or distribution" of sound recordings.[117]

These myriad changes in both radio and recording cultures had broad impacts for both forms. The relationship between the recording and film industries deepened. Links between phonography and cinema go back at least as far as Thomas Edison, who hoped that motion pictures would "do for the eye what the phonograph had done for the ear." Edison was not alone in his desire to link the two technologies. While smaller labels had found niches that now defined popular music, the major labels had managed not only to hire away talent, but also to invest in technological changes that became significant obstacles to smaller labels. The technological changes that came were "neither seamless nor inevitable" in the trajectory they eventually took.[118] Record labels small and large sought connections to film in the age of talkies—as they had before—through a variety of avenues, including sound-on-disc playback for film shorts and features. But in the talkies age, the relationships multiplied and deepened, as did connections between film and radio.

While radio had already influenced film in various ways by the mid-to-late 1920s, film studios and recording companies increasingly understood radio's promotional power.[119] This was made clear when Warner Bros.' Sam Warner purchased transmission equipment from Western Electric's Nathan Levinson to launch Warner-owned KFWB Los Angeles, which began broadcasting in March 1925.[120] In January 1928, NBC's flagship radio station, WEAF New York, took part in a coast-to-coast simulcast of the *Dodge Brothers Victory Hour* (to publicize Dodge's new Victory 6 automobile), "when 12,000 miles of wire make a nation-wide 'studio' possible," as the *New York Times* reported in its preview of the radio year to come.[121] Paul Whiteman appeared in the New York studio, while other stars chimed in via radio stations around the country: Al Jolson in New Orleans, Will Rogers in San Francisco, and Fred and Dorothy Stone in Detroit."[122] The *Variety* review noted that this bit of ballyhoo did its job despite its significant cost, in what the article called "a shrewd and effective hook-up with radio."[123] On March 29, 1928 another high-profile promotional network radio event took place on the *Dodge Brothers Film Stars Hour* when NBC (with technological help from AT&T) "set up a special, one-time link from Hollywood to Detroit, with a second New York to Detroit tie-up to allow WJZ and fifty-five NBC stations to switch back and forth from all three locations in the course of the broadcast."[124] The show featured Douglas Fairbanks, John Barrymore, Norma Talmadge, Charlie Chaplin, and D. W. Griffith, with Paul Whiteman and his orchestra providing the broadcast's musical cohesion. The broadcast also featured Dolores Del Rio, who sang the aforementioned title song from *Ramona*.

The changes across media industries imprinted in various ways, including on the careers of the era's most emblematic figures. In this cross-media environment, the celebrities who excelled were those who managed to cultivate an audience on records, through radio, and on the movie screen. Among these was perhaps an unlikely celebrity, but in an era when bandleaders were stars in their own right, none shone brighter than bandleader Paul Whiteman.

"Tickling Your Muscles, Stretching Your Soul": Paul Whiteman across Media

Implicit in the shifts in musical tastes was a rejection of Victorian morals and a move toward a liberatory modernity, an openness that thrilled some and frightened others. In its new popularity, one of jazz's best-known proponents was bandleader Paul Whiteman, the "King of Jazz," whose iconic plump face became most associated with the music in its new context. The bandleader never claimed to have invented jazz, though some mistakenly acted as though he had. Whiteman's 1926 book *Jazz*, written during the height of second-wave Ku Klux Klan activity in the United States, began with this sentence: "Jazz came to America three hundred years ago in chains," followed by an account of slaves' arrival to North American shores by Dutch traders, through which the slave industry "unwittingly assured a future for jazz."[125] *Jazz* is a fascinating and contradictory text, cowritten with newspaper journalist and future radio star Mary Margaret McBride. The chapter "Jazz in America" features Whiteman revisiting various "criticisms, constructive and otherwise": doctors who warned of jazz's various ill effects, a musician supposedly driven insane by it, the Nebraskan town that banned it, and so on.[126] "I took it where I found it," Whiteman explained of his involvement in jazz, "and I wish the preachers and club lady uplifters who put on sheets and pillow cases to go jazz-klanning wouldn't concentrate on me. I don't deserve it, really, nor the snorting editorials from Burma to Sydney, either. . . . All I did was to orchestrate jazz. If I had not done it, somebody else would have."[127]

This Whiteman passage reflects the sociocultural tensions that shaped his public persona. But his larger career was also symptomatic of sweeping changes in media industries during the late 1920s, in which discrete media industries drew ever closer together. By the time Paul Whiteman's *Jazz* was published, the bandleader's already formidable cultural reach had expanded dramatically. The bandleader's dominance across media was tied to at least four initiatives that bridged the turn of the decade: (1) Whiteman's move from Victor to Columbia Records from May 1928 to September 1931; (2) the bandleader's search for hot jazz credibility via new musicians and arrangers, a counterbalance to his long-standing mainstreaming agenda; (3) his Columbia-brokered deal to host the *Old Gold Radio Hour*, sponsored by Old Gold cigarettes; and (4) Whiteman's Universal film debut, *The King of Jazz*.

Within the recording industry, Whiteman's move from Victor to Columbia was the most symbolic. After eight years with Victor, the move to Columbia on May 12, 1928, "shocked the music world."[128] Whiteman resented the powerful head of Victor's "light music" division, composer and music director Nat Shilkret.[129] It was Shilkret and not Whiteman who had first choice on repertoire for upcoming recordings and who shared a desire to unite jazz and symphonic styles for the listening public, a strategy that mirrored Whiteman's own, which put the two in direct competition.[130] The deal Columbia offered Whiteman was difficult to refuse and perhaps unwise to offer. As Don Rayno notes, Columbia offered to "triple Victor's current payment of $25,000" to a $75,000 salary, as well as a percentage of the company's overall profits, recording royalties on Whiteman releases, and a board of directors' position on "British Columbia, the Columbia Gramophone (*sic*) Company."[131]

In competing with Victor, Louis Sterling's Columbia gambled and triumphed with their deal to sign Paul Whiteman, who had a steady stream of symphonic dance band hits throughout the decade. Russell Sanjek wrote, "Sterling temporarily revitalized his American Columbia company with a number of flamboyant business moves, among them the capture of Paul Whiteman from Victor with a guarantee of $175,000 a year. Labels of new releases by the King of Jazz bore a caricature of the bandleader in five colors."[132] A May 1928 four-page *Talking Machine World* advertisement announcing Whiteman's move to Columbia was extravagant in its attempt to transmit Columbia's power as well as Whiteman's prestige. It was a concerted marketing blitz: national newspaper advertisements in two hundred newspapers coast to coast, including ads made for the dealers' local press contacts; customized color labels for Whiteman releases (in ten-inch and twelve-inch formats) featuring the "potato head" Whiteman caricature on custom record sleeves; and "an impressive hook-up of radio stations," in which Whiteman's new Columbia releases were highlighted.[133] While Columbia had spent much of the decade recovering from its earlier financial difficulties, the Whiteman announcement was meant to assert its competitive muscle in a direct challenge to Victor.

The advertisement announced "Big News for Every Columbia Dealer," and it promoted Columbia Records' artists and technologies. It highlighted the new "Columbia Viva-Tonal and Kolster Radio Combination" on the first page, as well as blackface humorists Mack and Moran on the fourth page; their "Two Black Crows" minstrelsy skits had continued to sell (this ad touted *parts 7 & 8* of the series "You'll Never Make Canada with That Horse").[134] But Whiteman was the announcement that conveyed prestige. A two-page spread in the middle announced, "Paul Whiteman and His Orchestra Now Record for Columbia Exclusively," adding that "The artist who is in a position to choose, turns naturally, as does Whiteman, to the Columbia recording studios and laboratories, because he knows his records will be precisely what his music is."[135] In September's *Talking Machine World*, Columbia advertising manager W. H. Clock penned "How Columbia Sold Whiteman," which hailed Whiteman as a Columbia artist while bragging about sales of his first Columbia releases.[136]

Paul Whiteman's career can be understood, at least in part through an argument he made and remade, in his desire to link jazz's hotter variants with the symphonic music in which he was trained. This was certainly the idea behind Whiteman's 1924 "Experiment in Modern Music" Aeolian Hall performance and numerous other moments in the bandleader's career. Decades after the Aeolian Hall concert, Whiteman was making the argument in print, coauthoring *Records for the Millions* (1948), a book (and companion LP) on how to build a record collection, in which he played taste arbiter. At the same time, he sought credibility within more discerning jazz circles. Whiteman raided less financially successful bands for talented players, as he did with Jean Goldkette's band.[137] Through such talent acquisitions, the bandleader worked to increase his hot-jazz credentials in the late 1920s and early 1930s. He did so first by hiring white musicians that understood hot jazz and who had made it their own (Bix Beiderbecke, Mildred Bailey).[138] Later in the decade, Whiteman featured African American musicians in compositional roles, employing Don Redman and Duke

Ellington as composers.[139] But the larger public had relatively little interest in Whiteman's hot-jazz affiliations. It was the popularity of his symphonic jazz that attracted broader listeners and created opportunities within and beyond the recording industry. By the end of the decade, Whiteman would be involved in both film and radio in altogether new ways. His foray into film and his first coast-to-coast radio show were both launched via a coordinated intermedial strategy involving recordings, radio, and cinema.

In the age of the talkies, sound meant music too. The first half-decade of talkies was focused on film musicals, a focus that would last through the early years of the Depression. Whiteman's feature-length film debut was Universal Pictures' *The King of Jazz* (Anderson/Fejos, 1930), an early Technicolor film and a stage revue musical format reinvented for the screen. It was a stunning and ambitious film, broadly experimental in artistic and technological scope. It was also a deeply contradictory film that sought to honor the genre while diminishing its potential cultural impact through racial caricature and containment. The film remains an odd totem of the endemic racism of the 1920s United States. It is ostensibly framed around the relatively progressive racial attitudes of Whiteman as jazz king, but the film is largely absent of African Americans. It is also rife with jungle imagery and motifs, meant to reinforce jazz's supposed primitivism.

Whiteman and his band relocated from New York to Los Angeles to film the picture. But a draft was not yet complete on their arrival in Los Angeles, on a train dubbed "the Old Gold Special," after Whiteman's radio show sponsor, Old Gold Cigarettes. Old Gold's advertising agency had encouraged the sponsors to arrange for the Whiteman band to travel across the country in luxurious train accommodations, giving free concerts and making radio broadcasts along the route. The publicity value of the Whiteman band's trip was considerable, as it reached a majority of the U.S. population via radio during numerous public appearances.[140] On April 25, 1929, Universal's Carl Laemmle Jr. threw Whiteman and his band a party at "Whiteman Lodge" in Universal City, a home away from home for Whiteman and his band provided by the studio. Laemmle's party was a send-off for Whiteman and band's short West Coast tour before they returned to shoot *King of Jazz*.[141]

The *King of Jazz* was singular in its extravagant 1920s hullabaloo. The film benefited both from Universal's experiments in talkie-era sound reproduction and Technicolor technology. Emerging from an era in which musical revue films were relatively commonplace, this one still stood alone, released in 1930 alongside that the Oscar-winning Universal film *All Quiet*

Fig. 4. Paul Whiteman promoting Universal Studios' *The King of Jazz* with Carl Laemmle Jr. and Carl Laemmle, *Cinelandia y Film*, October 1929. (Billy Rose Theatre Division, the New York Public Library for the Performing Arts.)

On the Western Front (Milestone, 1930), a film that Fejos and Anderson satirized in *King of Jazz*.[142] Universal's attempts at big-budget or Oscar-contender films would lessen in the Depression-era 1930s, when the studio pivoted to the horror films for which they became known. But for the moment, Universal was interested in "quality pictures" of various kinds, and *King of Jazz* was Universal's attempt to compete with revue films like MGM's *Hollywood Revue of 1929* (Reisner/Cabane, 1929) and *Paramount on Parade* (Arzner/Brower, 1930).[143]

The final film was a giddy example of jazz-age excess that debuted in the first catastrophic months of the 1930s, as the country searched for equilibrium. After disappointing critical and audience responses to Los Angeles premieres, the film debuted on Broadway in April 1930. *King of Jazz* was one of two Universal Pictures prestige pictures produced by Carl Laemmle, Jr., who had been groomed to succeed his father at the studio. The other Universal prestige film that debuted was *All Quiet On the Western Front* (Milestone, 1930), an Oscar award-winning film, one of the most important of the decade, which garnered much greater critical and audience responses than *King of Jazz*.[144] Despite Whiteman's celebrity, Universal's *King of* Jazz was a poorly timed experiment in extravagance—though an invigorating, even dizzying one. As James Layton and David Pierce write, the movie's "infectious melodies, carefully choreographed ensembles and specialities, and grandiose sets" make for a "fantastic experience."[145] Whiteman and Universal's intermedial approach would retain importance, because of as much as in spite of difficult times. The recording industry's jitters had shifted the recording industry's best bets to those performers that could work across media platforms, as performers and their labels sought to survive in changing contexts. Entertainers diversified in ways that mirrored many of the largest companies' financial moves across media. Paul Whiteman was certainly not the only the entertainer seeking to ensure longevity through such diversification.

Transcending "Jazzmania": Louis Armstrong, Intermedial Star

If Paul Whiteman's career was near the center of the pop spotlight at decade's close, others who started on the periphery had transcended various borders to join him. One of the most important of these was Louis Armstrong. Armstrong's long-term importance far exceeds Whiteman's often underestimated contributions (perhaps most importantly, the popularization of jazz). It is impossible to characterize the importance of Armstrong in any simple way, because it is hard to imagine American music without him. There were already hints of this in the 1920s, when the formidable horn player followed opportunity to opportunity, moving from band to band and city to city. After surviving an impoverished and difficult childhood in New Orleans, he landed a stint with Kid Ory's band. He then moved on to riverboat gigs on the Mississippi with Fate Marable's band, followed by a spot in King Oliver's Creole Jazz Band in Chicago and his second marriage to pianist Lil Hardin (a hasty first marriage ended quickly). After a successful and relatively stable period in Chicago, he left on Hardin's urging for New York and Fletcher Henderson's band, only to return to Chicago roughly a year later. Back in Chicago, Armstrong formed his own band, still uncomfortable in leadership roles and unsure of his own skills despite considerable talents. Louis Armstrong and His Hot Five (and numerous variants for recording and performing purposes) established Armstrong's career as an increasingly formidable talent on his own terms. These recordings are now rightly the stuff of legend and drew on Armstrong's present and past (recording compositions by King Oliver, Kid Ory, Clarence Williams, and others). The recordings allowed a new vehicle for Armstrong's vocal and horn improvisations on tracks like "Heebie Jeebies" (Armstrong's version led to many covers) and the legendary "West End Blues." But while this second Chicago tenure is crucial to Armstrong's lasting importance in jazz circles, it was his 1929 return to New York that began his transformation into a celebrity beyond the confines of popular music, to borrow current parlance, an intermedial entertainer.

Initially, Armstrong's New York return looked to be more of the same, starting with a residency at Connie's Inn in Harlem (the Cotton Club's chief rival). What came next was more surprising: performing in *Hot Chocolates*, an off-Broadway play at New York's Hudson Theatre.[146] The show created by Fats Waller, Andy Razaf, and Harry Brooks was an all-black revue depicting Harlem life.[147] Edith Wilson originally sang the Waller-penned "Black and Blue," but when Armstrong adapted it, he left out much of the song's details about doomed romance. In Armstrong's hands it becomes poignant racial commentary, signaling Armstrong's indebtedness to Bert Williams, the vaudevillian whose masterful tragicomic approach addressed the wounds of racism to attentive listeners. In his *Hot Chocolates* review, Walter Winchell wrote, "Louis Armstrong, the most torrid of the horn-tooters and blue-shooters—his delightful specialties served to entice the bored-with-Broadway diversion-seekers—offers an endless routine of terrifically tempoed tunes."[148] Favorable reviews of *Hot Chocolates* led to a half-year Broadway run. After the show closed, Armstrong hit the road for dates in the Midwest and East Coast, appearing under his own name in what signaled a shift to an extensive touring schedule that he would continue for decades. A May 1930 *Variety* ad announced Armstrong as simultaneously a "Master of Modernism," "Creator of His Song Style, and, in the biggest type, "King of the Trumpet." As an industry trade paper, the ad targeted entertainment business insiders as much as those interested in an Armstrong performance or recording. Reviews culled from Winchell's *New York Daily Mirror* piece and a glowing review of a Detroit performance were included in the ad for further evidence. Meanwhile, Armstrong's recording career with Columbia Records was stronger than ever.[149]

In addition to his foray into theater, touring, and recording dates, Armstrong first appeared on film between 1930 and 1932. This was especially important given the recording industry's sharp downturn along with fewer opportunities for jazz players. He performed as himself in his first film, *Ex-Flame* (Victor Halperin, 1930) and perhaps a second (*Mixed Doubles*, late 1930 or early 1931) though neither is known to have survived. Tiffany Pictures, a Poverty Row production company founded by silent film star Mae Murray, produced both films. Armstrong then appeared in two 1932 shorts for Paramount that have survived: a Max Fleischer "Betty Boop" cartoon, *I'll Be Glad When You're Dead You Rascal You*, and a live-action Paramount short, *Rhapsody in Black and Blue*, in which Armstrong and band appear in an ill-conceived fantasy sequence.

The short begins and ends with Armstrong and band performing the song, which structures much of the screen action. Betty Boop, along with sidekicks Koko the Clown and Bimbo, find themselves in the jungle under attack from local tribesmen. As Armstrong and band perform the title song, one of the pursuing tribesmen's bodies is transformed into an animated rendering of Armstrong's head, which ascends into the sky. The animated head then transforms into Armstrong's head, courtesy of live-action footage embedded into the animation. Armstrong's head sings, "I'll be glad when you're dead" as the pursuit continues. The tribespeople meet their end at the top of a conveniently erupting volcano, after which we return once again to Armstrong and band.[150] In *Rhapsody in Black and Blue*, a companion 1932 short, Armstrong receives top billing. The film stars two uncredited actors: Sidney Easton as the husband who would rather listen to jazz records than mop the floor and Fanny Belle DeKnight as his long-suffering wife.[151] When the husband disobeys her request to wash the floor in favor of listening to an Armstrong 78, she knocks him unconscious with a mop. This leads to the Armstrong dream sequence, in which the band performs the title tune and "Shine," surrounded by suds and in jungle dress. After the husband enjoys Armstrong's performance, he asks where he is. "Why, you are in Jazzmania, and you are the king," his new servant replies.[152] When he comes to with his angry wife standing over him, he repurposes the short's title: "I'll be glad when I'm dead and gone," he says, knocking himself over the head with a vase. "I'll be more gladder," his wife responds, breaking the phonograph record over his head.[153]

Of the films that survive, Armstrong is not given much representative autonomy, as he worked within the era's racist iconography. Several authors have suggested that Armstrong's trust in manager Joe Glaser may well have led Armstrong into such endeavors with little question from the artist.[154] With King Oliver's late 1920s career decline (aggravated by health problems and poor business decisions that would leave him destitute at his death) and Glaser's admonitions to be an entertainer as much as a musician, Armstrong appears to have been convinced. And the leaner 1930s entertainment industry increasingly demanded it. However marginal these early film roles were, Armstrong must have considered acting an increasingly important part of his work. While in Denmark, he was asked to do another short film (sometimes mistakenly called his first), and in this film he is left to perform his music on stage, as represented in versions of "I Cover the Waterfront," "Dinah," and "Tiger Rag" for the camera. In the passport application Armstrong made ahead of his 1932 European tour, he listed his profession as "actor and musician."[155] These shorts gave way to more substantial acting roles, often overcoming a given script's racist shortcomings, throughout the mid-to-late 1930s, sometimes appearing with friend Bing Crosby: *Pennies from Heaven* (McLeod, 1936); *Artists and Models* (Walsh, 1937); *Every Day's a Holiday* (Sutherland, 1937); *Doctor Rhythm* (Tuttle, 1938); and *Going Places* (Enright, 1938).[156]

One would imagine that Armstrong would have been a natural on radio, and he was. But it would be a long wait for a role beyond his usual radio performances. Armstrong finally got the opportunity to be a radio host, though long after other similar musicians and bandleaders got their own shows. When crooner and radio star Rudy Vallee departed the United States for an extended summer stay in the United Kingdom, he encouraged both NBC and the sponsors to let Louis Armstrong host the immensely popular *Fleischmann's Yeast Radio Hour* in his absence.[157] That it took Armstrong this long to appear (albeit temporarily) as a host of his own radio show, due in part to Rudy Vallee's influence and largesse, is another sign of the strange paradoxes of the era, in which the exploitation of African American cultures were increasingly central to U.S. society, while entrenched institutional racism was furthered by the rising racism and xenophobia of the era. Radio excluded African Americans to a much greater degree than either sound recording or cinema. "Despite its tremendous need for personnel, however, the industry in its so-called Golden Age offered only limited opportunities for black men and women to develop. In fact, of all the popular arts, commercial radio possessed one of the most effective policies of discrimination along racial lines," noted Fred McDonald. He added that discrimination extended even to "black technical operators or actors, and it was evident in the strident stereotyping of black characters in actual broadcasts."[158] And, as William Barlow writes, "This sort of selective expropriation of African American culture by certain sectors of white society was nothing new—it had been going on since slavery; and, more often than not, the outcome was contradictory."[159] Battles for representation played out in local radio as much as network radio. In the late 1930s, African American representation on radio began to improve, often led by black entrepreneurs like Ed Baker of Detroit's WJIB, a radio man who started his own advertising agency to boost sponsorship on his station.[160] This underlines the extent to which black entrepreneurs in local radio had to build their own businesses outside the existing institutional structures in radio, which offered little or no support.

There is one last connection to make with Louis Armstrong that nicely illustrates his expanding range as an entertainer, as well as the artificial boundaries of genre as created by the industry and their easily permeable borders by like-minded musicians. While on the West Coast, Armstrong sat in on horn, with Lil Hardin-Armstrong on piano, for a Los Angeles recording session to record "Blue Yodel No. 9" and "Memories of You" with the biggest star in hillbilly music, Jimmie Rodgers.

Hillbilly and Race Artists across Media

Three years after his Victor recording sessions with Ralph Peer in Bristol, Jimmie Rodgers was in Los Angeles. While Rodgers, perhaps more than any one figure, is associated with the origins of country music aesthetics, his style always drew from a variety of sources, including jazz and blues. The divisions have always been the recording industry dividing audiences into market segments and not any real reflection of how music was made. In fact, Rodgers could as easily be understood as a bluesman first and foremost, as Nolan Porterfield suggests in his Rodgers biography.[161] This is borne out not only from listening to his records or surveying Rodgers' musical transformations over time. There were also blues players influenced by Rodgers who were surprised when they learned that Rodgers was white (Piedmont blues guitarist Jack A. Johnson is said to be one of several blues players who made the mistake). For several weeks in late June and early July of 1930, Rodgers recorded various tunes in Victor's Hollywood studios. Among the last songs recorded was the collaboration with Louis Armstrong and Lil Hardin-Armstrong, "Blue Yodel No. 9."[162] Jimmie Rodgers and the Armstrongs collaborated on a few songs in 1930 at RCA's new and relatively unused Hollywood studios, built largely for the company's Photophone sound-on-film technology.[163] This recording studio was a dramatic improvement from his 1929 sessions at the Atlanta Women's Club and only three years since his first recordings for Ralph Peer), during a stint that lasted nearly two weeks, in which Rodgers recorded a variety of material.[164]

Both Rodgers' and Armstrong's catalogs were already stylistically broad. The collaboration was not as farfetched as Jones and Chilton seem to suggest, given both the strong influence of blues and jazz in Rodgers' work as well as Armstrong's own interest in country music. Porterfield writes, "The divisions, more imagined than real, existed largely in the minds of naïve fans and feckless scholar-critics," and he argues for the universality of music across provisional and ad hoc genres devised for commercial purposes more than any aesthetic delineation.[165] "Blue Yodel No. 9," or "Standing on the Corner," is blues through and through. It doesn't take Louis or Lil Hardin-Armstrong's playing to bring that out. You can hear it in the improvisational feel of the take, Armstrong's trumpet playing along with what Dan Morgenstern described as Rodgers' "somewhat unpredictable bar lines," with Hardin-Armstrong's piano as rhythmic anchor.[166]

Like Armstrong, Rodgers had sought a balance of stage, phonograph, and radio work. Beyond his best-selling phonograph recordings, fans encountered Rodgers on the radio far more than during increasingly infrequent stage appearances due to the tuberculosis that would take his life in 1933 (he died at thirty-five years old).[167] His radio work had already begun. Asheville, North Carolina's WWNC radio went on the air in February 1927 and by April Rodgers was regularly heard on air with various musicians. These included Claude and Jack Grant with Jack Pierce as the Tenneva Ramblers, who famously fired Rodgers just before the Bristol recording sessions with Ralph Peer in October of the same year.[168] Rodgers would make radio appearances on and off throughout the rest of his career, sometimes using them to publicize one-off appearances and other times taking up a radio residency while trying to improve his failing health due to tuberculosis. In addition to Rodgers' WWNF appearances, he appeared on WTTF (Washington, D.C.), KMAC (San Antonio), KFYO (Abilene, Texas), and border radio station XED (Reynosa, Mexico, later changed to XEAW when purchased by radio quack Dr. John Brinkley).[169]

The film industry's voracious appetite for musical content began with musical shorts in the early talkies era. Rodgers performed three songs for a musical short, *The Singing Brakeman*, in a joint partnership release between Victor Records and Columbia Pictures, "Columbia-Victor Gems."[170] After appearing on stage with feature films like *Four Walls* (directed by William Nigh and starring John Gilbert and Joan Crawford, 1928) and *Why Be*

Good? (directed by William Seiter and starring Colleen Moore, 1929), perhaps Rodgers understood this as joining them if he couldn't beat them.[171] By January 1930, *The Singing Brakeman* was in American theaters. In the same summer of Rodgers' 1930 RCA Victor sessions with Louis Armstrong and Lil Hardin-Armstrong, "Peer took Jimmie to Hal Roach Studios for lunch with Mr. Roach and a visit to meet Laurel & Hardy," though Rodgers' health would not allow his film career to develop, while Peer's interest in film-music crossovers would continue for decades.[172]

Rodgers performed three songs in the short: "Waiting on a Train," "Daddy and Home," and "Blue Yodel No. 1 (T for Texas)." Keeping with the practice of many musical shorts of the time, a narrative setup was thought to help carry the songs in a filmic context. The scenario: Jimmie has come in from a brakeman's run on the railroad and is waiting for his next shift. Two women who seem familiar to him chat him up; the younger woman offers coffee and the older woman requests a song while knitting in her rocking chair. Despite the brief and stilted dialogue, viewers quickly got a sense of Rodgers the performer. He begins with an uncanny vocal approximation of a "high and lonesome and fading" train whistle sound, followed by a "vo-dee-o-doh" flourish meant to fill in for the cornet work by C. L. Hutchison that appeared on Rodgers' original Victor recording.[173]

Of course, Rodgers the brakeman was also Rodgers the vaudevillian, and to think of his approach to entertainment as anything other than hybrid is to misunderstand his life and work.[174] When Rodgers' own hybrid sensibility met Peer's focus on pop breadth, the eclectic lineups of the performers' recording sessions serve as further confirmation. As Barry Mazor writes, "There would be Peer-produced Jimmie Rodgers recordings featuring Hawaiian bands, jazz bands, country stringbands, pop orchestras, protocowboy and Texas swing contingents, and also those groundbreaking sessions with African-American musicians from Louis and Lil Armstrong to the Louisville Jug Band."[175] While now understood as a founder of the country music genre, Rodgers himself understood his own music in a much broader sense, both in terms of musical aesthetics and in terms of pop accessibility.

Like other successful artists initially targeted by record companies to fill geographic as well as generic niches, Jimmie Rodgers was soon an international star. Rodgers' recordings would see domestic release, via Victor or its Bluebird subsidiary, before being licensed to the Montgomery Ward chain stores for mail order (an important income source for both old-time and race recordings) as well as in-store sales.[176] Then, as Barry Mazor writes, "International Victor subsidiaries and such overseas labels as Regal-Zonophone and HMV took Jimmie's music across the English-speaking world and beyond."[177] In the U.K. and British colonial India, Rodgers' songs were on 78 in a variety of countries. Many of Rodgers' records appeared on HMV ("His Master's Voice," due to its ties with U.S. Victor), while London-based HMV subsidiary Zonophone released Rodgers' records for anglophone colonial markets in Africa. "In Jimmie's own day," Mazor writes, "Zonophone ads appeared in the black press, for black audiences, in segregated South Africa, featuring both his records and Zulu language discs, side by side."[178]

In the same Columbia–Victor Gems series in which Jimmie Rodgers appeared, Mamie Smith took her turn in front of movie cameras with *Jailhouse Blues* (1929), in which she performed the song in the film's title. In the short, she plays heartbreak for laughs, recounting the tale of a lover who is in jail as much as he is home (like Rodgers' *The Singing Brakeman*, Mamie Smith's *Jailhouse Blues* was credited to director Basil Smith). She was in between long-term OKeh contracts and was recording for both Victor (perhaps due to her previous work with Ralph Peer) and Ajax. She recorded the dialogue and music for the film at Victor's Camden studios for the single-reel short.[179] Although Smith was already a vaudeville star when "Crazy Blues" appeared, she too was working on extending her cross-media reach almost a decade later. Other female blues stars that had followed Smith's early 1920s success appeared on film. Ethel Waters appeared in films as early as 1929, with a musical appearance in the feature film *On with the Show!* In the film appearances that followed, Waters made the transition from musician to actor, a strategy that became a time-honored means of extending one's career. She never left her musical identity behind completely, as evidenced by her role in Vincente Minnelli's *Cabin in the Sky* (1943), in which she sings the title song while acting alongside Eddie "Rochester" Anderson in starring roles (along with Lena Horne, Louis Armstrong, Butterfly McQueen, and the Duke Ellington Orchestra). The career of Bessie Smith, "Empress of the Blues," would surpass that of her mentor "Queen" Mamie Smith (no relation), a Columbia Records performer who appeared in her own short. *St. Louis Blues* (1929) was a two-reel short directed by Dudley Murphy, who also directed Duke Ellington in *Black and Tan* the same year.[180] Murphy had assisted the artist Fernand Léger in his avant-garde film *Ballet Mécanique* (1924) five years before. By 1929, Smith was already known for the song "St. Louis Blues," which she made famous in a 1925 Columbia recording featuring Louis Armstrong on cornet. In this film short, she was called upon to act as well as sing: the film and song details a story of love gone wrong, in which our protagonist finds her man in the arms of another woman. It was Bessie Smith's only known film appearance before her death a decade later at forty-three years old.

Radio's Rise and the Phonograph's Decline

By 1925, retail record sales had fallen to half of what they were in 1921. "It was also the year the Victor Talking Machine Company failed to pay stock-holder dividends for the first time since 1901," Russell Sanjek wrote, "and Eldridge Johnson, founder and chief executive, seriously contemplated the disposition of all his holdings in the world's largest record company."[181] Despite important technological innovations—including the shift to electronic recording—that had begun to change the way records were recorded and how they sounded, listeners were slow to respond. Why buy a new phonograph machine when a family already owned one? How much better could the new recording technology be? And perhaps the more important question was: Why buy a phonograph (and all those records) when music on the radio was free? The recording industry held out hope that their new electrical recording technology might offer a higher quality sound reproduction that would boost business. As Eric Morritt writes, both Victor and Columbia pinned hopes on improved sound reproduction: "The Victor Talking Machine Company and the Columbia Phonograph Company began issuing electrically produced recordings to the public in mid-1925, and Victor's Orthophonic Victrola and Columbia's Viva-Tonal phonographs became available later that year." Both of these new machines were acoustic but designed to play electrically recorded discs (electrical playback would come later). While industry insiders agreed that electric recording was the preferred method, consumers did not buy the new machines in great numbers.[182]

The changes that electronic recording brought to the organizational structure of the industry, however, may have been even more important. Other recording companies were anxious to get into electric recording, though it wasn't easily attainable for many. Many companies sought to combine their mechanical phonographs with RCA's Radiola radios, one model of which was designed to fit in any number of cabinets and was used by various phonograph companies. The advertising manager for Sonora Records wrote an article for *Printer's Ink* called "Does the Phonograph Need a New Appeal?" in which he noted that phonographs could no longer be sold on features or the promise of the world's great artists.[183] With the realization that his company could compete on neither, he argued for paying more attention to women, who, according to the companies' in-house surveys, drove the all-important dance-record business. Sonora then catered to these buyers with ads that depicted a woman dancing with her boyfriend and then proposing marriage: "It's in in her home that he plans his.'"[184] But the smaller companies largely alternated between pitches such as these, based in their interpretations of social and cultural dispositions, or claims about technological improvements, which sometimes stretched the truth to make the sale. Each company had to decide whether to make deals for use of the technology while also deciding from the processes available. Recording engineers had to be retrained to use electric equipment, and the microphone was an altogether different recording device than the acoustic horn used previously. "Bell Telephone Laboratories," or Bell Laboratories, invented and owned the technology, which it in turn leased to Columbia (which first employed the technology in February 1925) and Victor (which made its first electronic recordings roughly a month later). AT&T's leases with Columbia and Victor were exclusive, so this left other companies to cut their own deals, as Brunswick did when it signed a deal with General Electric for its electronic recording technology in April 1925.[185] But not all companies wanted to lease the technology (or could afford to do so). The Edison Company once again held steadfast to its own acoustic process until it adopted the electronic standard in 1928. In the case of OKeh, the company reached an agreement with Columbia to gain access to the new technology by selling the company. An October 1926 *Talking Machine World* article announced the news: "By the terms of this deal the OKeh-Odeon record division of the General Phonograph Corp. is sold as a going concern to the Columbia Phonograph Co., Inc."[186] General Phonograph's Otto Heineman, president of OKeh Records in the United States, made a deal with Columbia Records that allowed him to head up Columbia's new OKeh division. Though Heineman may have had concerns about the future of the recording industry with the rise of radio (and despite dramatic successes in the race records market), the move seemed equally driven by OKeh's need for electric recording technology.

Meanwhile, the Gennett label entered into an agreement with J. O. Prescott in 1926. Prescott went to General Electric and hired away electrical engineer Robert Callen to convert Gennett's New York studios to electronic recording.[187] Many of the labels renamed record series to publicize the new technologies, such as OKeh's Truetone and Gennett's Electrobeam discs. Paramount's parent company, Wisconsin Chair, was reticent to spend the money for electronic recording and instead made agreements with Chicago's Marsh Laboratories as well as Gennett for use of their facilities. Longtime Paramount employee and scout Art Satherley cited this as his reason for leaving the company: "When Paramount wouldn't give electronic recording, I said 'I've worked here for fifteen years, I am going to walk on.'"[188] The Paramount label eventually opened a studio with electronic recording facilities at Grafton, Wisconsin, in the summer of 1930.[189] In the recording industry as in other media industries, the cost of adopting new technologies was not always welcome, but technological investment or partnership deals with other companies were made more often than not because it meant participating in changes, rather than trying to withstand them, in order to survive. The institutional machinations of this era point to industry consolidation into what economists call "mature oligopoly," a term that describes a given industry that is so centralized that any one change at a single company can directly affect other industry players. This certainly describes the events in the recording industry as it faced changes at the end of the 1920s and the beginning of the 1930s.

While sales had recovered from their mid-decade low point, profits from electronically recorded discs weren't what the record industry had hoped. The blame went to the radio business, which had grown each year during the 1920s. In a 1924 *Wall Street Journal* article titled "Phonograph Sales Have Declined" (and subtitled "Competition from Radio Affects the Industry and Reduces the Number of Manufacturers"), the author accounted for the downward slide. "While the general business depression in 1921 somewhat effected the output, the radio in 1922 really started phonograph sales on the toboggan."[190] The article noted the increase of phonograph-radio set sales, as phonograph companies sought to make deals with radio manufacturers. "Of the large companies, Victor Talking Machine Co. alone was able to withstand the growth of the radio without resorting to a reorganization or new financing."[191]

These events again mirrored those happening during roughly the same era in the film and radio industries, though the most dramatic mergers and acquisitions happened slightly earlier in the recording industry, since its technological innovations didn't have the same effect that talkies had on ticket sales in the film industry. Meanwhile, radio was still in its ascendancy as national networks formed and its social and commercial power expanded, so radio companies—most notably RCA—acquired other companies and made collaborative agreements far and wide. The radio industry's consolidation happened later as the Great Depression led to the attrition of all but a few central radio companies.

Amidst industry changes, Victor still seemed invincible, and executives at the company seemed confident in their ability to adapt to radio despite their resentment at its meteoric rise. The company was even more confident at its prospects when compared to those of its two competitors, the upstart Brunswick label and its longtime rival Columbia. As early as 1925, Columbia premiered its budget Harmony label to compete with more cheaply priced competitors, particularly those five-and-dime labels that posed an increasing threat by decade's end. "It was the coming of Louis Sterling" to Columbia Records, Victor composer and talent scout Nathaniel Shilkret remembered, "that changed everything."[192] Sterling restructured the company to his liking and turned his attention to competing on various fronts, not only against Victor but also against the company's smaller rivals. Columbia's aggressive moves against Victor fell short in overtaking that rival company's sales, even as that company was in the midst of transforming much of its phonograph manufacturing facilities for radio production. Thus, "in spite of the major commitment of its facilities to the manufacture of RCA radio phonographs," Russell Sanjek wrote, "Camden beat Sterling and all others, producing 34.5 million of the 65 million discs sold in 1929."[193]

Phonograph-radio combination sets became standard for RCA Victor in the mid-to-late 1920s as other record companies sought to cut deals with radio manufacturers. As early as 1925, Victor experimented with radio through the appearance of two of its most famous operatic stars, John McCormack and Lucrezia Bori, on WEAF New York. The appearance of two Victor artists not only signaled an embrace of radio by the company that now increasingly saw radio as a way to sell recordings and phonographs, but choosing McCormack and Bori also transmitted Victor's class status as the number one record company, famous for its Red Seal classical music recordings. In the pages of *American Mercury* magazine, Dane Yorke wrote a few years later that "it was whispered in the Victor organization that one very high executive of the company had actually refused to listen that night to the singing of his own stars."[194]

Tensions between sound recording and radio companies continued even after major recording companies were purchased by radio companies and are reminiscent in of earlier battles between songwriters, music publishers, and recording manufacturers. In fact, questions surrounding royalties for sound recordings remain a perennial issue (and are the subject of current debate with contemporary playback technologies). Recording companies worried that people would stop buying phonographs and sound recordings with radio's emergence. Radio companies wanted to use sound recordings to attract listeners more broadly, for its live content or its proprietary recorded transcription discs, but they were concerned about the threat of exorbitant royalty fees for the broadcast of sound recordings. While each had something to lose, both ultimately wanted to make a deal because the opportunities outweighed the risks of allowing radio to play sound recordings, or radio concerns about a royalty plan that would satisfy recording companies.

Even before the historic RCA Victor merger, the first Big Three recording label to experiment with radio partnership was Columbia Records. Arthur Judson, who had previously worked for Sarnoff at Victor, left to form United Independent Broadcasters (UIB), a small group of East Coast stations from which he wanted to create a viable network. Amidst mounting financial pressures, Judson formed a partnership with Columbia, which had been looking for a radio network opportunity of its own and launched what was (briefly) known as the Columbia Phonograph Broadcasting Company.[195] Columbia Phonograph's hope was that the radio network would be the perfect venue to publicize its recording artists and its latest releases. By September 1927, however (shortly after the network actually went on the air), Columbia was ready to sell its broadcasting system to UIB because of an inability to garner sufficient advertiser sponsorships for programming. Louis Sterling, the owner of Britain's Columbia Graphophone, who had heavily invested in U.S. Columbia, sought to unload the broadcasting venture.

A new Columbia Records owner was found in William Paley, who used his family's Philadelphia cigar manufacturing company to make the purchase. Paley, who had hoped to further his family's cigar business through radio and had convinced his father to bankroll his entry into the radio business, was not disappointed. Sales of his La Palina brand more than doubled during 1928, through radio advertising that led to a series of sponsored radio shows ("La Palina Concert" provided classical music; "La Palina Smoker" was a musical variety program).[196] "Sterling was bought out," Sanjek wrote, "and the network name was changed to the Columbia Broadcast System, eliminating any connection with the phonograph business."[197] Unable to make money from the sale of radios as RCA could do, Paley "understood what no one else did at the time," Tim Taylor writes, "that selling what one would now call 'content' was not the way to make money," and that "what broadcasters really sold were audiences."[198] Meanwhile, Columbia Phonograph Company's difficulties didn't end with the failed merger with UIB. Also in 1927, Sterling steered Columbia into an ill-advised deal with Federal-Brandes, Inc., who made Kolster radios, in hopes of marketing radio under the Columbia-Kolster brand. Tim Brooks writes, "Several companies had been fabulously successful in the mid 1920s by being first on the market with reliable and affordable home radios. However in 1927 everyone was jumping on the bandwagon, and by 1930 Kolster was in receivership."[199] While individual radio concerns failed, the industry as a whole continued to prosper. In the pages of the trade magazines, the term "radio star" found increased favorability, though many radio stars made recordings regularly. From various angles and in various ways, radio had bested the recording industry, especially in radio's favorable sound when compared to what Sanjek called the phonograph record's "scratchy, tinny reproduction of the original old Victrola and other talking machines," while increased competition by smaller companies with poorly made, inexpensive discs for five-and-dime labels also caused difficulties for industry leaders.[200]

Industry-wide, the drive to sign their talent to long-term contracts soon backfired amidst radio's rise. The competition led to deals that were out of step with the actual earning power of the artists, particularly since record prices themselves lowered due to budget label competitors, competition with radio, and a slowing economy. After a period of great experimentation, of trying a variety of new genres and subgenres, the industry once again returned its attention to fewer artists in favor of a larger-scale hit-driven approach. Popularity across media platforms—notoriety in radio or film—was especially useful in recording sales. While most bandleaders and artists would still have recorded for much less, since recordings were a great way to get people to live performances (where the best money for performers was often made) as well as increasing radio, theater, or even film opportunities, label competition and exuberance led to unwise speculation. Sanjek suggested that this atmosphere led labels into "signing too many unknown bands and too many unknown singers to record too many unknown songs."[201] The general effect that this new strategy had was to put increased pressure on fewer artists to perform. The industry's strategy of limiting the scope of their releases had worked in difficult times. Its reliance on fewer and fewer artists to deliver hits had left record companies more vulnerable to failure, locked into contracts that may or may not have paid off.

But individual hits by mainstream artists could not save companies from mounting debts as radio's encroachment led to an increasingly difficult time for all recording companies, even the untouchable Victor. As pressure mounted, Victor found itself dealing with the face of the new entertainment economy: RCA. As Alfred Chandler recounts, "Victor's senior managers had become convinced that they must move into the production of radios, particularly radio-phonograph combinations, if they were to retain market share. [David] Sarnoff convinced them to sell out instead."[202] And while Victor was the biggest phonograph company in the United States, there was a larger story about to emerge: David Sarnoff became president of the new RCA Victor just in time for the U.S. economy to fall right out of the sky.

CHAPTER FIVE

"The Trademark Dog's Stubby Tail"

Depression and Resurgence

"What's in 1930 for You?"

The RCA Victor partnership was still in its infancy—a deal was signed on March 15, 1929—when the stock market crashed on October 24, 1929.[1] "Wall Street Lays an Egg," read the terse *Variety* headline of November 2, one initial response that marked a catastrophe beginning to unfold. "The Depression," as Charles McGovern succinctly states, "was a crisis of capitalism that permeated American life."[2] "Two days later," Russell Sanjek writes, "the Edison company terminated all manufacture of cylinders, discs, and phonographs, retaining only the production of radios and dictating machines," while "RCA stock fell from 114 to 20 in the next few weeks."[3] Record sales immediately began to tumble, down just under 40 percent by the end of 1930.[4] The world had changed and relatively few had seen it coming. "Nobody paid attention to the overtones of financial trouble ahead," bandleader Vincent Lopez remembered, "even busboys were buying RCA stock on margin."[5] By 1931, Eddie Cantor had coauthored *Yoo Hoo! Prosperity*, his satiric take on the financial advice that had led so many to ruin.[6]

Before and after buying Victor, RCA considered a number of other scenarios, including a largely now-forgotten plan to purchase Victor's chief competitor, the financially struggling Columbia Records.[7] The primary reason seemed to be RCA's search for a bigger manufacturing capacity, which it got via Victor's Camden works. "RCA invested almost $11.5 million to convert the phonograph plant to radio production, Jefferson Cowie notes. "[David] Sarnoff proclaimed the South Jersey city to be the 'Radio Capital of the World' during rededication ceremonies in September 1930, and the move marked the beginning of RCA's dominance of the consumer electronics industry for over half a century."[8] In the months following the merger, Victor was already trying to comfort its retailers. But the *Voice of the Victor*'s first issue of the new year now had to speak to much larger events. Just a few months after the stock market crash, the cover of the January 1930 *Voice of the Victor* asked, "What's in 1930 for You?" while the first page featured "RCA's Victor's Message for 1930," signed by RCA Victor company president E. E. Shumaker.[9] The purpose of the article was to assuage fears any Victor dealers might have had about the finalization of the RCA Victor deal in late 1929. Most of Shumaker's statements address radio and emphasize the joining together of great companies and great minds, even while admitting to deeper motivations in discussing "the need of greatly expanded technical and research facilities."[10] "It is our privilege," the statement read, "not only to wish you, but *assure* you a prosperous New Year." The company made the assurance based on the new merger: "For, on January 1, the RCA Victor Company, Inc.—a unification of all the greatest forces in the home entertainment field—brings to bear upon the problems and the developments of the radio and general home entertainment field the full force of its powers."[11] But in addition to that, the text's dismissive tone about the recent economic collapse belies a bigger issue and a greater fear of economic depression: "In the light of recent experience, they saw, too, the need of powerful and immovable financial position, unaffected by any artificial economic disturbance."[12] The words here suggest the forced confidence necessary to sell a new company to those who could very well be affected adversely. But it also transmits the insecurity of those first months after the crash, when it was difficult to know whether to believe the worst or hope for the best. In Shumaker's position, and in addressing his retail-dealer readers, he had to convince them that all was well. A little over a month after the stock market crash, this was already becoming hard to do.

In his 1931 book *Sanctuary*, novelist William Faulkner evoked the emerging mood of the era on the sad main streets of a decaying and malevolent South. "The sunny air was filled with competitive radios and phonographs," he wrote. Describing the shops along the main streets of fictitious Jefferson, Mississippi, Faulkner captures a pervasive melancholy, as passersby were attracted by the mechanized sounds of jazz-age modernity in the form of "ballads simple in melody and theme, of bereavement and retribution and repentance, metallically sung, blurred, emphasized by static or needle—disembodied voices blaring from imitation wood cabinets or pebble-grain horn-mouths above the rapt faces, the gnarled slow hands shaped to the impervious earth, lugubrious, harsh and sad."[13] Faulkner had in mind a specific, southern milieu, but also pointed at larger societal realities that included—and far exceeded—economic concerns that were ubiquitous when the book was published. The sense of doom was pervasive, even if industry trade papers continue to chirp positive messages against all evidence. "The Great Depression was a defining event in American life; perhaps only the Civil War was more stressful and touched proportionally more people," William H. and Nancy K. Young note, as "national income fell in half," one in five Americans were unemployed, and many citizens made accommodations in every aspect of their lives.[14] Americans postponed marriages and grew gardens and sold apples, Lizabeth Cohen notes, while those employed workers organized into powerful unions that would transform American industries, from steel to meatpacking.[15] The recording and radio industries remained largely untouched by such organizing (that would change in time) but would be profoundly affected by the Great Depression.[16]

Even amid calamity, sound media had made music ubiquitous. That was irreversible. Shifting media materialities from phonography to radio (and later the jukebox) impacted how listeners thought about music as a cultural form, increasingly so through the 1930s. By the Depression's start, David Suisman writes, "the creation of a new musical culture was effectively complete," in that "a transformation had been affected in musical forms and the technologies and practices of making music and as such created a *new* musical culture," while also creating a "new *musical* culture, with more music everywhere."[17] But ubiquity did not necessarily mean variety, as media companies sold the same songs across platforms. The acceleration of "multi-media consolidations led to the simultaneous playing of a limited number of pop songs on movie sound tracks, radio broadcasts, and jukeboxes, saturating the media with hit songs" through which previous niche genres were absorbed into larger hybrid popular formulations.[18]

The pop songs of the era were, for at least a few minutes, a much-needed respite from everyday life. "Happy Days Are Here Again," written by Milton Ager and Jack Yellen, appeared in *Chasing Rainbows* (Reisner, 1930) and was recorded by the Ben Selvin, Jack Hylton, and Leo Reisman orchestras, then was further popularized as the theme song for Franklin Delano Roosevelt's 1932 presidential campaign. Fred Astaire sang "Puttin'

On the Ritz," while Marlene Dietrich sang "Falling in Love Again," and songs by Bing Crosby (with or without his Rhythm Boys), Rudy Vallee, Cab Calloway, Louis Armstrong, Duke Ellington, and Hoagy Carmichael were ubiquitous. But "hit records," on phonograph, radio, or in film, could hardly bolster a collapsing industry. "All branches of the economy suffered from the slump," Pekka Gronow and Ilpo Saunio write, but the recording industry was especially injured: "After all, records were a luxury."[19] Peter Martland's comparison between U.S. and U.K. sales between 1928 and 1938 reveals that "both the United States and British record industry experienced a ninety-percent loss of sales."[20]

It didn't take long for the gravity of the situation to sink in. Edison documents show that company had not been profitable for the previous half-decade, showing net losses exceeding $1,000,000 between 1924 and 1928.[21] While the Edison company had been in decline for more than a decade, the company's activities and Edison's personal involvement with them were closely followed. "Edison . . . had been quietly losing money on its records since 1912. The situation had long been shielded from investors and the public by cash infusions from Edison's profitable divisions."[22] A report from Edison's vice president Arthur Walsh mapped past losses and predicted a dire future.[23] Roland Gelatt wrote that Edison's post-stock-market-crash departure from the phonograph business—one that he played no small part in founding—"let loose a spate of pent-up resentment which record collectors had long been nurturing."[24] Gelatt quoted from a letter that appeared in the *Phonograph Monthly Review*: "Has Thomas A. Edison, Inc. proved to be a house of false alarms, a dispenser of disappointments, a mere 'promising' business? From the very first, Mr. Edison himself regarded the phonograph as merely a machine. He has never had any understanding of music: how could he hope to succeed in a musical venture?"[25] Edison's pulling out of the industry, at the same time many of its proponents were promising to stay the course, added insult to injury.

Columbia would survive but would lose much of what the company had been. Antitrust concerns connected to the merger of British Columbia with HMV Gramophone Company ("His Master's Voice," with ties to American Victor), and the formation of Electric and Musical Industries, Limited (EMI), led to the 1932 sale of American Columbia to Chicago's Grigsby-Grunow, makers of Majestic Radio.[26] The coming years would see the company sold and sold again.

The biggest and most productive labels at the start of the 1930s—RCA

Victor, Columbia, and Brunswick—appealed to radio and film companies in search of expanded production facilities or a means of satisfying their increasingly voracious need for music. But for Gennett, Paramount, and so many others, they had less to offer to radio and film companies. Those smaller labels—which had played such an important part in creating the foundational genres of modern American music—already faced two great difficulties by decade's close: the expensive transition to electronic recording and radio's increasing market dominance. For these labels, the great economic cataclysm would mean an end to what they had been doing, as they continued to search for ways forward. Even amidst dire forecasts and harsh realities, entrepreneurs sought new ways to reach consumers. The January 4, 1930, *New York Times* announced, "Unbreakable Discs Made of New Resin," a synthetic concoction by a Columbia University professor used to create flexible and inexpensive (and paper thin) recordings via the upstart Hit of the Week label, with new recordings released weekly and for less than most others.[27]

In September 1932, Dane Yorke wrote about the rise and fall of the U.S. recording industry in *The American Mercury*. Yorke employed Victor Records' Nipper, the company's iconic pet in ubiquitous ads, eternally listening for "His Master's Voice": "The stubby tail of the little trademark dog which the phonograph had made famous now quivered to the voice of a new master. The story of the phonograph, as a contained, independent industry, was over."[28] By the time Dane Yorke's "The Rise and Fall of the Phonograph" was published, it seemed to the author (and presumably his readers) that the phonograph had been beaten by the one-two punch of radio and the Depression. The collapse seemed total. After a decade or more of expansion and experimentation, a relentless search for niches of varying kinds—including the industry-altering genres of jazz, race records, and hillbilly (and affiliated subgenres)—it all seemed over. After a decade or more of the expanding breadth and depth of recording companies' recording efforts in the United States, Yorke wrote in terms of fallen empires: "Like ancient Gaul, the phonograph trade was for many years divided into three parts. Three great companies, Edison, Columbia, and Victor, held rigid control, and shared between them a sales income that rose to more than $125,000,000 a year."[29] Whatever the recording business had been, it was no more. But what had it become?

This chapter focuses on the U.S. recording industry's increasingly interconnected relationships across media platforms during the calamitous 1930s, as it negotiated an unprecedented economic depression that would further the process of consolidation begun before the crash. "The Great Depression period," Martland writes, "was a critical juncture for the recording business, destroying as it did pre-existing structures and at the same time laying the foundations for a new, and as it turned out, highly profitable industry that prospered for the next fifty years," during which many of the leading recording companies were purchased and assimilated into larger media companies.[30] But few could see such a future in the early 1930s. Amidst an unprecedented set of social and economic conditions, and after a breathtaking period of expansion, the recording industry further backed off from its speculative explorations into niche markets in favor of music most likely to reach wider audiences. This is an understandable response, as the stakes seemed to be the survival of the recording industry as a whole. And while the industry's path from genre expansion in the 1920s to its major companies becoming part of the radio industry is sometimes characterized as an Edenic fall from grace, that narrative misses the larger story detailed here: the reorganization and maturation of the recording industry as a whole. During an era in which the very survival of the business seemed to be at stake, recording companies de-emphasized previous genre trends and instead targeted that material they felt was likely to appeal to a broad cross-section of the public. This was especially true at those labels now owned by radio companies, who saw great value in labels' music holdings, but less immediate value in releasing speculative new material. These socioaesthetic decisions made by record companies, I will argue, were informed by new economic realities, which in turn quickly impacted the whole of American society in unprecedented ways. Companies had to reckon with larger forces, which sometimes meant selling out to other concerns or folding altogether. Other companies hoped to persist one way or another.

Articles in the trade press continued to champion recording companies alongside radio companies, while asserting that if sales had declined, it wasn't the recording companies' fault. Sure, the Depression had taken a toll, but the blame rested with retailers. Couldn't a positive outlook and a snappier window display change everything? The trade press suggested so. Behind these endless articles on how to attract the Depression-era consumer lay anxieties about struggling American industries and the U.S. government, which was in search of ways to exit the misery of the Depression's breadth and depth, especially with the rise of President Franklin Delano Roosevelt's New Deal policies that arose after his 1932 victory in the U.S. presidential election. "By the mid-1930s," Charles McGovern writes, "reforming everyday purchasing power had become a persistent minor theme of public discourse," while the government sought to encourage and protect consumers.[31] Recording companies sought consumers where they could find them, working with advertisers who increasingly sold consumption as Depression-era patriotism, while the trade press joined recording companies in imploring retailers to do the work for them.[32] The pages of *Talking Machine World* were full of admonitions when expectations were not met and placed an inordinate amount of responsibility and blame on retailers given the unprecedented social and economic contexts of the era.

"Business Is Tired of Long Faces": The Recording Industry's Depression-Era Discourse

Recording company rhetoric had long placed blame for flagging sales in the face of radio on retailers. First they argued for the phonograph's continued relevance amid radio's rise, and then made predictions about the phonograph's connections to radio, before pinning their hopes to the future of radio itself. In "Workers Should Profit in 1928," the titans of industry were asked about the future. While the first few companies were associated in one way or another with the recording industry's past (Brunswick, RCA Victor, Columbia), the rest of the quotes were from radio executives. RCA's David Sarnoff touted technological advancements, nationwide broadcasting, public interest, and the nascent work of the Federal Radio Commission.[33] Meanwhile, R. W. Jackson, general manager of Brunswick-Balke-Collender's Panatrope radio-phonograph division, sang the praises of the electric phonograph and predicted a bright future. "Nineteen hundred and twenty-eight, as other years, will be just what we make it by good merchandising, confidence, and thoughtful and vigorous work."[34]

But with the onset of the Depression, trade papers' messages to retailers carried increasing amounts of anxiety and even blame. Rhetoric targeted at retail dealers couched the economic downturn in psychological terms. This reflected economic liberalism's championing of individual autonomy while ignoring overwhelming social odds: the depression is in *you*, this line of reasoning went. If your sales are down, you finally have to blame yourself, don't you? Trade papers suggested that the fault lay in retailers' poor performances, while critiques masked as motivational columns became increasingly judgmental. A February 18, 1931, editorial in *Voice of the Victor*, "Business Is Tired of Long Faces," began with the parable of a Dr. Vickers, a physician who "waited and waited for patients who were slow to come," but who "bought a sleek chestnut horse and a shiny surrey," in which he sought out patients two or three times a day, after which "[h]e looked busy and business perked up."[35] The article was excerpted from the *Cincinnati Enquirer* advertisement printed by "Ralph H. Jones Company, advertising agents."[36] What was Dr. Vickers' secret, besides initiative and having the means to appear successful? While suggesting that young Dr. Vickers had become a rich man and a pillar of his profession, the article finally pins his success down to the topic of the article: "He knew his psychology, didn't he?"[37]

> The depression has been psychological, too. No need to deny *that*. . . . Just why, no one seemed to say. . . . But the fact remained—the depression was *contagious*. . . . Business has gotten tired of inactivity and inertia, of long faces and milk-toast prescriptions. It has had a revival meeting—*the revival of the fit*.[38]

The Victor Company was imploring its retail dealers to dig deep and give a little more, through framing depression as an illness caught by the least worthy among us. In this, Victor was selling a concept as old as American culture and older than the country itself: the puritan's notion that success comes to those who are good and failure is visited upon those who are bad.[39] The 1931 editorial concluded with the opinion that the Depression had finally been character building for the American spirit, optimistically writing about it in the past tense. "It was a good thing, after all, wasn't it? If you meet a man who doesn't think so, suggest that he buy a house, an automobile, a suit of clothes or a potted plant. He'll find that he can get a lot more of better things for his money today than he could have gotten last year."[40] It is hard to imagine a retail dealer masochistic enough to take this editorial to heart.

But the era would not be defined by the unfounded optimism in the pages of industry trade papers. It would be a 1930 Tin Pan Alley tune by E. Y. "Yip" Harburg and Jay Gorney, "Brother, Can You Spare a Dime?" that would define the era. Gorney based the song on a Russian lullaby, making it a rare minor-key Tin Pan Alley hit.[41] The song was recorded and popularized by some of the era's key vocalists; Rudy Vallee recorded it for Columbia in 1931 while Bing Crosby recorded it for Brunswick in 1932. The controversial song was almost struck from *Americana*, the Broadway show in which it appeared, and some politicians sought to ban the tune. Its lyrics became a template for subsequent songs on the Depression, detailing a man's Depression-era woes. After helping to build the country, after fighting for it, the song's protagonist begged for help from a country that had forgotten him:

> Once I built a railroad, I made it run, made it race against time.
> Once I built a railroad; now it's done. Brother, can you spare a dime?
> Once I built a tower, up to the sun, brick, and rivet, and lime;
> Once I built a tower, now it's done. Brother, can you spare a dime?[42]

But the crooners of the moment didn't encourage second thoughts on those dealers or companies who wanted out. The numbers were too disturbing. Roland Gelatt noted, "A total of six million records were sold in the United States during 1932, approximately six percent of the total record sales in 1927."[43] The Depression and radio seemed to be a one-two punch that might mean the end of the industry altogether. Gelatt goes further by suggesting "something intangible" in the way consumers turned away from records and phonographs, suggesting that the very idea of phonograph records fell out of favor as "a sudden disenchantment" took hold, "on a country-wide scale with phonographs, needles, records, and the whole concept of 'canned music.'"[44] Despite the advent of electronic recording and playback, recording companies had to fight the notion that the phonograph was a quaint machine of the past.

The Recording Industry in the Conglomerate Era

The lasting effect of the sale of Victor to RCA was devastating to many longtime proponents of the record industry, amidst cheery forecasts for the future in this royal wedding of the best of radio and the premier recording company. In fact, RCA was more interested in Victor's Camden complex and the company's manufacturing, distribution, and retail capabilities than in the record business. As Roland Gelatt wrote, "No sooner had the ink dried on the agreement giving RCA control of Victor than plans were set in motion to begin the manufacture of radio sets in the Camden factory. For despite the assurances about radio and phonograph joining together on equal terms, the management at RCA had no real interest in the talking machine."[45] "It took months of corporate fighting and intrigue for Sarnoff to get control," Russell Sanjek wrote, followed by his effort to "reorganize and consolidate the combined RCA Victor Company, of which he became president."[46] The new RCA Victor was clearly the most powerful and prestigious organization in the recording industry. Those loyal to the old Victor, however, had reason to believe the new company's interest in that facet of the record industry was marginal at best.

Record executives severely limited (and in some cases discontinued) their efforts in the race records market first, quickly putting an end to most rural scouting trips as a cost-cutting measure, bringing the broad expansion in recording American vernacular culture to a close. Sanjek wrote, "Only the hot dance bands—those of Duke Ellington, Fletcher Henderson, King Oliver, all on Victor—and certain white Red Nichols units" survived, linked as they were to the pop-jazz hybrid dance band craze, while most of the rural blues acts found their recording opportunities over."[47] Victor, in the first years of Depression-era RCA ownership, severely limited recording even among most of its bestselling stars, while its attempt to cut in to the discount records market via its Bluebird imprint faltered (oddly, the label held on to its prestige Red Seal classical music imprint, releases of which retailed for between two and three dollars).[48] Ralph Peer sought to protect his bestselling hillbilly artists, Jimmie Rodgers and the Carter Family, by essentially paying an advance, which might not ever be recouped (Rodgers received $250 a side, while the Carters received $100), while Peer's own Southern Music was once again by 1932 growing dramatically, with branch offices in "London, Paris, Mexico City, Havana, and Buenos Aires."[49] Only the recording careers of the most successful race artists survived, including an emerging urban blues pianist Leroy Carr, whose 1929 hit with Scrapper Blackwell for Brunswick's Vocalion imprint, "How Long How Long Blues," ensured him a recording career into the early 1930s. Unlike many of the rural blues players captured by Paramount, Carr had moved from Tennessee to Indianapolis as a child. Carr's style was ruminative, melancholy, and humorous in turn. Elijah Wald has described Carr as "an understated pianist with a gentle, understated voice," particularly compared to the rough country blues of the late 1920s.[50] His records far outsold those cherished by collectors now, in deeply mournful, reflective, affecting songs (his songs were also uplifting in turn, while he could also ham it up, taking his turn at Tampa Red–style hokum on occasion). Carr's importance has been underplayed, but he was one of the few race record performers to enjoy continued success through the worst years of the Depression (his career was cut short when he died from complications due to alcoholism in 1935). In the era, Carr's importance was largely relegated to the African American press, as race record stars rarely if ever appeared beyond the African American press or received mentions in recording industry trade journals. Meanwhile, in the increasingly important radio publications, those African American stars were largely relegated to depictions of minstrelsy, given renewed vigor from the radio success of white blackface comedians Freeman Gosden and Charles Correll's *Amos n' Andy* narrative radio series. On rare occasions, those African American artists most successful at mainstreaming appeared in the pages of *Radio Digest*, as the Mills Brothers, the Three Keys, and Cab Calloway did in a "Special Colored Supplement" titled "Radio's Dark Town Harmonizers," reifying the racist attitudes of the day.[51]

Meanwhile, the hillbilly market stayed surprisingly strong. As Russell Sanjek suggested, hillbilly found its continuing niche through chain stores and mail-order business, most importantly Sears & Roebuck. Money was still being made in "the 'hick disc' business, as *Variety* dubbed it, which had tripled since 1930."[52] Although sales declined in absolute numbers, hillbilly music actually increased its share of the overall record market during these years. "Surprising everybody but [Ralph] Peer and a few other talent scouts and field-recording supervisors, hillbilly records, generally unadvertised by the manufacturers, who left such exploitation to the artists themselves, enjoyed a 25 percent share of all popular-record sales."[53] Jimmie Rodgers' "Blue Yodel Number 1," the first of his series of blues-infused hillbilly recordings, was said to have outsold all of Victor's popular releases. There was now so much demand for hillbilly records that some New York companies, who found only a few suitable local artists available, used as many as a dozen different names for single performer in order to continue quick production.[54] Ralph Peer, who had guided Rodgers' career closely, continued to do so amidst difficulties of his own. As Russell Sanjek wrote, "Fellow recording supervisors were jealous of the partnership he enjoyed with Victor through the music house he had formed, Southern Music Publishing Company."[55] Under Peer's supervision, and in spite of the singer's failing health, Rodgers remained an important figure at RCA Victor. (Rodgers would finally succumb to tuberculosis on a trip to New York to record for RCA Victor in May 1933.)

In a January 1932 RCA Victor advertisement, Rodgers was the only hillbilly performer whose photograph was included. He was featured dressed formally in straw hat, suit, and bow tie, with no traces of his railroad man or cowboy guises.[56] His appeal as a recording artist had led Rodgers out of the hillbilly market and into the larger realm of popular song. Rodgers' popularity is clear in his inclusion in the ad alongside the most popular dance band leaders and singers of the day: Russ Columbo, Paul Whiteman (back on Victor after a two-year stint with Columbia), Mildred Bailey, Fred and Adele Astaire, Sylvia Froos, and Fred Waring. For recording artists in general, contracts were unlikely to be renewed, instead returning to a session-by-session payment schedule. For many, royalties for the recordings were no longer offered. To keep in the game, those artists fortunate enough to have recording offers took the work.

The larger labels continued to curtail their recording schedules. RCA Victor's Los Angeles studios were initially built to capitalize on film sound through their Photophone sound-on-film technology. They had recorded Jimmie Rodgers with the Armstrongs before the crash and had hoped to record West Coast talent, but the studios were rarely used for anything but film in the Depression. Early on in the RCA years, Victor's entire record distribution wing was "collapsed into RCA's product-merchandising operation," and by June of 1931 "an official order cut recording down to a single wax take per selection, unless it was defective."[57] Technological experimentation did not stop completely, but company executives watched expenditures carefully. Columbia experimented with its Radiograph, an inexpensive record player meant to be powered through any radio receiver, while Victor developed 33 1/3 rpm records, primarily geared to the classical music market so that recordings of their favorite compositions would not be interrupted as often by changing 78 rpm records. But these failed to fire the public's imagination in the midst of the Depression.[58] Even Victor's legendary advertising efforts, which had made the brand one of the most recognizable in the world, were greatly reduced during this period. As Russell Sanjek wrote, "The national newspaper and magazine advertising which had kept consumers apprised for years of new weekly Victor issues was dropped and replaced with stickers on music-store windows and pamphlets describing new products that retailers sent to their local mailing lists."[59] RCA Victor did, however, sign Paul Whiteman for $60,000 after two years on a Columbia contract. Although the cost may have seemed excessive in 1930 (it wouldn't have seemed so even a year earlier), Whiteman's return to the Victor label (albeit under RCA's control) had symbolic as well as economic importance.[60] RCA Victor and Columbia also manufactured transcription discs as the market for them grew. The World Broadcasting System was the biggest manufacturer of

16-inch 33 1/3 rpm discs, which provided prerecorded radio programs to play on air, but both RCA Victor and Columbia competed with their own ventures as well.[61]

Beyond RCA Victor's continued dominance, Brunswick was a notable success. As Edison declined through the 1920s and Columbia suffered in varying ways, Brunswick rose steadily to become one of the most successful labels of the era. As one of the few success stories of the late 1920s, Warner Bros. bought Brunswick in April of 1930, a moment in which Warner was optimistic about film/music cross-promotions such as their earlier success with Vitaphone shorts, sound-on-film technology, the *Warner Bros.' Vitaphone Jubilee Hour* radio show, and the prospect of a Brunswick record line featuring Warner stars.[62] But the Brunswick-Warner relationship would be short-lived, as the full impact of the Depression was felt across media industries. But as Victor was the largest record company of the three, company executives managed to ignore the changes happening all around them.

Talkies-era film's interest in sound recording led to a precrash buyout offer from Fox Pictures, which Brunswick turned down.[63] Shortly after, Warner Bros. bought Brunswick in April 1930, the same year in which recording industry veteran Jack Kapp began work at the company. At Brunswick, Kapp had some cross-promotional plans in mind and focused on using their Warner Bros. movie star roster to make recordings. Warner Bros., Russell Sanjek wrote, "had recently acquired a dozen major music houses, looked to ensure the public exploitation of those catalogues and their singing-star personalities, and so purchased the failing Brunswick business."[64] Through recording songs from motion pictures in great numbers (the general practice dates back to the beginnings of both the film and recording industries), Warner had hoped that a hit song might lead to greater ticket sales as well as record sales. As Brian Rust wrote, "Warner did its best to maintain the Brunswick image by signing up film stars they had under contract. What followed, in addition to the dance bands already established as Brunswick artists, were a batch of unusual records by Gloria Swanson, Noah Beery (Swanson's former brother-in-law, during Swanson's disastrous early marriage to Wallace Beery), as well as sides by Harry Richman and Al Jolson."[65] All four of these names was involved in film in some way, whether working on feature films or recording musical shorts, as Richman had done with Irving Berlin's "Puttin' On the Ritz" at the close of the 1920s (another song about jazz-age opulence popularized in the first days of the Depression). Warner had the right impulse to try to capitalize via celebrity cross-promotion, but none of these recordings became hits. As Rust writes, "The Depression had bitten deep. So in December 1931 Warner sold out to the newly formed American Record Corporation (ARC), which was already producing Banner, Cameo, Oriole, Perfect and Romeo records for the chainstore counters."[66] American Record Corporation's Herbert Yates had bought up various discount labels and had successfully turned them collectively into a powerful industry presence, via the ARC imprint. ARC's parent company was Consolidated Film. "Consolidated Film Industries," the *New York Times* announcement read, "has acquired through a new subsidiary the phonograph business of Brunswick Radio Corporation, a subsidiary of Warner Brothers Pictures, Inc.," to then be known as "Brunswick Record Corporation" (Warner kept Brunswick's radio business).[67]

At Brunswick, Jack Kapp was able to assemble an impressive roster of vocalists and help repair some of the managerial missteps under Warner Bros., while he focused more directly on music he felt would connect with dance fans. With Victor focused on radio and Columbia in trouble yet again, Brunswick moved ahead with an impressive list of recording artists: "In 1931 he [Kapp] was general manager for the Brunswick label, building an imposing roster—the Mills Brothers, the Boswell Sisters, Victor Young, Cab Calloway, Glen Gray, Duke Ellington, and his personal favorite, Guy Lombardo."[68] The Mills Brothers were a singular outfit that created a unique vocal group sound, mimicking the sound of "a small jazz band" but without using any orthodox instruments other than one guitar.[69] First-rate talent acquisition continued after American Record Company's purchase of Brunswick, with Guy Lombardo and His Orchestra relocating from Columbia; Duke Ellington's band, a growing favorite among dance fans from coast to coast; and the Boswell Sisters, a trio from New Orleans whose distinctive and inventive vocals placed them in heavy demand. Besides owning Brunswick, American Record Corporation owned a variety of other labels that it managed to purchase through the Depression, including Banner, Grey Gull, Pathé-Perfect's American subsidiary, and Vocalion. With this talent, Rust suggested, "the Brunswick roster looked unbeatable."[70]

The same page of the January 1932 issue of *Radio Retailing* noted the sale of Columbia to the Grigsby-Grunow Company (the Chicago manufacturer of Majestic radios) and the announcement that Herbert E. Yates' Consolidated Film Industries, Inc. (the East Coast film studio) had purchased Brunswick's record division. *Radio Retailing* reported both mergers in January 1932. Regarding the Columbia–Grigsby-Grunow announcement, the magazine noted "This action signalizes [*sic*] one of the most interesting consolidations yet witnessed in the radio industry."[71] But the company's struggles soon led Grigsby-Grunow to bankruptcy. *Radio Retailing* described the Brunswick-ARC merger in a more matter-of-fact way: "The Consolidated Film Industries, Inc., New York, has organized the Brunswick Record Corp. to take over the manufacture and sale of Brunswick, Melotone, and Vocalion records for the U.S., Canada, and certain foreign countries."[72]

Yates had even bigger plans. "It was Yates' intention," Russell Sanjek wrote, "to use British financing for the purchase of Columbia Records and its licensed Western Electric recording equipment from the creditors of the bankrupt Grigsby-Grunow Company—so that he could go into the electrical-transcription business—that started the move of the record business into a new direction."[73] Yates, who had a Poverty Row film processing lab, had also been busy "acquiring film companies unable to pay their processing bills," including Chesterfield, Invincible, Liberty, Majestic, Mascot, and Monogram, all operating under Republic Pictures by 1935, just as he had acquired faltering record companies.[74] Yates now sought to join with Jack Kapp, chief manager of his recording interests at Brunswick, and bandleader and singer Ted Lewis, "Mr. Entertainment," to buy Columbia from Grigsby-Grunow's creditors.[75] He then wanted to close Columbia's Bridgeport, Connecticut, manufacturing plant and move operations to his Scranton Button Company, which he had purchased in 1929 to manufacture releases by Banner, Cameo, Conquerer, Domino, Melotone, Vocalion, and other discount ARC labels. When Kapp and Lewis rejected the sale of Columbia's Bridgeport plant, and Yates' British financing fell through, Kapp departed ARC's Brunswick (followed by his new investor, Ted Lewis) for a new venture that would also have interests across media platforms, to be called Decca.

Meanwhile, conglomerated RCA Victor's dedication to the phonograph lagged in relation to radio manufacturing, which was no surprise given that RCA's main attraction to Victor was its unparalleled manufacturing capacity. The new company, with David Sarnoff as its new president, was using its Camden facilities to make radios almost exclusively, since the phonograph business and record business seemed unlikely to recover. By the early 1930s—in an unthinkable reversal—Herbert Yates' ARC, which specialized in five-and-dime store labels such as Banner, Domino, and others, surpassed even Victor in sales. "The general apathy prevailing at all record companies around Christmas-time of 1932 reflected the 40 percent decline in a single year," Russell Sanjek wrote, "to sales of $11 million at retail from total production of 10 million discs; American Record Company-

Brunswick, 6 million; RCA Victor, 3 million; the balance spread among Columbia and lesser firms."[76]

In the pages of the trade magazines, retail dealers of phonographs and radios were encouraged to investigate other, more practical items for the home: refrigerators and washing machines. In the September 1931 issue of *Radio Music Merchant* (the renamed *Talking Machine World*), an advertisement announced, "Columbia Presents: A New Profit Maker," a tilted drum with a hand crank and lid complete with dry cleaning fluid, the "Columbia Home Dry Cleaner."[77] This unlikely (and unwieldy) product line followed an industry shift toward the sale of other home appliances, as phonograph and radio companies looked for sales where they could be found. When discographer Brian Rust wrote about it later, he emphasized Grigsby-Grunow's manufacturing "Majestic radios, refrigerators, and washing machines," reinforcing the role of home appliances in attempts to survive economic catastrophe.[78] Deals that had previously been unthinkable now became commonplace.

Meanwhile, those radio dealers that also carried appliance lines found that their refrigerator sales helped them survive the Depression. Radio and refrigerator sales had long been joined in that they were both among the first electrical appliances in the American home. In a March 1932 *Radio Retailing* article, the headline screamed, "Refrigeration Saved My Shirt!" Radio dealer Charles Reynolds of Binghamton, New York, detailed his success in selling refrigerators, noting that it had helped immensely to have such a practical machine to sell in hard times.[79] At the same time, the article's anonymous author made sure to underline radio's continued importance, in a way reminiscent of the pro-phonograph rhetoric found in the pages of *Talking Machine World* (later changed to *Radio Music Merchant* and finally *Radio Merchant*). Reynolds, identified as a radio enthusiast since the "amateur operator" days, said that refrigerators would outsell radios in 1932, adding, "This will be a difficult year for all selling due to the general low-ebb of business but radio will be the harder-hit of the two because of its greater saturation, stiffer competition."[80] For many retailers, one appliance that sold was as good as any other, whether they were selling phonographs, radios, or refrigerators.

Grigsby-Grunow went bankrupt in November 1933, making Columbia "just another label in the ever-growing collection owned by the American Record Corporation," as Rust wrote.[81] It was startling to see how far the once great Columbia had fallen, for anyone who was still paying attention. It reflected not only the sorry state of the label but of the industry in general. "The low esteem in which this business was held," Roland Gelatt wrote, "was effectively underlined by a transaction consummated in 1934 between the receivers of the Grigsby-Grunow Company and the American Record Company. ARC bought it—the plant at Bridgeport, the entire catalog of discs built up slowly over three decades, the trademark, the affiliation with Columbia in Europe, everything—for $70,500." Louis Sterling had paid $2,500,000 for the same operation nine years earlier.[82] "Companies that once had occupied a respectable position in the business community," Roland Gelatt wrote, "were sold for whatever they could bring."[83]

After merging with Columbia back in 1926, OKeh had the imprimatur of a key U.S. recording company, with the financial backing of British Columbia. But with the departure of Ralph Peer to Victor in 1927, along with Columbia's sale to Grigsby-Grunow and then ARC in the early 1930s, OKeh's importance diminished. Tommy Rockwell, who had worked for both Columbia and OKeh previously, replaced Peer at Columbia-owned OKeh and went on to record Mississippi John Hurt in 1928.[84] While Rockwell's reputation as a financially predatory scout is worthy of further investigation, he did nurture OKeh-Columbia's biggest stars, Louis Armstrong and Bessie Smith and was convinced by Eddie Condon and Armstrong to allow recording with a "mixed group" session for the label.[85] Armstrong's "(What Did I Do to Be So) Black and Blue" appeared on OKeh in 1929, culled from his work on *Hot Chocolates*. But during the Depression, OKeh releases slowed considerably. Grigsby-Grunow first "temporarily suspended" any new OKeh releases in June 1932 and discontinued its lesser-known discount labels. A year later, Grisgsby-Grunow was in bankruptcy, along with both Columbia and OKeh.[86] Both labels would later find new life via ARC in the short term and CBS in the long term, but for now both labels were defunct.

Legendary Paramount talent scout Art Satherley had long asked Paramount Records to make a proper transition from acoustic to electrical recording, but by the time a new studio was built in Grafton, Wisconsin, it was too late in at least two ways: Satherley had long complained both about the quality of Paramount Records' material composition as well as New York Recording Laboratories (NYRL)'s poor facilities. Satherley had made his mind up to leave, briefly to QRS (the piano roll company), then to ARC, before finally building a long-term career at Columbia, where he retired in 1952.[87]

"Sales of Wisconsin Chair Company products dropped dramatically as the economy deepened," Alex van der Tuuk notes, as the company struggled to cover its payroll through the first few years of the Depression.[88] Recordings were made sporadically through the early years of the economic calamity. In addition to recording artists from nearby Milwaukee—van der Tuuk notes that one of the last sessions was with an award-winning choir from nearby Shorewood High School—Paramount released sides by Jabo Williams, Blind Blake, and the Mississippi Sheiks.[89] Some who had worked with Paramount in varying capacities blamed the company's slow transition to electrical recording and its inability to retool its works for radio production.[90] At the same time, mail orders fell off, and, even worse, its distribution agreement with St. Louis' Artophone collapsed as well (though that company continued to reissue Paramount (and Gennett) releases through mid-1930 via Artophone's Herwin label.[91] Agreements to supply Montgomery Ward releases from their Broadway label ceased, but the label continued to make recordings and even made studio improvements in the months before Paramount's closure in December 1933, amidst discord and accusations among the label's leadership.[92] Wisconsin Chair would reinvent itself once again, surviving several decades more before finally closing in 1954.[93]

"Gennett Records, a marginally profitable venture before the Depression, was swept up in the chaos," Rick Kennedy writes.[94] Starr Piano, Gennett's parent company, discontinued its Electrobeam label along with the company's phonograph manufacturing in December 1930, while its discount Champion label's hillbilly and race record releases lasted until 1934, when Gennett ceased operations as an active recording company.[95] Harry Gennett, Starr's president in Richmond, Indiana, had diversified the company into refrigerators and thought it best to focus future efforts there.[96] "When unemployment rates rose in the early in the early 1930s, many consumers stopped paying the installments on their pianos. Starr Piano was stuck with enormous quantities of unpaid merchandise all over the country with no way to recoup the losses."[97] Economic tensions led to discord between the three Gennett sons (Harry, Fred, and Clarence) who had inherited the company from patriarch Henry Gennett after his death in 1922.[98]

Those Gennett metal masters not sold for scrap by recording director Fred Wiggins found rerelease via Jack Kapp's Decca, which had bought Gennett's back catalog and made reissues in the United States and elsewhere of the label's most popular recordings.[99] Meanwhile, Hoagy Carmichael, with his Gennett days behind him and his Hollywood music and film career ahead of him, was working as a songwriter and plugger for Peer's Southern Music in New York City, commenting tersely about his boss:

"I thought Ralph Peer was a good businessman. He was, and he profited much more out of the deal than I."[100]

One corner of the Gennett company continued recording: its sound effects division run by Harry Gennett Jr., which began in 1928 with a recording of a train passing through Gennett's Richmond, Indiana, works in the Whitewater gorge.[101] On the Gennett Special and Speedy Q labels, the company released sound effects recordings for radio and film. They recorded steamboats on the Ohio River and croaking frogs on an Indiana lake, and record after record of common, everyday sounds using Gennett's mobile recording truck, which traveled from coast to coast.[102] These endeavors built on Gennett's tradition of spoken word recordings by William Jennings Bryan and James Whitcomb Riley, its 1923 canary recordings (with violin accompaniment) released on Gennett and as a souvenir record on the West's label (West's was said to be "Largest Manufacturer of the Most Complete Line of Bird Products in the World'), and its 1926 project with anthropologist J. Walter Fewkes for a series of Hopi Indian recordings at the Grand Canyon's El Tovar Hotel.[103] Gennett's sound effects recordings lasted through the 1940s (the Starr company finally dissolved after being sold to J. Solotken Company, an Indianapolis "scrap metals and paper salvage" concern, in 1952).[104]

"The Canine's Industrial Horoscope": Swing, Jukeboxes, and Resurgence

As if in answer to Dane Yorke's 1932 summation of the recording industry, H. S. Maraniss's 1937 article "A Dog Has Nine Lives: The Story of the Phonograph" described how record companies were once again on the rebound.[105] He referenced the once-great Victor label through its iconic advertising symbol, suggesting that "since this paper is to be about phonograph and phonographs are symbolized by the 'His Master's Voice' trademark, there appears to be considerable truth in that remark. In fact, if I know anything about that canine's industrial horoscope, it would seem that it was born with nine lives, only three of which has been spent so far."[106] While Maraniss had to bend his metaphor a bit awkwardly—a cat, not a dog, is said to have nine lives—he suggested that the phonograph industry's first three lives were spent (1) in its slowness in convincing people the phonograph was more than a toy; (2) through increased competition from radio in the 1920s; and (3) the effects of the Great Depression. "Record sales suffered severely, as did the sale of every other commodity," Maraniss wrote. But, he continued, "In spite of adverse conditions and the counter-attraction of radio, there were still large numbers of people who were fond of records, and some who were willing to pay money to prove it."[107]

Record companies saw the beginnings of a turnaround in 1933, buoyed by the growing sense of national optimism on various fronts, including Franklin Roosevelt's young presidency. Maraniss wrote, "By 1933 the RCA Victor Company had stopped talking about the plight of the record business and was aggressively doing something about it. The first thing it did was to revolutionize the process of recording again and to introduce the almost incredibly realistic 'Higher Fidelity' system."[108] The company also began advertising again in radio trade magazines. Gelatt adds that RCA's 1933 record sales "were purported to show a hundred percent increase over 1932." Gelatt recognizes that such an increase "hardly connoted a booming business," since the company's 1932 sales figures showed only five millions records sold, "or less than one tenth of Victor record sales in 1927."[109] But the rules had changed so fundamentally in the interim that this turnaround represented real hope for the record business. Brunswick also reported an increased demand for its recordings in 1933, suggesting the beginnings of an industry-wide re-emergence.[110]

Ralph Peer's varying interests had grown too large for even RCA Victor. In an attempt to further distance himself from the jazz, race, and hillbilly artists that had catapulted him to his position, he increasingly saw music publishing and popular music as his focus, while he also had developed "markets in Latin American pop, and in globalizing Hollywood tunes."[111] On his departure from RCA Victor, and much to the consternation of rival Eli Oberstein, Ted Wallerstein was named new head of the record division (Oberstein oversaw Victor's ongoing if scaled back interests in jazz, race records, and hillbilly releases).[112] Unlike others within the RCA operation, Wallerstein had a long and abiding interest in the record business, primarily with his work at Brunswick, beginning in 1922.[113] While the Victor brand was still the most recognized in the world, its advertising efforts had largely ceased in the early years of the Depression. In 1932, Wallerstein convinced RCA management that there was still money to be made in the record business. "For the first time since 1929," Bob Ault writes, "a steady flow of new record releases began reaching dealers' shelves, and RCA Victor began to publish catalogues describing the available records," adding "he had to fight the radio-oriented RCA to reinstate this important service."[114] Wallerstein had convinced RCA, Roland Gelatt writes, but there was the additional problem of the aging phonograph equipment in American homes: "Since 1930 the manufacture of phonographs in America had all but ceased."[115] RCA Victor knew firsthand, however, that radio sales had remained strong in comparison and that, as Gelatt notes, "twenty million American homes had one or more radio sets to which a record player could easily be hooked up. They presented a huge potential market, and RCA Victor began to create inexpensive phonograph players that could be plugged in to existing radio sets. While radio sales continued to dominate, companies began to reconsider their approaches to the phonograph market with budget-priced record players.

Wallerstein responded to calls for radio to program "good music" as a public service, a campaign launched by the Federal Communication Commission, which replaced the Federal Radio Commission via the Communications Act of 1934.[116] NBC was putting together its own orchestra led by Arturo Toscanini, which led to further opportunities via RCA Victor's resurgent Red Seal label (which would control the overwhelming majority of classical releases by 1937).[117] Though popular music was not necessarily Wallerstein's milieu, he successfully sought to refocus RCA Victor's pop music catalog along the mainstreaming lines of his old Brunswick colleague Jack Kapp at Decca.[118] He also was persuaded to follow jazz's latest incarnation, swing, bringing (or agreeing to bring via Eli Oberstein) Benny Goodman, Glenn Miller, Duke Ellington, Tommy Dorsey, and Artie Shaw to the label via Victor or its Bluebird subsidiary.[119] But just as Peer had done before him, Wallerstein would tire of dealing with RCA's leadership, which still did not see much future in phonography, instead favoring ongoing research in both radio and television. As John Hammond wrote in his remembrance of Wallerstein, "The record division was only a tiny part of the huge RCA complex and he had to report to men who had no faith in the future of the business."[120] Wallerstein would depart RCA Victor a few years later, and he would persuade CBS leadership to purchase the ARC stable of labels, including CBS' namesake, the once-great Columbia, a label he would spend the rest of his career reviving into a label of great influence and renown, surpassing perhaps even its own historical place in American culture.

Some have argued that Hammond wrote a "phonograph music" column under the pseudonym "Henry Johnson," in the arts section of *The New Masses*, a 1930s leftist publication.[121] In the April 1937 issue, the author was moved to denounce working conditions with the old Columbia Records plant, under ARC ownership since 1934. He noted, "So many criticisms have been made by record reviewers about the surfaces of certain records, the residue that is allowed to remain in the grooves of laminated discs, poor wearing qualities, etc., that sweatshop conditions among the factory pressmen and mixers are primarily to blame."[122] But given ARC's asking price for Columbia and associated ARC holdings, CBS decided to buy more than ARC's labels in 1938. "Originally," Russell Sanjek writes, "CBS intended only to retrieve the Columbia name by purchasing that record company. It soon succumbed to Yates' irresistible asking price of $700,000 for American Record, with all its assets and properties, a modest figure dictated by a year during which sales were about six million discs, one fifth of the total market. With the company came the Bridgeport factory where electrical transcriptions could readily be produced and Brunswick's new and remodeled studios."[123] William Paley noted his increased confidence in the phonograph market: "The popularity of home records has grown tremendously during the last few years . . . and there is every indication that future increases are assured. Technical advances in radio combinations, phonographs, record playing attachments, and in the discs themselves, have greatly improved the quality of music that records can reproduce in the home."[124]

With Peer gone from RCA Victor and Wallerstein heading up the label for the radio conglomerate, Eli Oberstein rose to power, albeit during a period in which the label had scaled back its recording budget so much that Oberstein was the sole recording supervisor for some sessions.[125] Oberstein wielded most influence on Victor's Bluebird subsidiary, through signing race and hillbilly artists such as the Delmore Brothers, Big Bill Broonzy, Ernest Tubbs, and Roosevelt Sykes and bringing swing-era favorites Benny Goodman, Tommy Dorsey, Glenn Miller, and Artie Shaw to RCA Victor. Feeling underpaid, Oberstein found money by less scrupulous (if routine) means. "Oberstein's life style demanded a much higher income, which allegedly came from a share in the cut ins, under-the-desk deals, and participation in music-business speculation at an insider's privileged price—all of which were regarded as normal practices at the time."[126] Nonetheless, Oberstein was known as "a shrewd judge of talent," who held "complete mastery of recording-studio techniques."[127] With Wallerstein's departure from RCA Victor, former Columbia scout Frank Walker (who had brought Bessie Smith, Clarence Ashley, and Gid Tanner & the Skillet Lickers to Victor's chief 1920s competitor) took over the chairmanship of all of RCA Victor's operations, after supervising RCA's transcription disc venture. "In early 1938, it came as a surprise when . . . Walker, in one of his first official moves, fired Oberstein without any public explanation."[128] Oberstein had hoped to take many of Victor's key performers, including Tommy Dorsey, Artie Shaw, and Glenn Miller, with him, but when his financial backing (said to be based in "organized gambling and other underworld activities") collapsed, United States Records stumbled along via additional backers before ending in bankruptcy.[129]

Meanwhile, Wallerstein's former Brunswick colleague, Jack Kapp, whose great success at Brunswick had made him a central figure in the recording industry's recovery, had found success on his own. The name of the label was Decca, which was launched as an American subsidiary of the successful British company. It was envisioned by Kapp and his associates as a major competitor from its inception. "A series of complicated business transactions took place as Kapp assembled a complete recording, merchandising, and distributing operation that could compete with Victor or American Record-Brunswick . . . he had already prepared by ensuring the services of nearly the entire Brunswick roster through personal contacts with him, rather than his employer [Herbert Yates] for all future recording work."[130] The new company found financial support from British Decca's Edward R. Lewis, who served as president with Kapp as chairman.[131] Roland Gelatt wrote that the new label's "*raison d'etre* lay in the conviction, shared by Kapp and Lewis, that good phonograph records did not need to be expensive."[132] "Ever since World War I," he continued, "the big three . . . had been charging seventy-five cents for their bread-and-butter staple, the ten-inch dance record," but this now seemed exorbitant in the economic downturn. Of course, budget labels had long been important to the record industry and had created a parallel industry analogous to Hollywood's Poverty Row. But because the barrier to entry into the recording industry has historically been much lower than that of film production, there were many competitors. Discount record labels were often sold through chain stores, retail outlets that battled both traditional shopkeepers and powerful department stores.[133] As Alfred Chandler has noted, "In the 1920s mail-order enterprises began to build chains of their own. By the 1920s, therefore, the chain store had become the fastest growing type of mass marketer and was becoming the standard instrument for mass retailing in the United States."[134] Record companies in search of niche audiences created labels for specific stores that featured off-brand releases, often rereleases of a given company's premier label offerings: Challenge (Sears), Banner (Kresge), Broadway (Montgomery Ward), and so on. ARC's strategy of purchasing the back catalog of failing record labels and then reissuing material on inexpensive discs (often a quarter, sometimes as low as a dime) made them an increasingly powerful entity. Victor ran its Bluebird subsidiary and Columbia used the once-great OKeh label for its cut-rate brand. Brunswick,

Kapp's former label, continued at seventy-five cents under ARC's control, while their budget-rate labels (Vocalion, Melotone, and many others) continued at a cheaper price. By the mid-to-late 1930s, the companies that controlled the U.S. business were ARC's stable of labels, Kapp's insurgent Decca, and RCA Victor.[135]

Kapp was not a risk taker in terms of his labels' musical repertoire, often banking on the safe and familiar to the point that he was derided for it by some. But Kapp's innovation was to realize that he could sell the *best* artists at an inexpensive price and still make money. "Jack Kapp's gambit," Roland Gelatt wrote, "was to offer the biggest personalities in popular music at thirty-five cents a record. He persuaded most of the Brunswick 'seventy-five-cents' artists to sign up with Decca, and in his first advertisements was able to announce exclusive contracts with Bing Crosby, the Dorsey Brothers, Guy Lombardo, Glen Gray, Fletcher Henderson, the Mills Brothers, and Arthur Tracy."[136] While this was important, the discount labels mentioned above had been undercutting major labels' prices for years. As William Howland Kenney writes, it was Kapp's knack for the 1930s pop mainstream (as well more than a decade of work with recording companies) that allowed Decca to flourish: "He championed a reorientation of an economically prostrate industry away from the relatively long-term preservation of 'immortal and timeless' recorded concert hall music that would surely sell eventually (but not necessarily soon), and towards short-run profits from the quick sale of the latest recordings of popular music."[137] And he did so with an emphasis on the cross-platform sale of music. It was this ethos, along with Bing Crosby's loyalty to Kapp, that led him to the new label, and in so doing Crosby "became the new company's leading light."[138] Kapp made sure that Decca's launch would garner attention. In a period in which record advertising had become increasingly rare, Decca ran advertisements to announce itself in newspapers around the country, exclaiming "Decca Scoops Music World." The text read, "Here they are—your favorite stars of radio, screen, and stage—in their greatest performances of instrument and voice! *Not* obsolete records, cut in price to meet a market, but the latest, newest smash hits—exclusively DECCA. Hear them *when* you want—as *often* as you want—right in your own home."[139]

The recovery was assisted by a new jazz subgenre known as swing. A favorite mid-1930s parlor game was to define the subgenre, with musicians, executives, and social theorists all taking part.[140] Most swing musicians favored the notion that swing was nothing new but had its roots in the hot jazz of the late 1910s and early 1920s. As players from this milieu found their way into the "sweet" and symphonic bands of the late 1920s (Paul Whiteman's band, Guy Lombardo's, and many others) they helped transform those bands accordingly. This hot vs. sweet distinction was an early iteration of popular music's seemingly endless authenticity debates, in which distinctions between genres and sub-genres are fought over by musicians, critics, and industry executives. These particular debates gave voice to a burgeoning collector community, as evident in Ulysses "Jim" Walsh's "Pioneer Recording Artists" column in *Hobbies* magazine circa the 1940s.[141] In Louis Armstrong's book *Swing that Music* (named after a tune that he performed on a Decca 78), he wrote, "Now I know there are a lot of people who will read this book who will say that 'swing' is just a new name for the same old jazz they've been hearing for many years and that I am trying to make it look like something new . . . but I cannot say too strongly to these people that there is all the difference in the world and if they will just try to understand it they will very soon be singing out when they tune in a band on their radios, 'That's swing!,' or 'That's not swing!' and will be able to tell at once."[142] If the subgenre was new, "the *basic idea* of swing music is not new. The swing idea of free improvisation by the players was at the core of jazz when it started back there in New Orleans thirty years ago."[143] It was "these swing men" who came up through sweet bands and made them swing.[144] Duke Ellington agreed after becoming associated with the subgenre via his Brunswick release "It Don't Mean a Thing (If It Don't Got That Swing)" (1932). He too emphasized its roots in the past: "Swing? Well, that's what they're calling it this year . . . But it all goes back to something just about as old—and as natural as the circulation of the blood." Swing was neither "highbrow" or "lowbrow," but "belongs to everybody, whether he lives on Lenox, Park, or Second Avenue, whether he plays or dances or just listens."[145]

Swing's appeal (as it had been with the hot jazz boom of the early 1920s and would be with rock 'n' roll in the 1950s) was with American youth. Like many youth cultures, swing came with its own mode of dress (defined by the zoot suit), dance (the Lindy Hop and Jitterbug), and slang (the "hep" talk that would eventually become so prevalent it was lampooned in Saturday morning cartoons). The country, which had endured a deeply traumatic period of economic depression, needed optimism and irreverence, which is what swing provided. As Arnold Shaw writes, "After four years of frustration, economic shortages, and gloom, the American people were ready to seize on something that would give them a lift. Swing *did*."[146]

To attract youthful filmgoers to movie theaters, exhibitors paired films with live performances from swing bands, as Benny Goodman's band did with *Every Day's a Holiday* (Sutherland, 1937), a Mae West vehicle in which Louis Armstrong appeared as himself.[147] Many critics dismissed the music, such as jazz aficionado Wilder Hobson, who described swing dismissively in 1939: "The 'swing' fad, which still continues as this is written, has largely been built on the commercially salable mixture of a certain amount of jazz playing" combined with "popular-melody-and-jazz orchestrations."[148] As Paul Whiteman's star faded, bandleader Benny Goodman was dubbed "the King of Swing." Goodman and his band, along with the Dorsey Brothers and Glenn Miller, all became household names. Duke Ellington's suave and sophisticated countenance over the exuberant music of his big band seemed to define the new approach. The bandleaders and vocalists relied on solid songwriting, and the period greatly benefited from great songs. In many ways, the 1930s represents the zenith for jazz as commercial music. As Ted Gioia writes, "Never again would popular music be so jazzy, or jazz music so popular."[149] During the period in which swing thrived as commercial music, American popular songwriting was also at the height of its powers; Harold Arlen's "Stormy Weather," Cole Porter's "Night and Day," and Hoagy Carmichael's "The Nearness of You" were all written during the era.

"Jazz was the nation's popular music in the Machine Age (1919–1945) because its driving, syncopated rhythms reflected the speeded-up tempo of life produced by industrialization in the American workplace and the mechanization of urban life," Joel Dinerstein writes. "When big-band swing became the nation's most popular music in the mid-1930s, it was both a morale boost for a nation in depression and a reflection of the relative freedom African Americans felt" in the cities of the north and west.[150] Within media industries, a variety of technological and economic shifts greatly assisted swing's rise as commercial music. Radio again played a crucial role in the popularization of swing (western and otherwise). As James Lincoln Collier writes, "The big swing bands entered into a symbiotic relationship with the radio stations. For radio, the bands playing in the restaurants and dance halls were a golden source of free programming; for the bands, radio was an equally golden fountain of free publicity."[151] Radio shows from New York, Chicago, and Los Angeles were regularly broadcast into American homes, as listeners not only learned the names of bandleaders but soloists in the band. A few of them became stars in their own right, such as drummer Gene Krupa from Benny Goodman's orchestra, who appeared in film, radio, and television into the 1960s. Swing brought the names of Billie Holliday and Count Basie, both recording for Columbia, to national prominence (they worked together early on). Holliday was among the acts to bring Milt Gabler's upstart Commodore Records to prominence when John Hammond allowed Holliday to record "Strange Fruit" for the label. The song was written by "Lewis Allan," a pseudonym for Jewish American teacher and activist Abel Meeropol. His inspiration might have been a Lawrence Beitler photograph of the horrific Marion, Indiana, lynching of Thomas Shipp and Abram Smith in 1930.[152] "Strange Fruit" remains one of the best-known American songs that made plain, with plaintive fragility, the horrors of American racism.[153] On December 23, 1938, swing joined the pantheon of African American musical culture via John Hammond's *From Spirituals to Swing* concert at Carnegie Hall, one of the country's most prestigious venues, synonymous with the performing arts' establishment in the United States. The night featured a sweeping overview hinted at in the evening's title, including performances by Sister Rosetta Tharpe, Sidney Bechet, Albert Ammons, Count Basie, and others. Robert Walser notes that African American musicians had been welcomed at Carnegie Hall previously, in James Reese Europe's "Clef Club" performances of 1912–14, a W. C. Handy tribute in 1928, and a performance by Benny Goodman's integrated band, also in 1938.[154] But Hammond's *From Spirituals to Swing* offered a sweeping overview of African American music traditions that recording companies had broadly represented for decades, making a de facto argument for the importance and vitality of the music and the people who made it in Jim Crow America. In the program notes, James Dugan and John Hammond wrote, "The music that will be presented . . . is rarely heard. To be sure, it is not rare, for the country is rich with it, but serious audiences have neglected it, and it has had to find its followers among uncritical groups."[155] This was still a radical notion in 1938, quite literally in that Hammond found funding via *The New Masses*' Eric Bernay (more circumstantial evidence of Hammond's substantive connections with the leftist publisher).[156] But its symbolic importance resonated far beyond that night at Carnegie Hall.

In addition to its presence on America's radios, swing also helped bring back coin phonograph technologies, which had been responsible for saving the nascent 1890s recording industry; they were renamed "jukeboxes," though they had never fully left. In fact, these companies made a variety of instruments that competed with the phonograph in different ways, particularly in the 1910s. "Coin-operated machines became a fixture of American life during the 1920s and 1930s, as jukeboxes became ubiquitous in taverns, restaurants, hotels, and other establishments."[157] The jukebox's resurgence meant greater record sales for labels big and small, joined together in the same nickel-in-the-slot technology. "The fortunes of the jukebox industry and the record industry became intertwined in this period," Kerry Segrave writes. "Sales of discs to juke operators boosted industry sales considerably while technological improvements in records, along with the 1934 introduction of inexpensive records, helped in the widespread rapid growth of the juke industry."[158] Five companies dominated the domestic jukebox market: Cincinnati-founded and North Tonawanda, New York–based Wurlitzer; Grand Rapids, Michigan's Automatic Musical Instrument (AMI); and the Seeburg, Rock-Ola, and Mills Novelty companies, all based in Chicago by the early 1930s.[159] These companies made an array of music media over the decades, including player pianos, nickelodeon pianos, orchestrions, electric pianos, pipe and electric organs, and other music media, much of which were geared for silent-film nickelodeon theaters and movie palaces. But the rise of late 1920s sound-on-film technologies represented a crisis that led these companies to jukeboxes, a closely related sideline to the machines they had already been producing.

Like later radio formats, Eric Weisbard writes, the jukebox represented "the tension between seeing formats as pandering or as deferring to popular taste . . . it could encompass many varieties *or* group a subset for a defined public," much as the record catalogs of the 1920s and early 1930s sought to do, but commonly without the immediacy of listening to a recording there and then.[160] In doing so, the jukebox prefigured the eclecticism of 1950s "Top 40" radio formats. Jukeboxes embraced both pop music's self-conscious performativity at the heart of American popular music since minstrelsy, while also trading in the politics of authenticity just beginning to emerge in the 1930s.[161] "Early in the Depression the record companies were not slow to understand the importance of the jukebox business," and "in many parts of the country hillbillies and blacks were more likely to hear their own recording artists on a jukebox than on a radio."[162] "Coin men," Chris Rasmussen writes, "became indispensable middlemen in the negotiation between producers and consumers during the 1930s and 1940s as they attempted to bridge the various interests and desires of listeners, location owners, musicians, phonograph manufacturers, record companies, and, of course, the coin-op industry."[163] If swing directly appealed to the youth market, a coin in a jukebox was within their means as various businesses rushed to add jukeboxes. As Roland Gelatt noted, "Coin phonographs in one form or another had been around for a long time, but like everything else to do with recorded music they had gone into a serious decline during the early years of the Depression. They came out of retirement with the end of Prohibition and—as juke-boxes—proliferated in bars, drugstores, and diners."[164] Domestic sales also picked up. "Sales of records started to recover in the mid-thirties," Gronow and Saunio write. "The worst of the slump was over. The increased record sales in the United States were partly due to jukeboxes, which after the end of Prohibition appeared in every little bar and restaurant."[165] By 1940, a *Variety* article asked, "How many nickels are tossed into coin phonograph machines every year is something the jukebox statistical department has never had time to figure out; it's too busy adding up profits." The article claimed that jukebox companies were making as much as $150,000,000 by 1940 (a conservative estimate).[166]

Roland Gelatt wrote, "Price cuts, jukeboxes, and intensive promotion by three highly competitive companies," RCA Victor, Decca, and a reinvigorated Columbia, once under CBS ownership by decade's end, "served to swell and quicken the phonograph's return to public favor."[167] The recording industry benefited from the popularity of swing in various ways, as social and economic changes helped re-establish the industry as an economic power. Decca benefited most, through its emphasis on smooth vocalists and subtle rather than emphasized syncopation. William Howland Kenney finds an aesthetic he links to Jack Kapp—and thus to many of the stars of the swing era—in the smoothed-out pop tones of the 1930s: Bing Crosby's vocals, Guy Lombardo's band, the soothing melodies of the Mills Brothers and Boswell Sisters, the refined voice of Ethel Waters, and the muted trombone of Tommy Dorsey.[168] The softer dance-band sound that Kapp had captured at Brunswick in the 1920s had now allowed him to continue his success via swing in the 1930s.

By the late 1930s, the two largest record companies, RCA Victor and Decca, battled for dominance. The third company to make up the Big Three, ARC (by which point included both the Brunswick and Columbia labels) lagged behind. At ARC, Columbia's name was used largely for dance bands, though the best material and performers were funneled to Brunswick as ARC's premier label. As Brian Rust wrote, "Brunswick was ARC's principal 'quality' label, Columbia very much an afterthought."[169] But Edward Wallerstein, Kapp's former associate at Brunswick, had risen through the ranks at RCA Victor and had his eyes on new opportunities. Rust writes that "in 1938 . . . Wallerstein . . . suggested to Williams S. Paley, president of the Columbia Broadcasting System, that the Columbia label be reactivated, to resume its place of honour in the record market."[170] Paley, who had experienced great success with Columbia Broadcasting, was anxious to see the old Columbia phonograph imprint reborn and had the money to do it. Paley purchased the company in 1938, making Wallerstein general manager, which allowed "a new and vigorous Columbia [which] took its place in the American record market from September 1939 with Benny Goodman, Kay Kyser, Count Basie, Mildred Bailey, Kate Smith, Gene Krupa, and Harry James among the popular stars."[171] In December 1940, RCA vice president and advertising director Thomas Joyce spoke before New York's Sales Executive Club, noting that RCA considered leaving the record business entirely in 1932. The article detailing Joyce's comments in *Advertising Age* noted that industry record sales in 1940 reached seventy-five million discs sold. "He predicted that 1941 sales would pass the 100,000,000 mark set in 1921, and added, 'We haven't scratched the surface yet.'"[172]

Unfinished Business: Copyright and Royalties

Given the Depression-era conglomerations detailed here, the reasonable question may be asked: Why were music publishers and radio companies still in conflict, often with sound recording companies caught somewhere in between? There are several answers to consider. The simple answer would be that although Victor and Columbia both joined conglomerates during this period, not all recording companies became part of larger conglomerates. The answer beyond this is much more complicated, in that the transformations of the era involved many different constituents, songwriters, music publishers, record companies, radio networks, local radio stations, and film studios, to name a few. If any subset of those were owned by the same parent company, it was no guarantee that the disparate parts felt especially tied to the whole. And during the tumultuous Depression years, media companies (especially recording companies) were sold and sold again, coming in and out of relationships with parent companies, as well as music publishers and radio interests sometimes owned by the same conglomerate. And while radio's varying interests were represented by the National Association of Broadcasters (NAB), networks could hardly control the practices of all their affiliates, let alone the many nonaffiliated local and regional stations that made use of transcription discs and commercial sound recordings in general. Even closer relationships between music publishers, recording companies, radio networks, and film studios could solve that. The larger transformation of the recording industry into multilayered configurations did not mean that all players newly gathered under a single parent company would play nicely together. It never does. Nor does conglomeration mean that larger entities, sometimes only partially housed within these new companies, would not continue on with concerns about financial equity for their constituents (and themselves). And with each changing technology and shifting context, old concerns and animosities were revived. Simmering tensions of the 1930s foreshadowed even larger conflicts between music publishers and media interests in the 1940s and beyond (up to the present day). These tensions spilled over into public conflicts throughout the decade, but now are seen as a preamble to even bigger battles in the 1940s, if they are remembered at all. But mapping at least some of the ongoing issues between varying constituents serves to give much needed context for later battles and underlines that these issues are never truly settled, especially given shifting media playback technologies that call up the same issues regarding earnings and equity again and again.

"Prior to 1932, ASCAP licenses required flat fees from radio broadcasters," John Ryan writes, and broadcasters "had continually fought ASCAP's right to do business."[173] But radio and music interests had "reached an uneasy truce by the early 1930s," as ASCAP had implemented a percentage-based plan that involved a flat "sustaining fee" similar to previous arrangements, but with a new "commercial fee" in addition.[174] This and other ASCAP provisions did not please broadcasters, but they played along in the short term. In addition, the Music Publishers Protective Association (MPPA) asked RCA Victor, Brunswick, and Columbia to print "Not Licensed to Broadcast" on each newly pressed record label, which in turn led to newfound concerns about the future of recorded music on radio from the National Association of Broadcasters (NAB).[175] "They were assured," Russell Sanjek wrote, "that the warning meant nothing, that property rights ended once a record was sold across the counter, a legal opinion vigorously opposed by manufacturers and many leading artists, particularly Paul Whiteman and Fred Waring."[176]

Bigger troubles were inevitable on various fronts. ASCAP's monopolistic structure necessitated certain fundamental approaches. "The logic of ASCAP's operations, particularly the logic of the blanket license, is the logic of monopoly: only by gathering all copyrighted compositions into its repertory could ASCAP give users a blanket license that would enable them to perform any composition without fear of a lawsuit," Paul Goldstein writes. "But monopoly breeds discontent—and not only in the U.S.

Congress and among the antitrust enforcers in the Justice Department," as newer members accused ASCAP of favoring its senior members.[177] Radio's 1930s dominance necessarily altered ASCAP's approach, based on the notion that radio broadcasts would soon surpass income from the restaurant, bar and nightclub businesses that had been their focus.[178] "As radio became more and more successful in the 1930s, ASCAP tried to capture a larger share of its revenues through increased license fees," due to recorded music's increased centrality on U.S. radio, as well as the extent to which airplay was "cutting into its members' other sources of income, notably record and sheet music sales."[179] The NAB responded by arguing that radio "boosted sales of sheet music and records" in what was in effect "free advertising."[180]

ASCAP was not the only organization unhappy with radio's use of recorded music. Under the leadership of Joseph Petrillo, the Chicago chapter of the American Federation of Musicians had battled for musicians' pay since Petrillo became president of the powerful Chicago chapter. "Prohibition killed the cabarets and beer gardens" and radio had not yet become the "source of profitable work" it would become later.[181] Petrillo noted, "Our protest is based on self-preservation."[182] Petrillo's Chicago union won battles against local radio (largely owned by the local press) in 1924 and again in 1927, in a fight against movie theaters firing musicians in favor of organs, in the same that year that sound-on-film "talkies" debuted. In 1931, Petrillo's Chicago union took on local hotels on behalf of musicians and won.[183] But it was in December 1936 that Petrillo announced that by February, "no union musician in the Chicago jurisdiction would be permitted to make any kind of recording for use in broadcasting."[184] This unorthodox approach gained national attention for the strike, and Petrillo argued at the AFM's 1937 Louisville convention that the tactics and campaign should be extended to a national AFM strike against radio interests. By February 1, 1938, the AFM had "control over the output of canned music in 589 stations—289 chain affiliates and 300 independents."[185] "New licensing agreements were approved at the musicians' annual convention in 1938, and were accepted by RCA Victor, Columbia, and Decca soon after. Before a recording session could take place, it had to be registered with the local AFM chapter, and union representatives looked in to verify the presence of paid-up members or that of union musicians in a number equal to noncard bearers," Sanjek wrote, noting that the AFM also now threatened to revoke licenses for those recording companies using nonunion labor on race records, after African American musicians protested that AFM policies were specifically ignored in this regard.[186] They were instead paid in one-off arrangements popularized by Ralph Peer in which "they bought all rights for a small sum, kept the publishing rights, and paid only a small or no mechanical royalty."[187]

By the late 1930s, ASCAP's already shaky public reputation faced further setbacks as detractors' efforts in denouncing the organization took hold. A 1937 editorial in *Broadcasting* magazine suggested that AFM was doing ASCAP's bidding, asking: "Is AFM Acting as the Catspaw of ASCAP?"[188] The editorial noted that another possible radio "crisis" loomed, a threatened AFM strike scheduled for September 16, 1937, "unless the broadcasting industry brings in an acceptable plan for reemployment of musicians."[189] The editorial added that the AFM's demands on fees from broadcasters, transcription, and phonograph record manufacturing "would spell the doom of many independent stations" and could "conceivably destroy the industry," noting the broadcasters' opinion that it was movie studios and not radio broadcasters that sent musicians to the unemployment lines.[190] The lead story in the next issue of *Broadcasting* reported on a meeting at New York's Waldorf Astoria hotel, in which more than 150 network affiliates (members of the Independent Radio Network Affiliates (IRNA) met to avoid the strike, while the NAB met in New York days before the strike was to start to address the issue.[191] The mid-September issue declared an agreement had been reached, in which broadcasters promised to "double their payrolls."[192]

By 1939, battle lines were drawn once again, when the original agreement expired. "The NAB is in the middle," the article noted, "secretly hoping that both the chains and IRNA will stand firm and persist in nixing the AFM demands but alarmed that one or the other will weaken, making complete surrender inevitable."[193] In January, *Broadcasting* reported that "a national calamity for radio—and the public—apparently has been averted. . . . But will the armistice be permanent? . . . We think not!"[194] This was the pattern of crises and tentative deals, amidst a series of proxy battles and internecine squabbles that would lead to the most famous battle of this era, the so-called "Petrillo Strike," named after Joseph Petrillo, now the public face of the American Federation of Musicians after he took over national leadership in 1940. Leaders of the NAB were deeply dissatisfied with ASCAP's 1939 increase in percentage-point payments and sought and implemented a plan to create a music licensing source all its own, known as Broadcast Music, Incorporated (BMI). They recognized that "success and profits depended almost entirely on an ability to control network plugging of selected songs," especially in an era when thirteen

Hollywood-affiliated Tin Pan Alley firms now controlled more than half of Tin Pan Alley's output.[195] On April 1, 1940, BMI began business, bolstered later that September by unanimous agreement at NAB's Chicago meeting.[196] BMI partnered with E. B. Marks, a Tin Pan Alley house unhappy with ASCAP's leadership, which provided them legitimacy in early operations. The radio networks did not renew their ASCAP licenses in 1941, and instead licensed with BMI, which necessarily relied on licensing the music that ASCAP had largely ignored: race records (a term that would fragment into "blues" and "rhythm and blues" by the 1940s) and hillbilly artists, as well as Latin American music and other subgenres, genres that had already received a boost from jukebox eclecticism. The Department of Justice's concerns about both ASCAP's and the nascent BMI's dominance in U.S. music licensing led to a dispute in which it was "alleged that each organization had unlawfully exercised market power through the aggregation of public performance rights in violation of Section 1 of the Sherman [Anti-Trust] Act" in 1941.[197]

On August 1, 1942, amidst the first full year of U.S. involvement in World War II, the American Federation of Musicians made good on a strike, this time against the nation's leading recording companies, in a dispute over royalty payments to musicians. The strike was contentious and though the AFM's voting members supported the strike, Petrillo was the focus of public animosity. In Charles Williams' "Mr. Petrillo's Hopeless War," he was described as a "cocky Chicago labor politician with a great disdain for public opinion," who had provoked "a Congressional investigation, by forbidding his musicians to make recordings for radio station and juke boxes."[198] As the strike went on through all of 1943 and most of 1944, labels turned to reissues and even all-vocal performances to sidestep the musicians' ban. The recording companies with the most to lose were those with the smallest back catalog of sound recordings. Decca's relative youth and agility had benefited the company through most of the 1930s, but the AFM strike had forced Decca (and its World Broadcasting System subsidiary) into a deal by September 1943.[199] The emergent Capitol Records label settled in October of the same year. By 1944, President Franklin Roosevelt himself got involved in a publicized effort to settle the dispute. He sent a telegram to Petrillo, noting that while the National War Labor Board had found the AFM strike had not "unduly impede[d] the war effort," the action "may encourage other instances of noncompliance" that might do so.[200] But the AFM would not budge and both RCA Victor and Columbia Records conceded to the AFM's demands on November 11, 1944, amidst mounting competition from Decca and other labels that had already settled with the union. But this agreement did not end tensions, and, though it is rarely discussed, another almost year-long strike occurred in 1948. The Taft-Harley Act was meant "to benefit workers and the economy by stimulating the organization of labor,"[201] but it called into question the legality of the AFM's musicians' union fund, which led to the strike. The strike was settled on December 14, 1948, at which point twelve RCA Victor popular and operatic recording artists joined in a New York studio to record "I'm Just Wild About Harry," a salute of Harry Truman, who helped bring the strike to a close.[202] The AFM's Petrillo and RCA Victor's Sarnoff posed for a photo at the event. These perennial conflicts have continued at various intervals up to the present day. In media industries' dispute over royalties, agreements are only temporary truces.

A Transformed Industry

Those record companies that emerged out of the convulsive transitions of the late 1920s and early 1930s found themselves in an industry that had fundamentally changed. The industry's narrowed focus on the biggest sellers (despite some missteps and inflated contracts) had generally seen them through. The merger between RCA and Victor Records in 1929 was the beginning of an increased relationship between the recording, radio, and film industries that would extend further with CBS's purchase of the Columbia imprint in 1938. In the new mediascape, stardom wasn't as simple as a hit record (though a hit record wouldn't hurt your chances). Companies increasingly looked for performers that could excel across formats to save the expense of all that scouting and recording.

By the mid-1930s and beyond, there were clear signs that the recording industry had found a new equilibrium, for some of the reasons described above as well as the shifting government policies and market conditions within U.S. culture. Within the recording industry, a period of great experimentation had passed, as had most speculative work on niche genres and commercial recording trips, which in turn limited the flow of various forms of American vernacular music. The recorded culture that was the result would have lasting impact, though many recordings were lost, the master discs thought to be of no value whatsoever and sold for scrap or considered trash. The expansion into hot jazz, race records, and hillbilly granted unprecedented access to recording for musicians whose music would otherwise be lost forever.

Many of the labels that had innovated in these areas were purchased by larger companies or seemingly disappeared as their parent companies returned to what they knew best. Radio's dominance allowed recording companies now owned by radio conglomerates to reinvent themselves with some economic security, given that radio had become a ubiquitous presence in American life. "By the end of the 1930s, radio was seriously undermining newspapers as the preferred source of news and film as the favorite form of diversion," while the medium's "heavy diet of popular music" through swing-based shows like *Make Believe Ballroom* were beloved by listeners and important to recording companies interested in promoting new and established artists.[203]

The recording industry's responses to the economic depression were telling, a reminder of its role as a material media industry with deep connections to appliance manufacturing and retailing. In the most difficult periods of the late 1920s and early 1930s, the recording industry sought a renewed focus on hardware. Labels large and small did so first through radio adapters for phonograph machines, then hybrid radio-phonograph machines, and finally through changing business models that led to manufacturing and retailing other appliances. In the case of the biggest labels, these changes came through sellouts and mergers. For the smallest labels, it meant an embrace of the emerging appliances of the era—refrigerators and the like—or a return to its main business lines such as furniture. In the toughest times, the most practical hardware (radios for news and entertainment, home appliances for the family) bested entertainment software in terms of sales.

The recordings of the era would go on to have impacts that no one would have predicted. While the music of the 1930s had a smooth, dreamy quality that belied the difficulties of the decade and provided a much-needed escape, it also had a power and beauty—a desperation, a visceral impact too—all its own. Repeating patterns of expansion, experimentation, economic decline, and consolidation would return again as new genres arose and the industry capitalized on them, but this was the first big transformation of the recording industry in terms of the restructuring of the recording industry and the emergence of the foundational genres of American popular music. It marked a dramatic industrial and technological shift that also changed the company's means of doing business as well as understanding the business itself. In this way, the recording industry grew from a business that quickly learned through dramatic bouts of trial and error to an increasingly multifaceted, sophisticated, and self-aware industry in which the manufacture of audio recordings and playback machines were mere parts of companies involved across media platforms and contexts.

I have sought to bridge media history (which traditionally ignores the recording industry) and cultural industry research (which largely analyzes contemporary trends), while encouraging attention to the recording industry in the former and historical research in the latter. I have examined a few of the key recording companies of the 1920s and 1930s and their actions during a great period of experimentation and expansion, through how they organized the emerging genres of the era through recording and scouting practices and through advertising and promotional materials. Through this, I have identified industry patterns that should resonate. While each new genre or technology comes with its own specific set of connotations and complications, industry behavior as depicted here leads me to invert a familiar saying: everything new is old again. My purpose in engendering such a response is not to encourage simple comparisons, but to recognize repeating patterns as they emerged in an era of tremendous importance to the U.S. recording industry. Like all histories, its ultimate purpose is to use history to inform the roots of the ever-shifting present.

The specific aim of this analysis is to contribute to an understanding of the U.S. recording industry during a crucial interval, when it went through a series of defining changes both in terms of genre formations, institutional cultures, and company ownership. I have attempted to capture the recording business in its transition from a powerful emerging industry into a mature oligopoly with deepening ties across media industries, technologies, practices and cultures. I've sought to do that not only by concentrating on the general trends in genre and mode of production, but also in tending to those at varying levels of the recording industry who influenced its products. I have focused on depicting genres as malleable, flexible categories that are constantly shaped and reshaped by industry as well as by artists and audiences. My task has been to convey some sense of these genres as they transformed in order to suggest their use by record companies amidst constantly changing social, cultural, and industrial dynamics.

Decades after these companies had searched for new niches and had recorded unprecedented amounts of American vernacular culture, those involved would be again of interest. For some, the interest would engender exhaustion, as in Fred Gennett's exasperation at how little he remembered about all that work, so many years ago.[204] Record executives, recording directors and engineers, talent scouts, they all sought to define what the work meant on their own terms. If some found it too difficult to remember ancient history, others like Ralph Peer dismissed what they had done as a mere means to an end. Still others saw what they had done as an act of citizenship, a kind of American humanism in which they had asserted that the people they had recorded and the music they made were important parts of American culture. The layers of myth from the retellings and the rediscoveries of those who made the music in various ways have clouded our understanding of their work and what it meant as years passed.

"We should now like to note that the largest repository so far of recordings of American Folk Music (AFM) has been established not by academic institutions or folk music associations," John Fahey wrote, "but by the combined efforts of the American recording industry."[205] Huge swaths of the thoughts, experiences, and feelings of American people were and are expressed in slick pop and in rough, niche sounds, and points in between, all potentially impactful, depending on the relative openness of one's ears. The realities of how the music was sometimes made—many marginal figures visited a studio once or maybe twice, paid a flat fee, perhaps were given a train ticket, never to be heard from again—is often how we remember this era. It happened too, as did so much more. That musicians were exploited too has made us especially (and rightly) critical of the recording industry, in our own often contradictory quests for winners and losers in the authenticity sweepstakes, a key part of music mythmaking. This study rejects any simple conceptions of this. What happened was more complicated. At the same time, the recording industry's place as one of the most maligned media industries is based in real evidence and good reason, amplified given the business's low barrier to entry and the marginal status of those who made the music. The social and cultural realities of American life in the 1920s and 1930s meant that those involved in the recording industry at various levels—musicians, scouts, executives, and so on—traversed in sometimes dangerous ways. Lost recordings from this era are still being sought out, discovered after being long forgotten, and reissued to play back once again as part of an ever-expanding constellation of ideas, emotions and experiences. What they represented is still being considered and reconsidered, mapped and remapped with each new rediscovery, and with each reissue of recordings from long ago, it becomes clearer and clearer that it was the commercial recording industry more than any other entity that was responsible.[206] We understand this germinal period in American music, media, and culture most clearly when we trace at least some of its intersecting histories represented here, when we hold them in hand all at once.

Notes

Introduction

1. Harry Charles, interviewed by Gayle Dean Wardlow, 1968. Gayle Dean Wardlow Collection. Center for Popular Music, Middle Tennessee State University (CPM-MTSU).

2. Allan Sutton, *A Phonograph in Every Home: The Evolution of the American Recording Industry, 1900–1919* (Denver: Mainspring Press, 2010), 286.

3. Roland Gelatt, *The Fabulous Phonograph, 1877–1977*. 2nd rev. ed. (New York: Macmillan and Company, 1977), 190–91.

4. Rick Kennedy, "Paramount Records," in *Little Labels, Big Sound: Small Record Companies and the Rise of American Music*, ed. Rick Kennedy and Randy McNutt (Bloomington and Indianapolis: Indiana University Press, 1999), 25.

5. Eugene F. Hartley, ed., Department of Commerce, Bureau of the Census, *Abstract of the Census of Manufactures*, W. M. Steuart, director (Washington, DC: Government Printing Office, 1919), 20.

6. Department of Commerce, Bureau of the Census, *Abstract of the Census of Manufactures*, 20, 204.

7. "Amateur Radio Restored," *Electrical Experimenter*, June 1919, in Hugo Gernsback, *The Perversity of Things: Hugo Gernsback on Media, Tinkering, and Scientification*, ed. Grant Wythoff (Minneapolis: University of Minnesota Press, 2016), online ed., https://manifold.umn.edu/projects/the-perversity-of-things.

8. Allan Sutton and Kurt Nauck, *American Record Labels and Companies: An Encyclopedia (1891–1943)* (Denver: Mainspring Press, 2000), xiii.

9. Sutton and Nauck, *American Record Labels and Companies*, xiv.

10. "Victor Co. Loses Suit," *Variety*, April 7, 1922, 7. By 1920, only Edison and Pathé continued to make vertical-cut recordings. See Allan Sutton, *A Phonograph in Every Home: The Evolution of the American Recording Industry, 1900–1919* (Denver: Mainspring Press, 2010), 282.

11. Sutton, *A Phonograph in Every Home*, 279–81.

12. Sutton, *A Phonograph in Every Home*, 279–80. As evidence, Sutton cites a decision from Victor to withhold any new releases in December 1918 just to catch up with back orders.

13. "Our National Victory Should Make Business Men Optimistic," *Talking Machine World*, January 15, 1919, 3. Media History Digital Library (MHDL).

14. Allan Sutton, *Recording the Twenties: The Evolution of the American Recording Industry, 1920–29* (Denver: Mainspring Press, 2008), 5–6.

15. The answer, as one would expect in an ad for the record company, was OKeh. See OKeh Records, "There Has Never Been Such a Boom in the Record Business," advertisement, *Talking Machine World*, April 15, 1919. 75 (MHDL).

16. "No Victor Records for August," *Talking Machine World*, July 15, 1919, 172 (MHDL).

17. Leonard DeGraaf, *Edison and the Rise of Innovation* (New York: Sterling Signature, 2013), 110–14.

18. DeGraaf, *Edison and the Rise of Innovation*, 114.

19. Karl Hagstrom Miller, *Segregating Sounds: Inventing Folk and Pop Music in the Age of Jim Crow* (Durham, NC: Duke University Press, 2010), 198–99.

20. Laszlo Schwartz, "The Phonograph Industry Has Become an Ideal Musical-Commercial Power," *Music Trade Review*, June 14, 1919, 37.

21. Pathé Records, "Advance August List of Pathé Records," *Music Trade Review*, June 14, 1919, 37. The practice of using telegraphic codes for record orders is also mentioned in the record collectors' periodicals. See Martin F. Bryan, "And We Wonder Why They Are So Common?" *New Amberola Graphic*, October 1999, 8–10. The article references a May 1919 Victor Advance List that similarly used telegraphic codes in the ordering process.

22. Kyle Barnett, "Record Men: Talent Scouts in the U.S. Recording Industry, 1920–1935," in *Making Media Work: Cultures of Management in the Entertainment Industries*, ed. Derek Kompare, and Avi Santo (New York: New York University Press, 2014), 114.

23. Kathleen Morgan Drowne and Patrick Huber, *The 1920s* (Westport, CT: Greenwood Press, 2004), 28.

24. T. M. Kando, *Leisure and Popular Culture in Transition*, 2nd ed. (St. Louis, MO: C. V. Mosby, 1980); Michael Brian Schiffer, *The Portable Radio in American Life* (Tucson: University of Arizona Press, 1992), 48; and Gilbert Seldes, *The Seven Lively Arts* (New York: Harper & Brothers, 1924), 60.

25. F. Scott Fitzgerald, *Tales of the Jazz Age* (New York: Scribner and Sons, 1922). Fitzgerald's postmortem on the era, "Echoes of the Jazz Age," is perhaps even more important in its early redefining of the era from the perspective of an altogether different United States in 1931. See F. Scott Fitzgerald, "Echoes of the Jazz Age," in *The Crack Up*, ed. Edmund Wilson (New York: J. Laughlin, 1945).

26. Robert S. and Helen Merrell Lynd, *Middletown: A Study in Contemporary Culture* (New York: Harcourt, Brace & World, 1929), 244.

27. Many of these events are recounted in Irving Settel, *A Pictorial History of Radio* (New York: Grosset and Dunlap, 1967), 35–72.

28. Settel, *A Pictorial History of Radio*, 115.

29. Settel, *A Pictorial History of Radio*, 118.

30. For a description and application of "circuit of culture" research, see Paul

du Gay, Stuart Hall, Linda Janes, Anders Koed Madsen, Hugh Mackay, and Keith Negus, *Doing Cultural Studies: The Story of the Sony Walkman*, 2nd ed. (Thousand Oaks, CA: Sage Publications, 2013).

31. Pekka Gronow and Ilpo Saunio, *An International History of the Recording Industry* (London: Cassell, 1999), 38.

32. Peter Tschmuck, *Creativity and Innovation in the Music Industry* (Berlin: Springer Verlag, 2012), 49–51.

33. Paul Starr, *The Creation of Media: Political Origins of Modern Communications* (New York: Basic Books, 2004), 1.

34. Starr, *The Creation of Media*, 1–2, 5.

35. Starr, *The Creation of Media*, 1–2.

36. Jonathan Sterne, *The Audible Past: Cultural Origins of Sound Reproduction* (Durham, NC: Duke University Press, 2003), 213–14.

37. Alfred Chandler, *Scale and Scope: The Dynamics of Industrial Capitalism*, with the assistance of Takashi Hikino (Cambridge, MA: Harvard University Press, 1994), 1, 212–13.

38. Alfred Chandler, *The Visible Hand: The Managerial Revolution in American Business* (Cambridge, MA: Harvard University Press, 1993), 3.

39. Chandler, *The Visible Hand*, 3.

40. For a full framing of the various dynamics that the modern business enterprise brought forth, see Chandler, *The Visible Hand*, 1–11.

41. Alfred Chandler, *The Essential Alfred Chandler: Essays Toward a Historical Theory of Big Business* (Boston: Harvard Business School Press, 1991), 71.

42. Chandler, *The Essential Alfred Chandler*, 71.

43. Chandler, *The Essential Alfred Chandler*, 215–16.

44. Chandler, *The Essential Alfred Chandler*, 219.

45. Scouts navigated varying social divides while pursuing business ends, in a period shaped and reshaped by women's suffrage, racism and xenophobia, temperance, and the eventual coming of the Great Depression. Barnett, "Record Men," 2014.

46. William Howland Kenney, *Recorded Music in American Life: The Phonograph and Popular Memory, 1890–1945* (New York: Oxford University Press, 2004), 24–43.

47. Geoffrey C. Bowker and Susan Leigh Star, *Sorting Things Out: Classification and Its Consequences* (Cambridge, MA: MIT Press, 2000), 1–2.

48. Bowker and Star, *Sorting Things Out*, 3.

49. David Brackett, *Categorizing Sound: Genre and Twentieth-Century Popular Music* (Berkeley: University of California Press, 2016), 6–11.

50. Brackett, *Categorizing Sound*, 10. Brackett's use of "articulation" is more akin to Bruno Latour's use of the term, as opposed to its use in cultural studies, as described by Stuart Hall, in which the term is used to identify points in which social formations connect in different ways, often in terms of cultural or political work. Articulation in Hall's work acts as a kind of antidote to traditional ways of thinking about sociocultural sites of power and ideologies. See Brackett, *Categorizing Sound*, 20. See also Bruno Latour, *Reassembling the Social: An Introduction to Actor-Network Theory* (New York: Oxford University Press, 2007); and Lawrence Grossberg, "On Postmodernism and Articulation: An Interview with Stuart Hall," *Journal of Communication Inquiry* 10, no. 2 (June 1986): 45–60.

51. Brackett, *Categorizing Sound*, 11–15.

52. Jennifer Lena, *Banding Together: How Communities Create Genres in Popular Music* (Princeton, NJ: Princeton University Press, 2014), 2–3.

53. du Gay, et al. *Doing Cultural Studies: The Story of the Sony Walkman*, xxx–xxi.

54. Brackett, *Categorizing Sound*, 19–26.

55. Thomas Schatz, "Film Studies, Cultural Studies, and Media Industries Studies," *Media Industries Journal* 1, no. 1 (2014), mediaindustriesjournal.org.

56. Simon Frith, *Performing Rites: On the Value of Popular Music* (Cambridge, MA: Harvard University Press, 1996), 16.

57. Keith Negus, *Music Genres and Corporate Cultures* (New York: Routledge, 1999), 20.

58. Hagstrom Miller, *Segregating Sounds*, 2.

59. Hagstrom Miller, *Segregating Sounds*, 2.

60. Hagstrom Miller, *Segregating Sounds*, 4.

61. Hagstrom Miller, *Segregating Sounds*, 7.

62. Hagstrom Miller, *Segregating Sounds*, 7.

63. Simon Frith, "The Industrialization of Culture," in *Music for Pleasure: Essays on the Sociology of Pop* (Cambridge, UK: Polity Press, 1988), 11–23.

64. Hagstrom Miller, *Segregating Sounds*, 11.

65. Lisa Gitelman, *Always Already New: Media, History and the Data of Culture* (Cambridge, MA: MIT Press, 2008), 10.

66. Gitelman, *Always Already New*, 7.

67. Gitelman, *Always Already New*, 8.

68. Latour, *Reassembling the Social*, 8–9.

69. Isaac Pitman, *Phonography; or Writing by Sound* (London: S. Bagster & Sons, 1942).

70. Raymond R. Wile, ed., *Proceedings of the 1890 Convention of Local Phonograph Companies* (Nashville, TN: Country Music Foundation Press, 1974), ix.

71. Wile, *Proceedings of the 1890 Convention of Local Phonograph Companies*, 9.

72. Thomas Edison, "The Phonograph and its Future," *The North American Review*, May 1, 1878, 527–36.

73. DeGraaf, *Edison and the Rise of Innovation*, 41.

74. Wile, *The 1890 Convention of Local Phonograph Companies*, xv.

75. Paul Israel, *Edison: A Life of Invention* (New York: Wiley & Sons, 1998), 278, 282.

76. Israel, *Edison: A Life of Invention*, 280–81.

77. Israel, *Edison: A Life of Invention*, 280; and Wile, *The 1890 Convention of Local Phonograph Companies*, xii.

78. Israel, *Edison: A Life of Invention*, 277.

79. Israel, *Edison: A Life of Invention*, 58.

80. Israel, *Edison: A Life of Invention*, 81–85.

81. Wile, *The 1890 Convention of Local Phonograph Companies*, xxxiv.

82. Wile, *The 1890 Convention of Local Phonograph Companies*, 163.

83. Israel, *Edison: A Life of Invention*, 299.

84. United States v. Motion Picture Patents Co., 225 F. 800 (D.N.J. 1915). Also see DeGraaf, *Edison and the Rise of Innovation*, 136–39; and Q. David Bowers and Kathryn Fuller-Seeley, *One Thousand Nights at the Movies: An Illustrated History of Motion Pictures* (Atlanta: Whitman Publishing, 2013), 124–26.

85. Michael Chanan, *Repeated Takes: A Short History of Recording and Its Effects on Music* (London: Verso, 1995), 7.

86. *Phonogram*, May 1900–June 1901.
87. "Automatic Music in the Home," *Phonogram*, June 1901, 26.
88. *ARSC Blog*, "The Independent Record Companies of the 1890s," by Mason Vander Lugt, arsc-audio.org.
89. *ARSC Blog*, "The Independent Record Companies of the 1890s."
90. Dave Laing, "A Voice without a Face: Popular Music and the Phonograph in the 1890s," *Popular Music* 10, no. 1 (1991): 7.
91. Paul Starr, *The Creation of Media*, 303.
92. Paul Goldstein, *Copyright's Highway: From Gutenberg to the Celestial Jukebox*. rev. ed. (Palo Alto, CA: Stanford University Press, 2003), 51–52.
93. Goldstein, *Copyright's Highway*, 51–52.
94. Edward Samuels, *The Illustrated Story of Copyright* (New York: Thomas Dunne Books, 2000), 34.
95. White-Smith Music Publishing Company, Appt., v. Apollo Company, 209 U.S. 1 (1908).
96. Samuels, *The Illustrated Story of Copyright*, 37.
97. Samuels, *The Illustrated Story of Copyright*, 37.
98. John Philip Sousa, "The Menace of Mechanical Music," *Appleton's*, September, 1906, 278–84. A *rapprochement* would come in 1923, when Sousa and Edison would meet for the first time. The meeting, by which time Sousa's earlier antagonisms had been largely set aside, was chronicled in the October 1923 edition of *Etude*. "A Momentous Musical Meeting," is a short but far reaching article in which the two men discuss about sound, music, and hearing itself.
99. Samuels, *The Illustrated Story of Copyright*, 37.
100. Katherine Spring, *Saying It with Songs: Popular Music and the Coming of Sound to American Cinema* (New York: Oxford University Press, 2013), 40.
101. Spring, *Saying It with Songs*, 55.
102. Goldstein, *Copyright's Highway*, 54–55. Katherine Spring has argued that ASCAP's emergence was in process before Victor Herbert's experience at Shanley's, based on an earlier account provided by Bennie L. DeWitt's research. See Spring, *Saying It with Songs*, 45.
103. Herbert v. Shanley Co., 242 U.S. 591 (1917).
104. Goldstein, *Copyright's Highway*, 57.
105. David Suisman, *Selling Sounds: The Commercial Revolution in American Music* (Cambridge, MA: Harvard University Press, 2009), 151.
106. Sutton, *A Phonograph in Every Home*, xi–v.
107. Édouard-Léon Scott de Martinville, Thomas Edison, Charles Sumner Tainter, Chichester Bell experimented with disc technologies to varying degrees. See Oliver Read and Walter L. Welch, "Advent of the Discs," in *From Tin Foil to Stereo: Evolution of the Phonograph*, 2nd ed. (Indianapolis: Howard W. Sams/Bobbs Merrill, 1959), 119–35. For Berliner's comparison of the gramophone with competing sound technologies, see Emile Berliner, *The Gramophone: Etching the Human Voice*, pamphlet (Berliner Gramophone Company, 1909), Library of Congress (LOC).
108. Sutton, *A Phonograph in Every Home*, 1.
109. Sutton, *A Phonograph in Every Home*, 3. See also Tim Brooks, "High Drama in the Record Industry," *ARSC Journal* 33, no. 1 (2002): 21–76.

110. Sutton, *A Phonograph in Every Home*, 6.

111. Sutton and Nauck, *American Record Labels and Companies*, 293–94.

112. Sutton and Nauck, *American Record Labels and Companies*, 293.

113. Leeds & Catlin Co. v. Victor Talking Machine Co., 213 U.S. 301 (1909).

114. Suisman, *Selling Sounds*, 153–54.

115. Suisman's full quote suggests that these upstart labels focused on sound recordings solely, but a number of them made their own "talking machines," at least for a time. One common way into the recording business for these companies was initially making cabinets for others, then making their own phonographs and cabinets, then finally making discs of their own. Suisman, *Selling Sounds*, 153–54.

116. Raymond Wile, "The American Graphophone Company and the Columbia Phonograph Company Enter the Disc Record Business, 1897–1903," *ARSC Journal* 22, no. 2 (Fall 1991): 207–9.

117. Sutton and Nauck, *American Record Labels and Companies*, 45.

118. Sutton and Nauck, *American Record Labels and Companies*, 45.

119. Sutton and Nauck, *American Record Labels and Companies*, 45.

120. Sutton and Nauck, *American Record Labels and Companies*, 45–46.

121. Suisman, *Selling Sounds*, 153.

122. Suisman, *Selling Sounds*, 21.

123. Carl Wilson, *Let's Talk About Love: Why Other People Have Such Bad Taste*, new and expanded ed. (New York: Bloomsbury, 2014), 58–59.

124. "Music Publishers Sued Here on Trust," *New York Times*, August 4, 1920, 24. The suit was later dismissed.

125. "Government Moves Against Publishers," *Musical Courier*, August 12, 1920, 5–17.

126. "How Popular Song Factories Manufacture a Hit," *New York Times*, September 18, 1910, 11. The article highlights two of Tin Pan Alley's leading songwriters of the time, Victor Herbert and Charles K. Harris.

127. Suisman, *Selling Sounds*, 58–79.

128. Suisman, *Selling Sounds*, 59.

129. "Last Week's Ten Best Sellers among the Popular Songs," *Billboard*, September 15, 1917, 14; in Spring, *Saying It With Songs*, 19.

130. Spring, *Saying It with Songs*, 16.

131. "Utilizing Heart Throbs as Trade Ticklers," *Talking Machine World*, March 13, 1919, 34 (MHDL).

132. Only two record companies—Edison and Federal—were still making cylinder recordings by 1920. See Sutton, *Recording the Twenties*, 4–5.

133. "Victor Self-Service Record Cabinet a Novel Method Adopted by Dealers," *Music Trade Review*, January 18, 1919, 31.

134. C. L. Ketcham, "Let the Customer Play the Records," *Music Trade Review*, February 22, 1919, 36.

135. Charles McGovern, *Sold American: Consumption and Citizenship, 1890–1945* (Chapel Hill: University of North Carolina Press, 2006), 23.

136. Michele Hilmes, "Nailing Mercury: The Problem of Media Industry Historiography," in *Media Industries: History, Theory and Method*, ed. Jennifer Holt & Alisa Perren (Malden, MA: Wiley-Blackwell, 2009), 21.

137. Hilmes, "Nailing Mercury," 21.

138. Hilmes, "Nailing Mercury," 21.
139. Hilmes, "Nailing Mercury," 22.
140. Christopher Anderson and Michael Curtin, "Writing Cultural History: The Challenge of Radio and Television," in *Media History: Theories, Methods, Analysis*, ed. Niels Brügger and Søren Kolstrup (Aarhus, DK: Aarhus University Press, 2002), 17.
141. Simon Frith, "Towards an Aesthetic of Popular Music," *Popular Music: Critical Concepts in Media and Cultural Studies: Vol. 4, Music and Identity* (London: Routledge, 2004), 36.
142. Simon Frith, *Performing Rites*, 15.
143. While I focus here on U.S. recording industry histories, these companies had transnational ties early on. One of the few books to map recording companies' international relationships is Gronow and Saunio, *An International History of the Recording Industry*.
144. Kenney, *Recorded Music in American Life*, xiii.
145. Media scholars have traditionally thought of recorded music as content for other media platforms, not as a platform itself. In this sense, recorded music has only been considered "mass" when used by radio, television, or film. I am indebted to Jim Buhler, who alerted me to these dynamics in conversations about recorded music and the study of media.
146. Pekka Gronow, "The Record Industry: The Growth of a Mass Medium," *Popular Music* 3 (January 1983): 53.
147. Gronow, "The Record Industry," 53.
148. See John Williamson and Martin Cloonan, "Rethinking the Music Industry," *Popular Music* 26, no. 2 (May 2007): 305–22; and Jonathan Sterne, "There Is No Music Industry," *Media Industries* 1, no. 1 (2014). mediaindustriesjournal.org.
149. See David Hesmondhalgh, "Bourdieu, the Media, and Cultural Production," *Media, Culture & Society* 28, no. 2 (November 9, 2010): 211–31; Keith Negus, "The Work of Cultural Intermediaries and the Enduring Distance Between Production and Consumption," *Cultural Studies* 16, no. 4 (November 9, 2010): 501–15; and Liz McFall, "What About the Old Cultural Intermediaries?: An Historical Review of Advertising Producers," *Cultural Studies* 16, no. 4 (November 9, 2010): 532–52.
150. See Pierre Bourdieu, "Structures and the Habitus," in *Outline of a Theory of Practice* (New York: Cambridge, UK: 1977), 72–95. For useful introductory discussions of Bourdieu's theories and their application in cultural research, see Karl Maton, "Habitus," and Patricia Thomson, "Field," in Michael Grenfell, *Pierre Bourdieu: Key Concepts*, 2nd ed. (Durham, UK: Acumen, 2012), 48–64, 65–80.
151. Bourdieu, *Outline of a Theory of Practice*, 72.
152. Pierre Bourdieu, *Distinction: A Social Critique of the Judgement of Taste* (Cambridge, MA: Harvard University Press, 1993), 164.
153. Negus, *Music Genres and Corporate Cultures*, 17.
154. McGovern, *Sold American*, 24–25.
155. van der Tuuk, *Paramount's Rise and Fall: The Roots and History of Paramount Records* (Denver: Mainspring Press, 2012), 133–36.
156. Elijah Wald, *Escaping the Delta: Robert Johnson and the Invention of the Blues* (New York: Amistad/Harper Collins, 2004), xiv.
157. Wald, *Escaping the Delta*, xiv.

Chapter 1

1. Fred Gennett to George Kay, July 19, 1962, John K. MacKenzie Collection, Indiana Historical Society (IHS).

2. John K. MacKenzie, "Gennettics," unpublished ms., John K. MacKenzie Collection, Indiana Historical Society (IHS).

3. Gennett's manufacturing of KKK recordings remains shocking and confusing, given that at least some of the Gennett family were reportedly troubled by the private-press recordings made for the Klan. Gennett employee Tag Taggart recounted that Clarence Gennett once discovered some made-to-order Klan recordings and told Taggart to "throw everything away connected with that," either to cover up their manufacturing them, because of his general distaste for the connection, or both. See the interview with Wilson "Tag" Taggart, John MacKenzie Collection, Indiana Historical Society (IHS).

4. Gennett made numerous indigenous peoples' recordings released on the Gennett imprint in and in conjunction with Fewkes, going so far as to release a catalog in which the releases were featured in a Gennett catalog that included descriptions of the recordings written by the anthropologist.

5. Rick Kennedy, *Jelly Roll, Bix and Hoagy: Gennett Records and the Rise of America's Musical Grassroots*, 2nd ed. (Bloomington and Indianapolis, Indiana University Press, 2013), 2–4.

6. Kennedy, *Jelly Roll, Bix and Hoagy*, 4–5.

7. Kennedy, *Jelly Roll, Bix and Hoagy*, 4–5

8. Kennedy, *Jelly Roll, Bix and Hoagy*, 5–6.

9. Kennedy, *Jelly Roll, Bix and Hoagy*, 6–7.

10. Kennedy, *Jelly Roll, Bix and Hoagy*, 8–9.

11. Klein adds that Starr Pianos regularly won awards, including Nashville (1897), St. Louis (1904), Alaska-Yukon (1909), and Panama (1915). See Andrew Klein, "A History of the Starr Piano Factory," *Talking Machine Review*, February 1983, 1798.

12. Rick Kennedy and Randy McNutt estimate that there were fewer than 25 manufacturers in 1914, compared to more than 150 three years later. See Kennedy, *Little Labels, Big Sound*, 25.

13. The phonograph's transformation from public novelty to domestic furniture was in no way a simple transition, due in part to gendered debates surrounding media technologies in the home. The phonograph's domestication displaced the piano as the dominant home entertainment device, giving phonography a dominant influence in U.S. homes during the 1910s and 1920s, before radio rivaled the phonograph as a source for domestic entertainment. See Holly Kruse, "Early Audio Technology and Domestic Space," *Stanford Humanities Review* 3, no. 2 (Autumn 1993): 6–7; and Barnett, "Furniture Music: The Phonograph as Furniture, 1900–1935," *Journal of Popular Music Studies* 18, no. 3 (November 2016): 307–13.

14. While not offering a conclusive decision as to who might have been at fault,

Kennedy suggests that, Gennett's hard-driving and ambitious style notwithstanding, Jesse French's account is less trustworthy, given his unscrupulous past and a competing interest in the form of the Krell-French Company in Newcastle, Indiana. Much of Kennedy's account is based on "a compilation of the depositions and supporting exhibit documents, including news article and correspondence" that Kennedy culled from the Richard Gennett estate through his son Fred Gennett. See Kennedy, *Jelly Roll, Bix and Hoagy*, 9–10; 257fn.

15. Wilson "Tag" Taggart, interview, John K. MacKenzie Collection (IHS).

16. Sutton and Nauck, *American Record Labels and Companies*, 371–72.

17. The Victrola series was introduced in response to Columbia' low-priced phonograph line. The Victrola IV, which sold for $15.00 in 1911, was an immediate success. See Robert Baumbach, *Look for the Dog: An Illustrated Guide to Victor Talking Machines* (Woodland Hills, CA: Stationery X-Press, 1981), 70.

18. Library of Congress; "History of the Cylinder Phonograph," loc.gov.

19. Read and Welch, *From Tin Foil to Stereo*, 207.

20. Sutton and Nauck, *American Record Labels and Companies*, 86–87.

21. Alan P. Merriam and Fradley H. Garner, "Jazz—The Word," *Ethnomusicology* 12, no. 3 (September 1968): 389–93.

22. See Ted Gioia, *The History of Jazz* (New York: Oxford University Press, 1997), 36. Gioia notes that those active musicians both black (Buddy Bolden, Bunk Johnson, Joe "King" Oliver, Mutt Carey), Creole (Jelly Roll Morton, Sidney Bechet, Kid Ory, Freddie Keppard), and white (Nick LaRocca, Sharkey Bonano, and Papa Jack Laine) transcended racially segregated groups in such close proximity.

23. Lizabeth Cohen, *Making a New Deal: Industrial Workers in Chicago, 1919–1939*, 2nd ed. (Cambridge, UK: Cambridge University Press, 2008), 155.

24. Kenney, *Recorded Music in American Life*, xiv.

25. Kenney, *Recorded Music in American Life*, 5–10.

26. The next noteworthy discovery via recording would be even more significant for Carmichael: "In 1924, Carmichael heard the Wolverines with Bix Beiderbecke. They were the first band of white and northern Midwesterners." See Marshall Stearns, *The Story of Jazz* (New York: Oxford University Press, 1972), 175. Oddly, Victor would not do much to follow up on the ODJB recordings, instead staying with the traditional dance bands of the era.

27. The article mentions that the wife punched firefighters when they arrived and argued that it was her fire and she could take care of it. "Don't Try to Burn a Jazz Record," *Talking Machine World*, December 15, 1920. And there's a musical coincidence in the husband's name, Robert Johnson, evoking the blues legend who later had his own version of phonograph blues.

28. "The Appeal of Primitive Jazz," in *Jazz in Print (1856–1929): An Anthology of Collected Writings of Selected Early Readings in Jazz History*, ed. Karl Koenig (Hillsdale, NY: Pendragon Press, 2002), 119–20.

29. "The Appeal of Primitive Jazz," 119–20.

30. Maureen Anderson, "The White Reception of Jazz in America," *African American Review* 38, no. 1 (Spring 2004): 135.

31. W. Francis Gates, "American Folk Music: A Negligible Quality," *The Etude*, January 1918, 8.

32. Defining and assessing jazz became a parlor game of sorts for the magazine. See R. M. Goodrod, "Conquering the Jazz Craze of Young Pianists," *The Etude*, February 1934, 82; Isham Jones, "American Dance Music Is Not Jazz," *The Etude*, August 1924, 526; and Gustave Klemm, "The Jargon of Jazz," *The Etude*, August 1934, 155.

33. Ernst Felix Tschudi, "The Immortals Protest Against Jazz," *The Etude*, August 1924, 670.

34. "Where the Etude Stands on Jazz," *The Etude*, August 1924, 515.

35. "Where Is Jazz Leading America?" *The Etude*. August, 1924, 517–18, and September, 1924, 595–96.

36. Herman Gray, *Producing Jazz: The Experience of an Independent Record Company* (Philadelphia: Temple University Press, 1988), 4.

37. The choice of these two acts is due in part to the amount of archival material available. When record enthusiasts such as John MacKenzie interviewed those involved with record production during the 1920s and 1930s, their questions were largely dictated by their tastes. In MacKenzie's case, this meant a focus on jazz artists who were most influential in collector circles, such as King Oliver. Fortuitous sources in *Talking Machine World* and other trade papers allow for a larger cross-section of recorded jazz artists of the era.

38. By the time of Wiggins' death, there was no mention of his managing Starr's Gennett recording division, which both illustrates the Gennett label as a sideline for Starr Piano's operations as well as how distant their musical activity was by the late 1940s. "Fred Wiggins Dies Suddenly at Elks Club," *Richmond Palladium-Item & Sun-Telegram*, August 10, 1948. Curiously, no known photograph of Wiggins has ever been discovered, despite researchers' best efforts.

39. See George W. Kay, "Those Fabulous Gennetts!: The Life Story of a Remarkable Label," *Record Changer*, June 1953, 4–13.

40. Rick Kennedy, "The Man Who Made the Label a Legend," *Starr-Gennett Foundation, Inc.* newsletter, 2002.

41. Kay, "Those Fabulous Gennetts!" 4–13.

42. Kennedy, *Jelly Roll, Bix and Hoagy*, 74–75.

43. Kennedy, *Jelly Roll, Bix and Hoagy*, 73.

44. Kennedy, *Jelly Roll, Bix and Hoagy*, 73–74.

45. See extensive research on the Friars Society Orchestra/New Orleans Rhythm Kings in Richard Sudhalter, *Lost Chords: White Musicians and their Contribution to Jazz, 1915–1945* (New York: Oxford University Press, 1999). While honoring Sudhalter's contributions to scholarship on individual musicians, Gerald Early has rightly argued that jazz history would have been much better served by an integrated history that would have teased out complexities missing in Sudhalter's sprawling history. See Gerald Early, "White Noise and White Knights: Some Thoughts on Race, Jazz, and the White Jazz Musician," in *Jazz: A History of America's Music*, ed. Geoffrey C. Ward and Ken Burns (New York: Knopf, 2000), 324–25.

46. Suzanne Fischer, "Friars Society Orchestra/New Orleans Rhythm Kings," *Starr-Gennett Foundation, Inc.* newsletter, 2010.

47. The New Orleans Rhythm Kings are said to have made history on July 23, 1923, when they backed up Jelly Roll Morton at a session at Gennett's Richmond studio, with Ezra Wickemeyer as recording engineer. Several takes of "Mister Jelly Lord" and other tunes were cut in what is purported to be the first interracial recording session, though Morton regularly rejected the notion of his Creole heritage, insisting that all his ancestors were French in origin. Morton cut many records for Gennett. The February 2, 1924, issue of *The Music Trade Review* reported that "The demand for the Gennett records of all Mr. Morton's hits has exceeded the supply of these popular records." See Howard Reich and William M. Gaines, *Jelly's Blues: The Life, Music, and Redemption of Jelly Roll Morton* (Cambridge, MA: Da Capo Press, 2003), 89–90, 94.

48. See Kay, "Those Fabulous Gennetts!" 7.

49. Much debate has surrounded exactly why Oliver sent for Armstrong. Ted Gioia outlines several of the most likely possible scenarios: (1) Oliver's appreciation for Armstrong's talent; (2) Oliver's desire to improve the overall skills of the Creole Jazz Band through Armstrong's presence; (3) Oliver's own sense of his declining skills. There is likely more than a little truth to all these possibilities, but we're finally left with conjecture. See Gioia, *The History of Jazz*, 46–51.

50. Rick Kennedy outlines the North End/Goose Town's history in the second edition of *Jelly Roll, Hoagy and Bix*, in an account largely based on a first-person interview with Teresa Braun, a descendant of the Italian American family that ran a neighborhood grocery beginning in 1918. See Kennedy, *Jelly Roll, Bix and Hoagy*, 210–12.

51. Accommodations in the North End/Goose Town neighborhood included a private residence where black travelers could stay, which was said to double as a brothel. A whites-only brothel (with no overnight boarding accommodations) was also located in the neighborhood. See Kennedy, *Jelly Roll, Bix and Hoagy*, 210–12.

52. Gennett recording ledgers, Starr-Gennett Foundation. (SGF)

53. Austin Lescarboura, "At the Other End of the Phonograph," *Scientific American*, August 31, 1918, 164.

54. Kennedy, *Jelly Roll, Bix and Hoagy*, 21–22.

55. Kennedy, *Jelly Roll, Bix and Hoagy*, 22. I am also indebted to Charlie Dahan for emphasizing the significance of this move. Dahan and Malcolm Rockwell have long been at work on a discography of Gennett and related labels. See Charlie B. Dahan and Malcolm Rockwell, "Starr Piano/Gennett Records Discography," *ARSC Journal* 39, no. 2 (Fall 2008): 292–94.

56. Tom Tsotsi, "Gennett-Champion Blues," *78 Quarterly* 1, no. 3 (1988): 51.

57. Wilson "Tag" Taggart, interview, John MacKenzie Collection (IHS).

58. Kay, "Those Fabulous Gennetts!," 5.

59. Kay, "Those Fabulous Gennetts!," 5.

60. Interview with Rena Clarke, interview, John MacKenzie Collection (IHS).

61. Gioia, *The History of Jazz*, 49–50.

62. *The Columbia Gypsy*, excerpt from a 1971 letter from Robert P. Gennett to British jazz enthusiast, on Gennett descendent Judith Gennett's blog.

63. Excerpt from a Gennett instructional sheet, given to performers before recording. Reprinted in Frederick Ramsey Jr. and Charles Edward Smith, eds. *Jazzmen* (New York: Harcourt, Brace and Company, 1939), 73.

64. "'Making Record No Easy Job' says Pittsburgh Musicians by Deppe (Himself)," *Pittsburgh Courier*, October 20, 1923. Reprinted in *Jazz Advertised, 1910–1967, Vol. 1: Out of the New England Press, 1910–1934*, ed. Franz Hoffmann (Berlin: Franz Hoffmann, 1997), 69. See also Rick Kennedy's discussion of the Lois Deppe's Serenaders in Kennedy, *Jelly Roll, Bix and Hoagy*, 82–83.

65. "Making Record No Easy Job," 69.

66. "Making Record No Easy Job," 69.

67. Gioia, *The History of Jazz*, 45–46. Gioia notes that of all those famous names of New Orleans jazz, a list that would certainly include Louis Armstrong, Sidney Bechet, Jelly Roll Morton, and King Oliver, only those who left the city were able to gain larger recognition. All four of these musicians recorded for Gennett, including Oliver.

While Chicago did deliver for many on its economic promises, and its thriving entertainment scene offered an exciting if tentative interracial social and creative environment in the city's South Side clubs, a larger backlash soon dashed hopes. Chicago's 1919 race riots were followed by a decade in which racial intolerance and xenophobia in the United States found new cachet in a society in which dramatic changes were taking place.

68. Louis Armstrong, "Joe Oliver Is Still King," *Record Changer*, July–August 1950, 10–11.

69. Lil Hardin-Armstrong, "Lil and Louis: Satchmo and Me (Excerpts)," in *The Louis Armstrong Companion*, ed. Joshua Berrett (New York: Schirmer Books, 1999), 38. Gennett engineer Ezra Wickemeyer would later claim to have suggested the musicians' placement. See Kennedy, *Jelly Roll, Bix and Hoagy*, 62–73.

70. Hardin-Armstrong, "Lil and Louis," 38.

71. Hardin-Armstrong, "Lil and Louis," 38.

72. Larry Gara, *The Baby Dodds Story* (Los Angeles: Contemporary Press, 1959), 69.

73. Nat Shapiro and Nat Hentoff, *Hear Me Talkin' To Ya: The Story of Jazz as Told By the Men Who Made It* (New York: Dover Publications, 1966), 84.

74. This was the case with Hoagy Carmichael's "Stardust," which later became an American pop music standard, but made little impact for Gennett in its initial release.

75. Kay, "Those Fabulous Gennetts!," 6.

76. MacKenzie, "Recording and Record Manufacturing at the Starr Piano Company," John MacKenzie Collection. (IHS)

77. MacKenzie, "Recording and Record Manufacturing at the Starr Piano Company," John MacKenzie Collection (IHS).

78. MacKenzie, "Interview with Harold Soule," John MacKenzie Collection (IHS).

79. MacKenzie, "Interview with Florence Gennett," John MacKenzie Collection (IHS).

80. MacKenzie, "Recording and Record Manufacturing at the Starr Piano Company" (IHS).

81. MacKenzie, "Recording and Record Manufacturing at the Starr Piano Company" (IHS).

82. MacKenzie, "Interview with Joseph E. Geier," John MacKenzie Collection (IHS).

83. MacKenzie, "Recording and Record Manufacturing at the Starr Piano Company" (IHS).

84. MacKenzie, "Recording and Record Manufacturing at the Starr Piano Company" (IHS).

85. MacKenzie, "Recording and Record Manufacturing at the Starr Piano Company" (IHS).

86. MacKenzie, "Recording and Record Manufacturing at the Starr Piano Company" (IHS).

87. MacKenzie, "Recording and Record Manufacturing at the Starr Piano Company" (IHS).

88. In the pre–World War II recording industry, few stores specialized solely in records. Instead, music was sold in furniture stores (where phonographs were often purchased), musical instrument retailers, department stores, chain stores, and—for a time—even funeral homes (further capitalizing on the connection to the furniture industry). A critical history of record retailing and the rise of the record store remains to be written.

89. S. G. Marden, "Phonograph Books Have New Motif," *Printers Ink Monthly*, April 20, 1920, 14.

90. MacKenzie, "Interview with Wilson 'Tag' Taggart," John MacKenzie Collection (IHS). Taggart worked with Niles in the advertising department, then moved to the purchasing department and later the production department. Taggart started with Starr in 1914, left to serve in the U.S. Army during World War I, and returned to the company in 1919. He left in 1924 around the time that Fred Wiggins returned from Chicago to become director of the company's recording division.

91. MacKenzie, "Interview with Wilson 'Tag' Taggart," John MacKenzie Collection (IHS).

92. Starr Piano Company advertisements, *Talking Machine World*, February 15, 1925, and *Talking Machine World*, September 15, 1925 (LOC).

93. *Gennett Records*, June 1922 Catalog (SGF).

94. *Gennett Records*, June 1922 Catalog (SGF).

95. *1925 List of All Lateral Gennett Records*—Complete Catalog No. 5, Richmond, IN: Starr Piano Company, January 1, 1925 (SGF).

96. Mervyn Cooke, *The Chronicle of Jazz* (New York: Abbeville Press, 1998), 55.

97. Paul Whiteman and Mary Margaret McBride, *Jazz* (New York: J. H. Sears & Company, 1926), 87.

98. Don Rayno, *Paul Whiteman: Pioneer in American Music, Vol. 1: 1890–1930* (Lanham, MD: Scarecrow Press, 2012), 80–81.

99. Rayno, *Paul Whiteman: Pioneer in American Music*, 80–81.

100. Rayno, *Paul Whiteman: Pioneer in American Music*, 80–81.

101. Gertrude Swift, "What Do You Think of Jazz?" *Voice of the Victor*, December 1924, 10. Camden County Historical Society (CCHS).

102. Swift, "What Do You Think of Jazz?" 10.

103. Whiteman and McBride, *Jazz*, 115.

104. Neil Leonard, "The Impact of Mechanization," in *American Music: From Storyville to Woodstock*, ed. Charles Nanry (New Brunswick, NJ, Transaction Books, 1972), 50.

105. Gioia, *The History of Jazz*, 77.

106. William Howland Kenney, *Chicago Jazz: A Cultural History, 1904–1930* (New York: Oxford University Press, 1993), 141.

Chapter 2

1. Ronald Foreman Jr., "Jazz and Race Records, 1920–1932: Their Origins and Their Significance for the Record Industry and Society" (PhD diss., University of Illinois, 1968), 35, http://search-proquest-com.libproxy.bellarmine.edu/pqdt-global/. A $250 phonograph would have been one of the more expensive models marketed to the middle class, while tabletop models sold for as little as $15. High-end models included "the Chippendale," which sold for $600.

2. Foreman Jr., "Jazz and Race Records, 1920–1932," 35.

3. "Broome Special Phonograph Records" advertisement, *The Crisis*, January 1920, 159, in *Mainspring Press Record Collectors' Blog*, Allan Sutton, "The First Race Records Advertisements, 1918–1920," https://www.78rpm.club/record-labels/broome-special-phonograph-records/

4. Sutton notes that the Booker T. Washington oratory was likely a personal series recording made via Columbia. It appears that Broome pasted new labels over the original discs. See Sutton, "The First Race Records Advertisements, 1918–1920." For more on Broome, see Sutton and Nauck, *American Record Labels and Companies*, 27.

5. Foreman Jr., "Jazz and Race Records, 1920–1932," 39–40.

6. Foreman Jr., "Jazz and Race Records, 1920–1932," 34.

7. See Donald Clarke, *The Rise and Fall of Popular Music* (New York: Viking/Penguin Books, 1996), 135; and Patrick Huber, "Black Hillbillies: African-American Musicians on Old-Time Records, 1924–1925," in *Hidden in the Mix: The African American Presence in Country Music*, ed. Diane Pecknold (Durham, NC: Duke University Press, 2013), 23.

8. Foreman Jr., "Jazz and Race Records, 1920–1932," 39.

9. See Tim Brooks, *Lost Sounds: Blacks and the Birth of the Recording Industry* (Urbana: University of Illinois Press, 2004), 13–49, 105–48, and 192–215.

10. See Lynn Abbott and Doug Seroff, *Out of Sight: The Rise of African-American Popular Music, 1889–1895* (Jackson: University Press of Mississippi Press, 2002), xii–xvi.

11. Abbott and Seroff, *Out of Sight*, 40.

12. Sanjek, *Pennies from Heaven*, 62.

13. Sutton and Nauck, *American Record Labels and Companies*, 284.

14. Leroi Jones (Amiri Baraka), *Blues People: Negro Music in White America* (New York: Harper Perennial, 1999), 81.

15. Richard Middleton, *Pop Music and the Blues: Its Relationship and Significance* (London: Gollancz, 1972), 33.

16. Two useful studies provide good places to start: Daphne Duval Harrison's *Black Pearls: Blues Queens of the 1920s*, (Brunswick, NJ: Rutgers University Press, 1988); and Angela Y. Davis, *Blues Legacies and Black Feminism: Gertrude "Ma" Rainey, Bessie Smith, and Billie Holliday* (New York: Vintage Books, 1998).

17. Jones (Baraka), *Blues People*, 82.

18. Foreman Jr. "Jazz and Race Records, 1920–1932," 56–57.

19. This time, instead of using Fred Hager's house band ("the Rega Orchestra" of the previous release), Bradford's band ("the Jazz Hounds") played, suggesting an increasing confidence on behalf of Hager and OKeh of Smith's abilities.

20. Perry Bradford, *Born with the Blues* (New York: Oak Publications, 1965), 48.

21. Kenney, *Recorded Music in American Life*, 117.

22. Kenney, *Recorded Music in American Life*, 117.

23. While the *Defender* reporter notes Smith's recording as the first from an African American woman, there were a number of predecessors. Smith is generally believed, however, to be the first African American woman to record in the blues idiom. See Foreman, Jr., "Jazz and Race Records, 1920–1932," 56–57.

24. Foreman Jr., "Jazz and Race Records, 1920–1932," 57.

25. Foreman Jr., "Jazz and Race Records, 1920–1932," 56.

26. Foreman Jr., "Jazz and Race Records, 1920–1932," 58.

27. Foreman Jr., "Jazz and Race Records, 1920–1932," 59.

28. The matter is further complicated by my intention not to denigrate those African American performers, like Bert Williams and many others, who showcased their talents within the context of the minstrel tradition, which for many offered the only viable access to entertainment professions for many years.

29. "OKeh Records: To Hear Is to Buy!" advertisement, *Talking Machine World*, October 15, 1920. (LOC).

30. Bradford, *Born with the Blues*, 28.

31. David Wondrich, *Stomp and Swerve: American Music Gets Hot, 1843–1924* (Chicago: Chicago Review Press, 2003), 161.

32. Drowne and Huber, *1920s*, 193.

33. Foreman Jr., "Jazz and Race Records, 1920–1932," 59.

34. Sanjek, *Pennies from Heaven*, 64.

35. Kenney, *Recorded Music in American Life*, 110.

36. Suisman, *Selling Sounds*, 205.

37. Suisman, *Selling Sounds*, 206.

38. Jitu Weusi, "The Rise and Fall of Black Swan Records," *Red Hot Jazz* (blog), http://redhotjazz.com/blackswan.html. See also Suisman, *Selling Sounds*, 209–10.

39. Madeline G. Allison, "The Horizon," *The Crisis*, March 22, 1922, 221–22; and "104,628.74 Paid for 'Black Swan' Records," *New York Age*, January 28, 1922, 6.

40. "Distribution System of Black Swan Phonograph Records," *The Crisis*, March 1922, 221.

41. "104,628.74 Paid for 'Black Swan' Records," *New York Age*, January 28, 1922, 6; and "Fletcher Making Tour," *New York Clipper*, February 7, 1923, 18.

42. Art Satherley, interview by Norm Cohen, G. Earle, with Ken Griffis and Bill Ward, June 12, 1971, Southern Folklife Collection, University of North Carolina-Chapel Hill (SFC-UNC).

43. Sutton and Nauck, *American Record Labels and Companies*, 22.

44. Foreman Jr., "Jazz and Race Records, 1920–1932," 75.

45. Sutton and Nauck, *American Record Labels and Companies*, 21.

46. "Caruso's Voice" advertisement, *The Crisis*, October 1921, 236, in Allan Sutton, *Race Records and the American Recording Industry, 1919–1945* (Denver: Mainspring Press, 2016), 38.

47. Theodore Kornweibel Jr., "An Economic Profile of Black Life in the Twenties," *Journal of Black Studies* 6, no. 4 (June 1976): 307.

48. Norman Kelley, *R&B: Rhythm & Business: The Political Economy of Black Music* (New York: Akashic, 2002), 14.

49. Kelley, *R&B*, 15.

50. Kelley, *R&B*, 15.

51. Ross Laird and Brian Rust, *Discography of OKeh Records, 1918–1934* (Westport, CT: Praeger, 2004), 4–5.

52. Ralph Peer, interview, Lillian Borgeson, 1958 (SFC-UNC).

53. Hagstrom Miller, *Segregating Sounds*, 20.

54. Lizabeth Cohen, *Making a New Deal: Industrial Workers in Chicago, 1919–1939*, 2nd ed. (New York: Cambridge University Press, 2008), 104–5.

55. Cohen, *Making a New Deal*, 105.

56. Robert M. W. Dixon and John Godrich, "Recording the Blues," in *Yonder Comes the Blues: The Evolution of a Genre*, 2nd ed., ed. Paul Oliver, Tony Russell, Robert M. W. Dixon, John Godrich, and Howard Dye (Cambridge, UK: Cambridge University Press, 2001), 261.

57. Daphne Duval Harrison and John Cowley, "'Classic' Blues and Women Singers," in *New Blackwell Guide to Recorded Blues*, ed. John Cowley and Paul Oliver (Oxford, UK: Blackwell, 1996), 125–155.

58. Foreman Jr., "Jazz and Race Records, 1920–1932," 70.

59. Bradford, *Born with the Blues*, 116.

60. Sutton, *Race Records and the American Recording Industry, 1919–1945*, 23.

61. Sutton, *Race Records and the American Recording Industry, 1919–1945*, 70–77.

62. "New Victor Records—A Special List of 'Blues'" advertisement, *Chicago Defender*, August 4, 1923; and "Victor Records by Race Artists" advertisement *Talking Machine World*, September 15, 1923, 70, in Sutton, *Race Records and the American Recording Industry, 1919–1945*, 77.

63. Sutton and Nauck, *American Record Labels and Companies*, 219–20.

64. Victor Records Advance Lists, June 8, 22, and July 6 and 20, 1928 (CCHS).

65. Victor Records Advance Lists, June 8, 22, and July 6 and 20, 1928 (CCHS).

66. Jeff Todd Titon, *Early Downhome Blues: A Musical and Cultural Analysis*, 2nd ed. (Chapel Hill: University of North Carolina Press, 1994), 205.

67. Titon, *Early Downhome Blues*, 200.

68. Lerone Martin, *Preaching on Wax: The Phonograph and the Shaping of Modern African American Religion* (New York: New York University Press, 2014), 5–8.

69. Irene Jackson-Brown, "Developments in Black Gospel Performance and Scholarship," *Black Music Research Journal* 10, no. 1 (Spring 1990): 36.

70. Dixon and Godrich, "Recording the Blues," 275.

71. Dixon and Godrich, "Recording the Blues," 276.

72. Titon, *Early Downhome Blues*, 200.

73. Titon, *Early Downhome Blues*, 202.

74. Huber, "Black Hillbillies: African American Musicians on Old-Time Records," in *Hidden in the Mix: The African American Presence in Country Music*, ed. Diane Pecknold (Durham, NC: Duke University Press, 2013), 23. See also Hagstrom Miller, *Segregated Sounds*, 157–86.

75. Lizabeth Cohen, *Making a New Deal*, 22.

76. Lizabeth Cohen, *Making a New Deal*, 22.

77. C. Vann Woodward, *The Strange Career of Jim Crow*, 3rd ed. (New York: Oxford University Press, 1974), 99–100.

78. Woodward, *The Strange Career of Jim Crow*, 200–201.

79. OKeh Race Records Catalog, June 1924, Jim Walsh Collection, Recorded Sound Reference Center, Library of Congress, Washington, D.C. (LOC).

80. OKeh Race Records Catalog, 1925, Jim Walsh Collection, Recorded Sound Reference Center, Library of Congress, Washington, D.C. (LOC).

81. OKeh Truetone Records Catalog, 1926 and 1927, Jim Walsh Collection, Recorded Sound Reference Center, Library of Congress, Washington, D.C. (LOC)

82. Eric Lott, *Love & Theft: Black Minstrelsy and the American Working Class* (New York: Oxford University Press, 1995), 139–41.

83. Doris Witt, "Soul Food: Where the Chitterling Hits the (Primal) Pan," in *Eating Culture*, ed. Ron Scapp and Brian Seitz (Albany: State University of New York Press, 1998), 260–68. Witt's larger argument links soul food with long-standing societal prejudices against both black femininity and homosexuality.

84. It is difficult to know why the OKeh catalogs' visual rhetoric changes in 1926. One tempting (but inconclusive) explanation is a shift in ownership. In October 1926, OKeh's Otto Heinemann agreed to sell his company to Columbia.

85. OKeh Truetone Records Catalog, 1926, Jim Walsh Collection, Recorded Sound Reference Center, Library of Congress, Washington, DC (LOC).

86. Duval Harrison, *Black Pearls: Blues Queens of the 1920s* (New Brunswick, NJ: Rutgers University Press, 1990), 50.

87. Tony Russell, "Blacks, Whites, and the Blues," in *Yonder Comes the Blues: The Evolution of a Genre*, 2nd ed., ed. Paul Oliver, Tony Russell, Robert M. W. Dixon, John Godrich, and Howard Dye (Cambridge, UK: Cambridge University Press, 2001), 155.

88. Titon, *Early Downhome Blues*, 205.

89. Titon, *Early Downhome Blues*, 205.

90. Stephen Calt, *I'd Rather Be the Devil: Skip James and the Blues* (Chicago: Chicago Review Press, 1994), 218.

91. Barnett, "Record Men," 113–15.

92. The historical similarities between Paramount and Gennett are striking. Both parent companies (Wisconsin Chair Company and Starr Piano Company, respectively) were involved in furniture production, both companies suffered catastrophic fires related to the use of varnish, and both were companies located in small towns. Both entered the record business via the "hardware" side of the business, first through the manufacture of phonograph cabinets, then phonographs, and finally phonograph records.

93. van der Tuuk, *Paramount's Rise and Fall*, 21–24.

94. van der Tuuk, *Paramount's Rise and Fall*, 26.

95. van der Tuuk, *Paramount's Rise and Fall*, 26–28; and Sarah Filzen, "The Rise and Fall of Paramount Records," *Wisconsin Magazine of History*, 82, no. 2 (1998): 109–10.

96. van der Tuuk, *Paramount's Rise and Fall*, 25.

97. van der Tuuk, *Paramount's Rise and Fall*, 43.

98. van der Tuuk, *Paramount's Rise and Fall*, 43.

99. Dixon and Godrich, "Recording the Blues," 263.

100. Dixon and Godrich, "Recording the Blues," 264.

101. "Local Firm Takes Over New York Phonograph Company," *Port Washington Herald*, April 2, 1924, 1. Port Washington Historical Society, Port Washington, Wisconsin (PWHS).

102. "Any Woman's Blues," *Printers' Ink*, May 29, 1924, 69–70. The article emphasized vaudeville blues, borrowing the title of Ida Cox's version of "Any Woman's Blues," made famous by Bessie Smith.

103. van der Tuuk, *Paramount's Rise and Fall*, 54.

104. William Barlow, *"Looking Up at Down": The Emergence of Blues Culture* (Philadelphia: Temple University Press, 1989), 3.

105. Art Satherley, interview by Ed Kahn, June 14, 1969. Ed Kahn Collection, Southern Folklife Collection, University of North Carolina-Chapel Hill (SFC-UNC).

106. Titon, *Early Downhome Blues*, 207.

107. Titon, *Early Downhome Blues*, 217.

108. Titon, *Early Downhome Blues*, 217.

109. Negus, *Music Genres and Recording Cultures*, 1996, 61–62.

110. Barry Mazor, *Ralph Peer and the Making of Popular Roots Music* (Chicago: Chicago Review Press, 2015), 1. Later Peer would contact folklorist Alan Lomax in a 1933 letter questioning Lomax regarding the copyright claim regarding "Home on the Range" by William and Mary Goodwin. Lomax believed his father was the first to publish the song in his book collection *Cowboy Songs and Other Frontier Ballads* (originally published in 1911). Peer later commented, "They hear one of these songs, and it's always 'legendary.' And I've always taken the position, 'Well, I don't know whether it's legendary or not. Now you *prove* it; I've got a contract on it." See Mazor, *Ralph Peer and the Making of Popular Roots Music*, 156.

111. Titon, *Early Downhome Blues*, 201.

112. Ralph Peer, interview, Lillian Borgeson, 1958 (SFC-UNC).

113. Titon, *Early Downhome Blues*, 201.

114. Titon, *Early Downhome Blues*, 201.

115. WCC cofounder Frederick Dennett is said to have been involved in the Paramount business until his death from heart attack in 1920 while at a Grafton, Wisconsin, trolley stop waiting for his ride back to Port Washington. See van der Tuuk, *Paramount's Rise and Fall*, 41, 47.

116. van der Tuuk, *Paramount's Rise and Fall*, 26.

117. van der Tuuk, *Paramount's Rise and Fall*, 43.

118. van der Tuuk, *Paramount's Rise and Fall*, 41. Van der Tuuk's evidence for this

comes from the late John Steiner, a record collector and Paramount enthusiast, who after John Bostwick's death found five hundred Paramount test pressings in the company president's attic.

119. van der Tuuk, *Paramount's Rise and Fall*, 41.

120. Barlow, "*Looking Up at Down*," 292.

121. van der Tuuk, *Paramount's Rise and Fall*, 56.

122. Barlow, "*Looking Up at Down*," 294–95.

123. van der Tuuk, *Paramount's Rise and Fall*, 64.

124. Stephen Calt, "Paramount—Part Two: Anatomy of a 'Race' Label," *78 Quarterly* 1, no. 4 (1989): 13.

125. van der Tuuk, *Paramount's Rise and Fall*, 66.

126. Dixon and Godrich, "Recording the Blues," 264.

127. Alex van der Tuuk, "Aletha Dickerson: Paramount's Reluctant Recording Manager," *Vintage Jazz Mart*, http://www.vjm.biz/new_page_18.htm.

128. Undated letter from Mayo Williams to John Steiner (ca. 1947), John Steiner Collection, Chicago Jazz Archive, Regenstein Library, University of Chicago, Chicago, Illinois (CJA-UC).

129. van der Tuuk, "Aletha Dickerson: Paramount's Reluctant Recording Manager."

130. Jacqueline Najuma Stewart, *Migrating to the Movies: Cinema and Black Modernity* (Berkeley: University of California Press, 2005), 117.

131. Stewart, *Migrating to the Movies: Cinema and Black Modernity*, 116.

132. Gayle Dean Wardlow, "Initial Inquiries and Encounters," in *Chasin' that Devil Music: Searching for the Blues*, ed. Edward Komara (San Francisco: Miller Freeman Books, 1998), 48.

133. John Steiner, Art Laibly notes (undated), John Steiner Collection (CJA-UC).

134. Later, Art Laibly denied in phone conversations with Gayle Wardlow that he ever supplied his artists with any type of liquor before or during a session. Wardlow also interviewed Paramount executive H. C. Speir, who said he didn't remember any liquor at Grafton. Speir suggested, however, that "talent recorded better" with "a tonic" or a drink to relax them. According to Wardlow, blues performer "[Ishmon] Bracey was emphatic on several occasions that both he and [Tommy] Johnson had access to bonded whiskey at their Grafton session and Johnson had a drinking problem during the session." Bracey added that "Tommy was supposed to make more songs than I was. But he stayed high most of the time. Mr. Layman [Laibly] told me to record what I could." See Wardlow, "Initial Inquiries and Encounters," 51–52.

135. Gayle Dean Wardlow, "The Talent Scouts: H. C. Speir (1895–1972)," *78 Quarterly* 1, no. 8 (1993): 13.

136. Gayle Dean Wardlow, "Legends of the Lost (The Story of Henry Speir)," *Blues Unlimited* 31 (March 1966), John Steiner Collection. (CJA-UC)

137. Gayle Dean Wardlow, "Godfather of Delta Blues: H. C. Speir," in *Chasin' that Devil Music: Searching for the Blues*, ed. Edward Komara (San Francisco: Miller Freeman Books, 1998), 133.

138. Titon, *Early Downhome Blues*, 213.

139. Titon, *Early Downhome Blues*, 212.

140. Titon, *Early Downhome Blues*, 212.

141. Dixon and Godrich, "Recording the Blues," 271.

142. Titon, *Early Downhome Blues*, 207.

143. If Art Satherley's account is to be believed, the unusual vocal phrasing may have been due in part to Jefferson's blindness and his need for lyrics of new songs to be whispered to him during takes. Satherley claims to have whispered the lyrics to a song (he doesn't specify which one) during a Paramount recording session. See Art Satherley, interview by Norm Cohen et al., June 12, 1971 (SFC-UNC).

144. Art Satherley, interview by Norm Cohen et al., June 12, 1971 (SFC-UNC).

145. Art Satherley, interview by Norm Cohen et al., June 12, 1971 (SFC-UNC).

146. If it's tempting to consider just how much music was recorded due to happenstance, when musicians and scouts both happened to be at the right place and the right time, it's also worth considering how much music from this era was missed due to too many scouts who only traveled to a much-traveled locale in search of replicating the last hit song.

147. Titon, *Early Downhome Blues*, 214.

148. Titon, *Early Downhome Blues*, 214.

149. Wardlow, "Godfather of Delta Blues: H. C. Speir," 130.

150. Paul Swinton, "A Twist of Lemon," *Blues and Rhythm* 121, 7, http://www.bluesandrhythm.co.uk/documents/121.pdf.

151. van der Tuuk, *Paramount's Rise and Fall*, 131–33.

152. van der Tuuk, *Paramount's Rise and Fall*, 132–36.

153. John Szwed, "Negro Music: Urban Renewal," in *Our Living Traditions: An Introduction to American Folklore*, ed. Tristram Potter Coffin (New York: Basic Books, 1968), 273–74. The title of Szwed's article, which wryly compares the nationalization of blues to urban renewal, has taken on different resonances over time as reassessments of urban renewal have rightfully pointed to the often damaging impacts such initiatives have had to African American communities.

154. John Szwed, "Negro Music: Urban Renewal," 274.

155. Hagstrom Miller, *Segregating Sounds*, 188.

156. Hagstrom Miller, *Segregating Sounds*, 189.

Chapter 3

1. Jeffrey Lange, *Smile When You Call Me a Hillbilly: Country Music's Struggle for Respectability, 1939–1954* (Athens: University of Georgia Press, 2004), 23.

2. *Handbook of Texas Online*, s.v. "Alexander Robertson." http://tsha.utexas.edu/handbook/online/articles/RR/froam.html.

3. Rodger Harris, "Wars of the Confederacy, Wars with the Indians, Wars with the 'Fiddle and Bow': The Life of Henry C. Gilliland," Oklahoma Historical Society, http://www.ok-history.mus.ok.us/folk/gilliland.html.

4. Bill C. Malone, *Country Music, U.S.A.*, rev. ed. (Austin: University of Texas Press, 1997), 35.

5. Malone, *Country Music, U.S.A.*, 35.

6. This by no means is to suggest that jazz was solely understood as modern

music; as previously noted, common discussions in the white press renewed racist critiques of black culture via jazz as primitivism.

7. Huber, "Black Hillbillies," 56.

8. Cohen, *Making a New Deal*, 133–34.

9. Gene Wiggins, "John Carson: Early Road, Radio, and Records," *Journal of Country Music* 8, no. 1 (May 1979): 27.

10. For a relatively recent example of this assertion, see Paul L. Tyler, "Hillbilly Music Re-imagined: Folks and Country Music in the Midwest," *Journal of American Folklore* 127, no. 504 (Spring 2014), 159–90. Tyler opens the article with this: "Sectionalism has worked all manner of mischief in American history and culture, exposing the difficulty of making one out of many."

11. Tyler, "Hillbilly Music Re-imagined," 159–60.

12. D. K. Wilgus, "The Hillbilly Movement," in *Our Living Traditions: An Introduction to Folklore*, ed. Tristram Potter Coffin (New York: Basic Books, 1968), 263.

13. While I agree with Kenney's assertion here, I'd also add that power in the North was dispersed and uneven. The upstart labels of the industrial Midwest did not have the same depth and breadth as those better-established East Coast labels like Victor and Columbia. But in comparison to those players (coming from the rural and urban South as well as from the rural and urban Midwest), the power differential between label executives and engineers vs. the musicians cutting the records was clear. See Kenney, *Recorded Music in American Life*, 135–36.

14. Malone, *Country Music, U.S.A.*, 31.

15. Cecil Sharpe, *Folksongs from the Southern Appalachians* (Oxford, UK: Oxford University Press, 1974).

16. Wilgus, "The Hillbilly Movement," 267.

17. Malone, *Country Music, U.S.A.*, 31.

18. Malone, *Country Music, U.S.A.*, 31.

19. Malone, *Country Music, U.S.A.*, 31.

20. Malone, *Country Music, U.S.A.*, 31.

21. Malone, *Country Music, U.S.A.*, 31.

22. Clifford Doerksen, *American Babel: Rogue Radio Broadcasters of the Jazz Age* (Philadelphia: University of Pennsylvania Press, 2005), 72.

23. Doerksen, *American Babel*, 72.

24. *Consider Fifty Million Customers: A Study of the Rural Section of the American Market* (Spencer, IN: Farm Life, ca. 1920s), 39, 44.

25. "Rural Residents Average Well As Music Lovers," *Talking Machine World*, December 15, 1920, 66.

26. "Much later," Gene Wiggins writes, "Brockman thought the resultant recordings were the first commercial recordings made in the South. This was not quite true. For instance, Brunswick had been in Atlanta to record Sacred Harp music in September, 1922." See Wiggins, "John Carson: Early Road, Radio, and Records," 30.

27. Wiggins, "John Carson: Early Road, Radio, and Records," 36.

28. Jeffrey Lange, *Smile When You Call me a Hillbilly*, 23.

29. Barlow, "*Looking Up at Down*," 194–95.

30. Wiggins, "John Carson: Early Road, Radio, and Records." Different ac-

counts have Brockman jotting down a bit more than this, but it's safe to say that Brockman would have jotted down Carson's name.

31. OKeh expedition notice, *Atlanta Journal*, June 15, 1923, 4, in Green, "Hillbilly Source and Symbol," Archie Green Collection, https://exhibits.lib.unc.edu/exhibits/show/hillbilly_music/green_hillbilly (SFC-UNC).

32. Lange, *Smile When You Call Me a Hillbilly*, 23.

33. "Fiddlin' John Carson OKeh Records Popular in Trade," *Talking Machine World*, April 15, 1925, 56.

34. Lange, *Smile When You Call Me a Hillbilly*, 23.

35. Wiggins, "John Carson: Early Road, Radio, and Records," 30.

36. Wiggins, "John Carson: Early Road, Radio, and Records," 60.

37. Wiggins, "John Carson: Early Road, Radio, and Records," 60.

38. Ralph Peer, quoted in Greil Marcus, *The Old, Weird America: The World of Bob Dylan's Basement Tapes* (New York: Picador/Henry Holt, 2001), 120–21. Gene Wiggins disputes Peer's account: "Pressings were made from two different takes. The one issued in the smallest quantity, apparently, carried the words 'Recorded in Atlanta.' Probably both takes were recorded there. There is the story of John being called to New York to re-do the record. Peer thought he was, but Polk Brockman is certain this did not happen." See Wiggins, "John Carson, Early Road, Radio, and Records," 30.

39. Country music history has given less attention to Samantha Bumgarner and Eva Davis' 1924 Columbia Records recording session, which made them one of the first women to record old-time music sound recordings. Banjoist Bumgarner's musical career in her native North Carolina took her up through the folk music revival of the 1950s (she died in 1960). Much less is known about fiddle player Eva Davis. See Robert K. Oermann, "Samantha Bumgarner," in *The Encyclopedia of Country Music*. Paul Kingsbury, ed. (New York: Oxford University Press, 1998): 65–66, and "Samantha Bumgarner: Newspaper Highlights (1924–1960)," *Mainspring Press* blog (78records.wordpress.com).

40. Malone, *Country Music, U.S.A.*, 38.

41. Malone, *Country Music, U.S.A.*, 38.

42. Walter Darrell Haden, "Vernon Dalhart: His Rural Roots and the Beginnings of Commercial Country Music," *ARSC Journal*, 3, no. 1 (Winter 1970/71): 6.

43. "Henry Whitter, OKeh Artist, Real Hill Country Type," *Talking Machine World*, May 15, 1925, 35. In the same issue, in an article touting OKeh's latest "Blue Book of Blues" catalog, a new "old time music" catalog is mentioned (LOC).

44. Richard Peterson, *Creating Country Music: Fabricating Authenticity* (Chicago: University of Chicago Press, 1997), 75.

45. John K. MacKenzie, *Gennett Records of Old Time Tunes: A Catalog Reprint*, JEMF Special Series, No. 6 (Los Angeles: John Edwards Memorial Foundation at the Folklore and Mythology Center, University of California at Los Angeles, 1975).

46. MacKenzie, *Gennett Records of Old Time Tunes.*

47. *OKeh Truetone Old Time Tunes*, catalog, 1926. Jim Walsh collection, Recorded Sound Reference Center (LOC).

48. Malone, *Country Music U.S.A.*, 55.

49. Wilgus, "The Hillbilly Movement," 267.

50. Wilgus, "The Hillbilly Movement," 263.

51. "Hill Billies' OKeh Records," *Talking Machine World*, April 15, 1925, 50d. Just below this article, Ralph Peer's scouting trips in the South are recognized: "R. S. Peer on Extended Trip." "Among other activities, he has had the pleasure of being a judge at an Old-Time Fiddler's Convention in Knoxville, Tennessee."

52. Archie Green, "Hillbilly: Source and Symbol" https://exhibits.lib.unc.edu/exhibits/show/hillbilly_music/green_hillbilly (SFC-UNC).

53. Michele Hilmes, *Radio Voices: American Broadcasting, 1922–1952* (Minneapolis: University of Minnesota Press, 1997), 105.

54. Ward Seeley, "Will the Great Artists Continue?" *The Wireless Age*, June 25, 1923, 22.

55. Sutton and Nauck, *American Record Labels and Companies*, 78.

56. Seeley, "Will the Great Artists Continue?" 22.

57. Elena Razoglova, *The Listener's Voice: Early Radio and the American Public* (Philadelphia: University of Pennsylvania Press, 2011), 3.

58. Wilgus, "The Hillbilly Movement," 268.

59. Wilgus, "The Hillbilly Movement," 266–67.

60. Wayne Daniel, "Old Time Fiddlers' Contests on Early Radio," *JEMF Quarterly* 17, no. 63 (Fall 1981): 159.

61. Cohen, *Making a New Deal*, 138.

62. Cohen, *Making a New Deal*, 138.

63. Wilgus, "The Hillbilly Movement," 268.

64. Diane Pecknold, *The Selling Sound: The Rise of the Country Music Industry* (Durham, NC: Duke University Press, 2007), 15.

65. Susan Smulyan, *Selling Radio: The Commercialization of American Broadcasting, 1920–1934* (Washington, DC: Smithsonian Institution Press, 1996), 23.

66. Smulyan, *Selling Radio*, 26.

67. Peterson, *Creating Country Music*, 100.

68. "Hill-Billy Music," *Variety*, December 29, 1926, 1, in Peterson, *Creating Country Music*, 68–69.

69. Peterson, *Creating Country Music*, 67.

70. Peterson, *Creating Country Music*, 70.

71. Peterson, *Creating Country Music*, 70.

72. Wilgus, "The Hillbilly Movement," 265.

73. Charles Wolfe, "Nashville and Country Music, 1926–1930: Notes on Early Nashville Media and Its Response to Old-Time Music," *Journal of Country Music* IV, no. 1: 4.

74. Wolfe, "Nashville and Country Music, 1926–1930," 3.

75. Steve Craig, "'The Farmer's Friend': Radio Comes to Rural America, 1920–1927," *Journal of Radio Studies* 8, no. 2 (2001): 330–46.

76. Jerome D. Traver and Joel M. Maring, "*Stand By*: Journalistic Response to a Country Music Radio Audience," *JEMF Quarterly* 14, no. 71 (Autumn 1983): 150.

77. Malone, *Country Music, U.S.A.*, 70.

78. Charles Wolfe, *A Good Natured Riot: The Birth of the Grand Ole Opry* (Nashville: Country Music Foundation/Vanderbilt University Press, 1999), 9.

79. Chet Hagan, *Grand Ole Opry* (New York: Henry Holt and Company, 1989), 5.

80. Wolfe, *A Good Natured Riot*, 9.

81. Nashville *Tennessean*, December 27, 1925, in Wolfe, *A Good Natured Riot*, 11.

82. WSM Press Release, in Wolfe, *A Good Natured Riot*, 12.

83. WSM Press Release, in Wolfe, *A Good Natured Riot*, 12.

84. Nashville *Tennessean*, December 27, 1925, in Wolfe, *A Good Natured Riot*, 21.

85. Chet Hagan, *Grand Ole Opry*, 24.

86. Wolfe, *A Good Natured Riot*, 21.

87. Wolfe, *A Good Natured Riot*, 21–22.

88. Richard Peterson and Paul DiMaggio, "The Early Opry: Its Hillbilly Image in Fact and Fancy," *Journal of Country Music* 4 (1974): 39–40.

89. William R. McDaniel and Harold Seligman, *Grand Ole Opry* (New York: Greenburg, 1952), 20.

90. Peterson and DiMaggio, "The Early Opry," 41.

91. Peterson and DiMaggio, "The Early Opry," 41.

92. Peterson and DiMaggio, "The Early Opry," 41.

93. Peterson, *Creating Country Music*, 80.

94. Wolfe, "Nashville and Country Music, 1926–1930," 2–16.

95. Peterson and DiMaggio, "The Early Opry," 43.

96. Peterson and DiMaggio, "The Early Opry," 43.

97. In Peterson and DiMaggio's "The Early Opry" article, they cite what may be the most outlandish tongue-in-cheek crafting of the hillbilly origins narrative. KMPC Beverly Hills announcer Glen Rice claimed to find "a group of wild musicians" in 1928, up in the Santa Monica mountains. They were a group of Los Angeles musicians that Rice dubbed "The Beverly Hill Billies." See Peterson and DiMaggio, "The Early Opry," 43.

98. Charles Wolfe, "The Legend that Peer Built," in Charles K. Wolfe and Ted Olson, *The Bristol Sessions: Writings about the Big Bang of Country Music* (Jefferson, NC: McFarland, 2005), 19.

99. Eric Morritt, "Early Sound Recording Technology and the Bristol Sessions," in Charles K. Wolfe and Ted Olson, *The Bristol Sessions: Writings about the Big Bang of Country Music* (Jefferson, NC: McFarland, 2005), 7.

100. John Lilly, "Jimmie Rodgers and the Bristol Sessions," in Charles K. Wolfe and Ted Olson, *The Bristol Sessions: Writings about the Big Bang of Country Music* (Jefferson, NC: McFarland, 2005), 65.

101. Mazor, *Ralph Peer and the Making of Popular Roots Music*, 71–72.

102. Mazor, *Ralph Peer and the Making of Popular Roots Music*, 74–75.

103. Although no scout had been through the area, musicians from the region had traveled to studios elsewhere to record for Gennett, OKeh, Vocalion, Brunswick, Victor, and other labels. For a more detailed list of artists, see Wolfe, "The Legend that Peer Built," 21.

104. Wolfe, "The Legend that Peer Built," 18.

105. Wolfe, "The Legend that Peer Built," 18.

106. Wolfe, "The Legend that Peer Built," 19.

107. This is not to say that the economic hardships of Appalachian life, or rural life in general, disappeared with the old hill culture. That poverty is still very much with us.

108. Malone, *Country Music, U.S.A.*, 21.

109. Kenney, *Recorded Music in American Life*, 140.

110. Mark Zwonitzer and Charles Hirshberg. *Will You Miss Me When I'm Gone?: The Carter Family and Their Legacy in Music* (New York: Simon & Schuster, 2002), 101.

111. Zwonitzer and Hirshberg. *Will You Miss Me When I'm Gone?*, 112.

112. *More and More and More*, New Victor Records Southern Series (order blank and advance list), July 20, 1928 (CCHS).

113. *National Listing for Record Number 21245*, letter from Victor Talking Machine Company to its retail dealers, May 24, 1928 (CCHS).

114. *From Singing Brakeman to Popular Exclusive Victor Artist* and *National Listing for Record Number 21245*, letter from Victor Talking Machine Company to its retail dealers, May 24, 1928. (CCHS)

115. The letter also mentioned another early release by Tal Henry and His North Carolinians, "Lonesome" (a foxtrot) and "I'd Trade My Air Castles For a Love Nest and You" (a waltz). See *Subject: Early Shipment August 3rd Records*, letter from Victor Talking Machine Company to Victor wholesalers, July 11, 1928 (CCHS).

116. The musical short can be seen on *Times Ain't Like They Used to Be: Early Rural & Popular American Music* (Yazoo DVD).

117. Mazor, *Ralph Peer and the Making of Popular Roots Music*, 144.

118. *New Victor International Records*, Victor Advance Lists, August and September 1928 (CCHS).

119. Mazor, *Ralph Peer and the Making of Popular Roots Music*, 124.

120. Mazor, *Ralph Peer and the Making of Popular Roots Music*, 124.

121. Mazor, *Ralph Peer and the Making of Popular Roots Music*, 124.

122. Mazor, *Ralph Peer and the Making of Popular Roots Music*, 125–26.

123. Mazor, *Ralph Peer and the Making of Popular Roots Music*, 126.

124. Mazor, *Ralph Peer and the Making of Popular Roots Music*, 127–28. Peer's later scouting work that took him to Mexico, beginning in 1928, would have profound impact and deserves much more attention than it receives here or elsewhere. Peer's deals from OKeh to Victor to RCA Victor shifted several times, reflecting the record man's increased emphasis on popular song and publishing and on the international market for music. See Mazor, *Ralph Peer and the Making of Popular Roots Music*, 160.

125. Smulyan, *Selling Radio*, 24.

126. Smulyan, *Selling Radio*, 24.

127. "West Virginia Hills Are in the Bronx, Says Barn Barnum," *Billboard*, March 4, 1939, in Nick Tosches, *Country: Living Legends and Dying Metaphors in America's Biggest Music*, rev ed. (New York: Charles Scribner's Sons, 1985), 157.

128. Joli Jensen, *The Nashville Sound: Authenticity, Commercialization, and Country Music* (Nashville: Vanderbilt University Press, 1998), 160.

129. Stuart Hamblen and Suzy Hamblen, interview with Douglas B. Green, Los Angeles, 1974, Country Music Foundation Oral History Project, Country Music Hall of Fame (CMHF).

Chapter 4

1. Alexander Russo, *Points on the Dial: Golden Age Radio Beyond the Networks* (Durham, NC: Duke University Press, 2010), 2.

2. Writing about early cinema, Gaudreault notes, "I believe it is essential for anyone interested in understanding how the cinema came to be and how it became an autonomous media institution to take this alternate route through the intermedial relationships of cinema at its conception. The question of the 'intermediality' of early cinema has been for some time on the agenda for research in this field." André Gaudreault, "The Diversity of Cinematographic Connections in the Intermedial Context of the Turn of the 20th Century," in *Visual Delights: Essays on the Popular and Projected Image in the 19th Century*, ed. Simon Popple and Vanessa Toulmin, trans. Wendy Schubring (Trowbridge, UK: Flick Books, 2000), 14.

3. "On With the Dance," *Voice of the Victor*, July–December, 1928, 8–9, 13 (CCHS).

4. "On With the Dance," *Voice of the Victor*, July–December, 1928, 8–9, 9 (CCHS).

5. Spring, *Saying It With Songs*, 39.

6. Spring, *Saying It With Songs*, 40.

7. Timothy Taylor, *The Sounds of Capitalism: Advertising, Music, and the Conquest of Culture* (Philadelphia: University of Pennsylvania Press, 2012), 44.

8. S. R. Winters, "An Interview with Secretary H. C. Hoover," *Radio News* 6, October 1924, 568.

9. Spring, *Saying It With Songs*, 21.

10. J. Fred MacDonald, *Don't Touch That Dial: Radio Programming in American Life, 1920–1960* (Chicago: Nelson-Hall, 1979), 13.

11. "Radio v. Phonograph," *Time*, October 4, 1926, 38.

12. Razoglova, *The Listener's Voice*, 34.

13. "Thos. A. Edison, Inc., Not to Enter Radio Field," *Talking Machine World*, January 15, 1925, 182.

14. "Tommy's Only Foolin'," *Radio Digest*, October 15, 1926, 14.

15. "Al Jolson Signs Contract to Record Exclusively for the Brunswick Co.," *Music Trade Review*, February 21, 1924, 60.

16. "Al Jolson Signs Contract to Record Exclusively for the Brunswick Co.," 60.

17. McGovern, *Sold American*, 48–49.

18. As Michele Hilmes notes, nonpayment of artists, particularly for phonograph recordings, would soon become an issue. See Hilmes, *Radio Voices*, 51–52.

19. "Radio Helps to Sell Your Victor Records," *Voice of the Victor*, February 1924, 24. RCA Victor Collection (CCHS).

20. "Radio Helps to Sell Your Victor Records."

21. David Ewen, *All the Years of American Popular Music* (Englewood Cliffs, NJ: Prentice Hall, Inc.,), 283.

22. Ward Seeley, "Yes, the Radiofan Buys Records!" *The Wireless Age*, August 1923, 23–26.

23. McGovern, *Sold American*, 53.

24. P. H. Pumphrey, "Choosing the Program Idea," *Broadcast and Broadcast Advertising*, 1931, 40, in Taylor, *The Sounds of Capitalism*, 26.

25. As Roland Gelatt wrote, "They proposed to pay approximately $28,000,000 for his 245,000 shares of Victor stock and to pay another $12,000,000 for the remaining shares held by various individuals (Emile Berliner among them). . . . On December 7, 1926, the agreement was signed." See Gelatt, *The Fabulous Phonograph*, 246.

26. Gelatt, *The Fabulous Phonograph*, 247.

27. Sanjek, *Pennies from Heaven*, 11–12, 66.

28. Mazor, *Ralph Peer and the Making of Popular Roots Music*, 129.

29. Mazor, *Ralph Peer and the Making of Popular Roots Music*, 129.

30. Gelatt, *The Fabulous Phonograph*, 247–48.

31. Sanjek, *Pennies from Heaven*, 71.

32. Sanjek, *Pennies from Heaven*, 69.

33. "A Glance into the Trade Future," *Talking Machine World*, January 1928, 34b.

34. "Workers Should Profit in 1928," *Talking Machine World*, January 1928, 76; and Spring, *Saying It With Songs*, 39.

35. Sanjek, *Pennies from Heaven*, 62–64.

36. Sanjek, *Pennies from Heaven*, 64.

37. Sutton, *Recording the Twenties*, 255.

38. DeGraaf, *Edison and the Rise of Innovation*, 119.

39. Kennedy, *Jelly Roll, Bix and Hoagy*, 163–65.

40. Kennedy, *Jelly Roll, Bix and Hoagy*, 127–28.

41. Schiffer noted that portable radio technologies did not go away with the arrival of domestic AC radio platforms but went through its own technological and cultural designs and redesigns over time. See Michael Brian Schiffer, *The Portable Radio in American Life* (Tucson: University of Arizona Press, 1992), 49.

42. Schiffer, *The Portable Radio in American Life*, 51.

43. Schiffer, *The Portable Radio in American Life*, 51.

44. Schiffer, *The Portable Radio in American Life*, 90.

45. Razoglova, *The Listener's Voice*, 55.

46. Michele Hilmes, *NBC: America's Network* (Berkeley: University of California Press, 2007), 7.

47. Taylor, *The Sounds of Capitalism*, 6. Taylor's vital research misses an opportunity in starting with radio and advertising. There is a largely unexamined history of advertising via phonograph in the era previous to and concurrent with radio's rise. One example: the NuGrape Twins' 1926 Columbia recording, "I Got Your Ice Cold NuGrape," a transcendent performance in service of selling a mundane product: grape soda.

48. Sanjek, *Pennies from Heaven*, 86; See also Ted Cox, *Sold on Radio: Advertisers in the Golden Age of Broadcasting* (Jefferson, NC: McFarland, 2008), 85.

49. Sanjek, *Pennies from Heaven*, 86.

50. Allison McCracken, "Real Men Don't Sing Ballads: The Radio Crooner in Hollywood, 1929–1933," in *Soundtrack Available: Essays on Film and Popular Music*, ed. Pamela Robertson Wojcik and Arthur Knight (Durham, NC: Duke University Press, 2001), 105. See also her book-length analysis of singing, masculinity, and culture, *Real Men Don't Sing: Crooning in American Culture*. (Durham, NC: Duke University Press, 2015).

51. McCracken, "Real Men Don't Sing Ballads," 106. While by the mid-1930s crooners had slid into a comfortable familiarity, McCracken sees this initial concern over crooning to constitute a "history that reveals larger concerns about the mass media's power to define and shape social life, in this case the sound of white masculinity." Vallee's career began in New York City, where he and his band (the Yale Collegians) had a residency at the city's Heigh-Ho Club. His initial recording attempts were disappointing (he released his first record, "Japanese Sunset," through Columbia's Personal Record Department in 1921; later that summer, he was rejected by Edison). Though it took Vallee and his band a number of years to garner a reputation at the Heigh-Ho Club, Victor eventually offered them the chance to record in 1928, and he recorded for the company into the next decade.

52. Sanjek, *Pennies from Heaven*, 117–18.

53. Sanjek, *Pennies from Heaven*, 66.

54. Spring, *Saying It With Songs*, 46.

55. Spring, *Saying It With Songs*, 46.

56. Sanjek, *Pennies from Heaven*, 78.

57. William M. Randle Jr., "A History of Radio Broadcasting and Its Social and Economic Effect on the Entertainment Industry" (PhD diss., Case Western Reserve University, Cleveland, OH, 1966), 368, in John Ryan, *The Production of Culture in the Music Industry: The ASCAP-BMI Controversy* (Lanham, MD: University Press of America, 1985), 31.

58. Ryan, *The Production of Culture in the Music Industry*, 32.

59. Ryan, *The Production of Culture in the Music Industry*, 34–35.

60. Samuels, *The Illustrated Story of Copyright*, 43.

61. Sanjek, *Pennies from Heaven*, 84.

62. Sanjek, *Pennies from Heaven*, 86.

63. Spring, *Saying It With Songs*, 49.

64. Spring, *Saying It With Songs*, 50–51.

65. Spring, *Saying It With Songs*, 51–53.

66. Spring, *Saying It With Songs*, 53.

67. Spring, *Saying It With Songs*, 53–54.

68. Some recording industry talent scouts of the era, including Paramount's Art Satherley, would refer to signing talent to exclusive contracts as "tying them up." See Barnett, "Record Men," 129.

69. "Exploit-O-Grams: 'Reckless Romance,'" *The Film Daily*, July 2, 1928, 7 (MHDL).

70. "Exploit-O-Grams: 'Chicken A La King,'" *The Film Daily*. July 8, 1928, 8 (MHDL).

71. Spring, *Saying It With Songs*, 53–60.

72. Spring, *Saying It With Songs*, 54.

73. Spring, *Saying It With Songs*, 59.

74. Spring, *Saying It With Songs*, 54.

75. Spring, *Saying It With Songs*, 54.

76. Sanjek, *Pennies from Heaven*, 106.

77. James P. Kraft, "Musicians in Hollywood: Work and Technological Change in Entertainment Industries," *Technology and Culture* 35, no. 2 (1994): 290.

78. Spring, *Saying It With Songs*, 61.
79. Spring, *Saying It With Songs*, 62–63.
80. Spring, *Saying It With Songs*, 106.
81. Spring, *Saying It With Songs*, 39.
82. Steve Wurtzler, *Electric Sounds: Technological Change and the Rise of Corporate Mass Media* (New York: Columbia University Press, 2007), 1.
83. Gelatt, *The Fabulous Phonograph*, 219.
84. Gelatt, *The Fabulous Phonograph*, 222.
85. Gelatt, *The Fabulous Phonograph*, 223.
86. Benjamin L. Aldridge, *The Victor Talking Machine Company*, ed. L. Frederic Bayh, David Sarnoff Library, College of New Jersey, http://www.davidsarnoff.org/vtm.html (DSL-CNJ).
87. M. C. Rypinski, "Will the Jobber Survive in Radio?" *Electrical World*, January 20, 1925, 1368.
88. "Will the Phonograph Industry Stage a Sensational Comeback?" *Sales Management*, November 14, 1925, 639.
89. A. W. Hilton, "How Victor Is Forcing Sales in Cities and Trade Centres," *Printers' Ink*, March 19, 1925, 17.
90. Hilton, "How Victor Is Forcing Sales in Cities and Trade Centres," 18.
91. Hilton, "How Victor Is Forcing Sales in Cities and Trade Centres," 18.
92. Hilton, "How Victor Is Forcing Sales in Cities and Trade Centres," 17.
93. Will the Phonograph Industry Stage a Sensational Comeback?" *Sales Management* 9, November 14, (1925), 639.
94. Hilton, "How Victor Is Forcing Sales in Cities and Trade Centres," 17.
95. "Will the Phonograph Industry Stage a Sensational Comeback?" 639.
96. Sutton, *Recording the Twenties*.
97. "Phonograph and Radio Industries Pool Their Interests," *The Magazine of Wall Street*, January 2, 1926, 424.
98. "Phonograph and Radio Industries Pool Their Interests," 424.
99. "Phonograph and Radio Industries Pool Their Interests," 425.
100. Victor advertisement, *McCall's Magazine*, September 1927, 3, in Emily Thompson, "Machines, Music, and the Quest for Fidelity: Marketing the Edison Phonograph in America, 1877–1925," *The Musical Quarterly* 79, no. 1: 131–71. See also Thompson's *The Soundscape of Modernity*, Cambridge, MA: The M.I.T. Press, 2004.
101. Advertisements in *Talking Machine World*'s October 15, 1925 issue: Victor Orthophonic, 18; Brunswick Panatrope, 6; Canvassing article, 8.
102. "Microphones, Part 2: The Electrical Era," Audio Engineering Society, http://www.aes-media.org/historical/html/recording.technology.history/microphones2.html.
103. Shawn VanCour, *Making Radio: Early Radio Production and the Rise of Modern Sound Culture* (New York: Oxford University Press, 2018), 71.
104. VanCour, *Making Radio*, 73, 82.
105. Wurtzler, *Electric Sounds*, 3.
106. Russo, *Points on the Dial*, 78.
107. Russo, *Points on the Dial*, 79.

108. Russo, *Points on the Dial*, 81.

109. Competing technologies included Fox's Movietone, Lee DeForest's Phonofilm, and the lesser known Pallo Phonophone, from General Electric. See Russo, *Points on the Dial*, 83.

110. Russo, *Points on the Dial*, 85.

111. Russo, *Points on the Dial*, 88.

112. Michael Biel, "The Making and Use of Recordings in Broadcast Before 1935" (PhD diss., Northwestern University, 1977), 593–94.

113. Russo, *Points on the Dial*, 97–98. For a listing of Federal Radio Commission orders from this era, see "Index to General Orders of the Federal Radio Commission as Published in the Radio Service Bulletin," which originally appeared in *Radio Service Bulletins*, August 31, 1931, https://www.fcc.gov/sites/default/files/radio-service-bulletin-FRC-general-orders-list.pdf.

114. Russo, *Points on the Dial*, 97–101.

115. Michele Hilmes, *Hollywood and Broadcasting* (Urbana: University of Illinois Press, 1990), 49.

116. Susan Schmidt Horning, "Chasing Sound: The Culture and Technology of Recording Studios in America, 1877–1977" (PhD diss., Case Western Reserve University, 2002), 52. See also her book, *Chasing Sound: Technology, Culture, and the Art of Studio Recording from Edison to the LP.* Baltimore: Johns Hopkins University Press, 2013.

117. Horning, "Chasing Sound."

118. Wurtzler, *Electric Sounds*, 68. There is a much larger story to tell regarding the film industry's interest in music publishing in the talkies era. Cory Messenger's "Calling the Tune: Hollywood and the Business of Music" maps the importance of music publishing as well the recording industry across the twentieth century, but chapters 2 and 3 on the film industry's purchases in music publishing during the 1920s and 1930s are especially useful. See Cory Messenger, "Calling the Tune: Hollywood and the Business of Music" (PhD diss., Griffith University, 2011), https://research-repository.griffith.edu.au

119. Film-radio cross-promotional partnerships have not always been recognized in either film or radio studies. Michele Hilmes debunked the notion of a 1932 ban on film stars performing on radio, an anecdote that she suggests became a misleading "metonym for the entire relationship" between radio and film, which was increasingly close during the 1930s, marked by both collaborations and conflicts during the era. See Michele Hilmes, "The 'Ban' That Never Was: Hollywood and the Broadcasting Industry in 1932," *The Velvet Light Trap* 23 (Spring 1989): 39–47.

120. Hilmes, "The 'Ban' That Never Was," 33–34.

121. "Radio Review of 1927 and Hopes for the New Year," *New York Times*, January 1, 1928, 12.

122. Wurtzler, *Electric Sounds*, 33–34.

123. "The Real Publicity Value of the Dodge Radio Hour," *Variety*, January 11, 1928, 53.

124. "The Real Publicity Value of the Dodge Radio Hour," 56. See also "Movie Stars Go On Air," *St. Petersburg Times*, March 25, 1928, 9.

125. Whiteman and McBride, *Jazz*, 3.

126. Whiteman and McBride, *Jazz*, 137–58.
127. Whiteman and McBride, *Jazz*, 20.
128. Rayno, *Paul Whiteman*, 189.
129. Rayno, *Paul Whiteman*, 189.
130. Rayno, *Paul Whiteman*, 189–90.
131. Rayno writes, "When Victor realized that Whiteman wasn't bluffing and that he intended to jump ship, they panicked and offered him $100,000 per year to stay. But it was too late. His mind was made up. Columbia celebrated by taking out large ads in some two hundred newspapers nationwide. All Victor could do now was to book Whiteman for as many sessions as possible in order to stockpile a large body of records for release after the bandleader's departure." Rayno, *Paul Whiteman*, 189–90.
132. Sanjek, *Pennies from Heaven*, 71. Record collectors reference these Whiteman artist-label releases by calling them "potato head" labels, after the somewhat unflattering likeness of Whiteman. Other Columbia artists (and at least one of those for Columbia's OKeh subsidiary) had artist labels as well, including Rudy Vallee, Ted Lewis, and Seger Ellis. None of these, however, had the lavish five-color printing that appeared on Whiteman's discs. For pictures and descriptions of these releases, see Mike Sherman and Kurt Nauck, *Note the Notes: An Illustrated History of the Columbia Record Label, 1901–1958* (New Orleans: Monarch Record Enterprises, 1998).
133. "Paul Whiteman Signs to Record Exclusively for Columbia Co.," *Talking Machine World*, May 1928, 34–99 (LOC).
134. "Big News for Every Columbia Dealer" advertisement, *Talking Machine World*, May 1928. The multipage advertisement is unpaginated but appears between pages 18 and 19 (LOC).
135. "Big News for Every Columbia Dealer" (LOC).
136. W. H. Clock, "How Columbia Sold Whiteman," *Talking Machine World*, September 9, 1928, 24 (LOC).
137. Whiteman hired away Bill Challis, Frank Trumbauer, Bix Beiderbecke, Joe Venuti, and Tommy and Jimmy Dorsey, all from Jean Goldkette's Detroit-based band. See Joshua Berrett, *Louis Armstrong and Paul Whiteman: Two Kings of Jazz* (New Haven, CT: Yale University Press, 2004), 87.
138. As Beiderbecke's health declined due to alcoholism and pneumonia, to the point that he could no longer perform, Whiteman kept him on the payroll. Berrett, *Louis Armstrong and Paul Whiteman*, 87–88.
139. Berrett, *Louis Armstrong and Paul Whiteman*, xiv.
140. Rayno, *Paul Whiteman*, 226–31.
141. See Paul Whiteman researcher and fan Albert Haim's "Paul Whiteman in Los Angeles: The Summer of 1929," posted to the Bixography discussion forum, http://www.network54.com/Forum/27140. Haim's post includes a reproduction of a Western Union telegram to theater/film critic and film producer Mark Hellinger and his wife Gladys. The telegram is from the private collection of fellow Whiteman aficionado Neil Lipes. http://www.network54.com/Forum/27140/message/1399144397/Paul+Whiteman+in+Los+Angeles-+the+Summer+of+1929. See also James Layton and David Pierce's detailed re-appraisal of Whiteman's feature

film debut, *King of Jazz: Paul Whiteman's Technicolor Revue*. (Severn, MD: Media History Press, 2016).

142. Bernard Dick, *City of Dreams; The Making and Remaking of Universal Pictures* (Lexington: University Press of Kentucky, 1997), 84–85.

143. Dick, *City of Dreams*, 77–78, 85.

144. James Layton and David Pierce, *King of Jazz: Paul Whiteman's Technicolor Revue*. Severn, MD: Media History Press, 2016.

145. Layton and Pierce, *King of Jazz: Paul Whiteman's Technicolor Revue*. Severn, 2016: 182-187.

146. Waller and his coauthors had originally developed the revue for Connie's Inn, where it originally debuted in 1926. See Aberjhani and Sandra L. West, "Hot Chocolates," *Encyclopedia of the Harlem Renaissance* (New York: Facts on File, 2003), 157–58.

147. Berrett, *Louis Armstrong and Paul Whiteman*, 97.

148. Walter Winchell, "On Broadway," *The New York Mirror*, March 14, 1930.

149. Sanjek, *Pennies from Heaven*, 117.

150. *I'll Be Glad When You're Dead and Gone You Rascal You*, directed by Dave Fleischer (Paramount, 1932). Short Film.

151. *A Rhapsody in Black and Blue*, directed by Aubrey Scotto (Paramount, 1932). Short Film.

152. *A Rhapsody in Black and Blue*.

153. *A Rhapsody in Black and Blue*.

154. William Howland Kenney, "Negotiating the Color Line: The Autobiographies of Louis Armstrong," in *Jazz in Mind: Essays on the History and Meanings of Jazz*, ed. Reginald T. Buckner and Steven Weiland (Detroit: Wayne State University Press, 1991), 38–59. See also Krin Gabbard, *Jammin' at the Margins: Jazz and the American Cinema* (Chicago: University of Chicago Press, 1996), 215.

155. Joshua Berrett, *Louis Armstrong and Paul Whiteman*, 114.

156. Thomas Brothers, ed., *Louis Armstrong in His Own Words* (New York: Oxford University Press, 2001), 147; and Ben Alexander, "'For Posterity': The Personal Audio Recordings of Louis Armstrong," *American Archivist* 71, no. 1 (Spring-Summer, 2008): 64.

157. Joseph Willems, *All of Me: The Complete Discography of Louis Armstrong* (Oxford, UK: Scarecrow Press, 2006), 103.

158. MacDonald, *Don't Touch that Dial*, 327.

159. William Barlow, "Black Music on Radio in the Jazz Age," *African-American Review* 50, no. 4 (Winter 2017): 325.

160. William Barlow, *Voiceover: The Making of Black Radio* (Philadelphia: Temple University Press, 1999), 94.

161. Nolan Porterfield, *Jimmie Rodgers: The Life and Times of America's Blue Yodeler* (Urbana: University of Illinois Press, 1992), 259.

162. Porterfield, *Jimmie Rodgers*, 409.

163. Sanjek, *Pennies from Heaven*, 117.

164. For many years, whether Louis or Lil Hardin-Armstrong played on these tracks was unknown. There is still some conjecture on how Rodgers and the Armstrongs got together for these sessions; biographers Max Jones and John Chilton

called their collaboration one of "jazz's unsolvable riddles." See Jones and Chilton, *Louis: The Louis Armstrong Story* (Boston, Little, Brown, 236). More recently, Porterfield dismisses this: "One may assume that Rodgers's very brief association with the great Satchmo was merely a natural result of Peer's acquaintance with both performers. Armstrong only recently moved to the West Coast and was anxious to get whatever work he could find," while noting Ralph Peer's boast that he "invented Louis Armstrong" while at OKeh points to Peer as the likely go-between that brought the two artists together. (See Porterfield, *Jimmie Rodgers*, 258). In Louis Armstrong's remembrance, the story was simpler: "I'd been knowing Jimmie for a *long* time. We met one morning, and he said 'Man, would like sing some blues with me?' And I said, 'Okay, daddy, sing some blues, and I'm gonna play behind you.'" This quote from Louis Armstrong's 1970 television appearance on the *Johnny Cash Show* comes from Barry Mazor, *Meeting Jimmie Rodgers: How America's Original Roots Music Hero Changed The Pop Sounds of a Century* (New York: Oxford University Press, 2009), 7. For the complete list of songs Armstrong and Armstrong recorded with Rodgers, see Porterfield, *Jimmie Rodgers*, 404–9.

165. Porterfield, *Jimmie Rodgers*, 404–9.

166. Dan Morgenstern, liner notes to *Louis Armstrong: Portrait of the Artist as a Young Man, 1923–1934* (Columbia-Legacy, 1994) (CD), in Patrick Huber, "Black Hillbillies," 56–57.

167. It is amazing to think of the depth and breadth of Rodgers' impact given his short, six-year career.

168. Lilly, "Jimmie Rodgers and the Bristol Sessions," 57.

169. As so often happened for Rodgers, one connection led to another: Rodgers knew XED owner Will Horwitz, a Houston theater owner. See Porterfield, *Jimmie Rodgers*, 270, and Mazor, *Meeting Jimmie Rodgers*, 96.

170. Porterfield, *Jimmie Rodgers*, 211.

171. Rodgers was involved in "vaudefilm," a short-lived practice that involved "four to seven live acts, often traveling as a unit . . . along with the latest movie release." Mazor argues that the practice probably served to make movie fans more than retain vaudeville fans: "In the end, as the Depression sapped movie attendance, vaudefilm would be used as one more, no-cost-added attraction, like the regular dishware giveaways or bingo games." So to learn of Rodgers' foray into film may seem like an unlikely turn of events for those only familiar with his recording career, but it also reflected the increasing importance of cross-platform "tie-ups" as stars capitalized on emerging and transitioning media forms. Though this was the only filmic appearance of Rodgers' short life, Mazor rightly notes, "His potency as a screen personality . . . is evident." See Mazor, *Meeting Jimmie Rodgers*, 91–98.

172. Mazor, *Ralph Peer and the Making of Popular Roots Music*, 66–168.

173. Porterfield, *Jimmie Rodgers*, 163, 397.

174. In Peterson's virtuosic *Creating Country Music*, he notes, "To see the affinities between the music performed by whites and blacks one need go no further than the recorded work of Jimmie Rodgers." But the book's keystone contribution is his descriptions of Jimmie Rodgers' various personas: brakeman, vaudevillian, cowboy. See Peterson, *Creating Country Music*, 196, 33–52.

175. Mazor, *Meeting Jimmie Rodgers*, 87.

176. Mazor, *Meeting Jimmie Rodgers*, 89.

177. Mazor, *Meeting Jimmie Rodgers*, 89.

178. Mazor, *Meeting Jimmie Rodgers*, 100, 102–3. Decades later came a now famous story that illustrates Rodgers' cultural reach: On a 1950 recording trip, ethnomusicologist Hugh Tracey and his colleagues recorded the Kipsigis and Nandi tribes on either side of what was known as the Kavirondo Valley in Kenya (then part of British colonial East Africa), who had been influenced by the recordings of the artist they knew as "Chemirocha." The musical performances Tracey and his group recorded "were simple, unpretentious and pentatonic," he wrote in his notebook, adding that they "experienced the full charm of their little wishbone-shaped lyres played songs of unusual poetic value." Tracey's framing of the Kipsigis' music in the rhetoric of authenticity then common to ethnomusicology was later undermined when it was understood that the songs featuring the mythic "Chemirocha" written by members of the tribe bore the influence of Jimmie Rodgers' recordings, particularly his voice and distinctive yodeling style. This discovery in turn was distorted into descriptions of the Kipsigis as a kind of cargo cult. But the enduring importance of Tracey's discovery is the tremendous cultural circulation of such recordings and their subsequent roles in influencing music making.

179. Mamie Smith, "Jailhouse Blues," Victor Records, *Discography of American Historical Recordings* (University of California-Santa Barbara), https://adp.library.ucsb.edu/index.php/matrix/detail/800023812/MVE-50603-Jailhouse_blues.

180. Donald Crafton, *The Talkies: American Cinema's Transition to Sound, 1926–1931* (Berkeley: University of California Press, 1997), 403.

181. Sanjek also notes that while Johnson's eventual decision to sell his shares was one he regretted, he avoided a dramatic devaluing of his holdings, "for sales in general continued to slip, eventually dropping by 60 percent in a single year, to $16.9 million in 1931." See Sanjek, *Pennies from Heaven*, 117.

182. Morritt, "Early Sound Recording Technology and the Bristol Sessions," 11–12.

183. F. W. Schnirring, "Does the Phonograph Need a New Copy Appeal?" *Printer's Ink*, October 28, 1926, 41–42.

184. Schnirring, "Does the Phonograph Need a New Copy Appeal?" 42.

185. David Wilkins, "Brunswick Goes Electric (1921–1926)," *Mainspring Press* (blog), http://www.mainspringpress.com/lightray.html.

186. "Columbia Co. Buys OKeh-Odeon Record Division of General Phonograph Corp.," *Talking Machine World*, January 1928, 18, in Ross Laird and Brian Rust, *Discography of OKeh Records, 1918–1934* (Westport, CT: Praeger, 2004), 5.

187. Robert J. Callen, "The Education and Tribulations of a Precursory Disc Recording Engineer," presented at the fortieth American Engineering Society (AES) Convention, April 27–30, 1971, Los Angeles, CA, http://www.aes.org/aeshc/docs/callen.education.and.tribulations.pdf.

188. Art Satherley, interview by Norm Cohen, G. Earle, with Ken Griffis and Bill Ward, June 12, 1971, Southern Folklife Collection, University of North Carolina–Chapel Hill (SFC-UNC).

189. van der Tuuk, *Paramount's Rise and Fall*, 154. Changes in the recording industry with the coming of electrical recording in this period bear some resemblance to

events in the film industry with the coming of synchronized sound, though the economic outcomes differed. Cecil B. De Mille wrote this in his introduction to Lester Cowan for the Academy of Motion Picture Arts and Sciences' *Recording Sound for Motion Pictures* (1931). "Two or three years ago the motion-picture industry visited the conjurers of the East. The electrical witch doctors made a brew—a strong one. It contained, among other ingredients, the horn of a radio, the needle of a phonograph, and some radio dust. All of these elements were moistened with a producer's tears and made into a record. The record was cremated in an electrical laboratory, the ashes wrapped into a film of celluloid and placed on the doorstep of the motion-picture industry." Just as Columbia and RCA Victor had lead the way in the recording industry, Warner Bros. and Fox were the innovators in the film industry. Film studios had to each make their deal for the new sound technology, though most had greater resources than those available to most record companies. The coming of sound not only enabled radio's rise and restructured the recording industry, it also restructured the film industry and concentrated its power, as Douglas Gomery has noted. See Douglas Gomery, *The Coming of Sound* (New York: Routledge, 2005), 6. The motion picture industry boomed with the coming of talkies and studios scrambled to adapt. In the pages of film exhibition magazines, they referred to the transition to sound as a "stampede." See Arthur James, "Sound? Well, Listen!," *Exhibitors' Film* Daily, December 1928, 1. A similar dynamic had occurred in the recording industry several years earlier with the coming of electrical recording, although the change from acoustic to electrical recording did not represent a full-scale sensory shift as it did in the film industry as it shifted from a sight-dominant medium to one in which synchronous sound increasingly mattered. The recording industry, after all, had been providing us sound all along. And while performance styles as well as recording styles would shift dramatically—most significantly with the introduction of the microphone—electrical recording contributed to a slight recovery for some companies while contributing to the demise of others. In general, the brief pre-Depression boom that "talkies" brought to the motion picture industry in 1928 and 1929 was not reflected in the recording industry's fortunes.

190. "Phonograph Sales Have Declined," *Wall Street Journal*, May 16, 1924, 6. The earlier recession is not well remembered, overshadowed by the dramatic Depression to come.

191. "Phonograph Sales Have Declined," 6.

192. Nathaniel Shilkret, *Nathaniel Shilkret: Sixty Years in the Music Business*, ed. Niels Shell and Barbara Shilkret (Lanham, MD: Scarecrow Press, 2005), 66.

193. Sanjek, *Pennies from Heaven*, 71.

194. Dane Yorke, "The Rise and Fall of the Phonograph," *The American Mercury*, September, 1932, 11.

195. Legal agreement between Columbia Phonograph Company, United Independent Broadcasters, and Columbia Phonograph Broadcasting System (undated), A.F.R. Lawrence Collection, Rodgers and Hammerstein Archives of Recorded Sound, New York Public Library (RH-NYPL).

196. Robert Toll, *The Entertainment Machine: American Show Business in the Twentieth Century* (New York: Oxford University Press, 1982), 53. For a listing of early radio shows in context, featuring time of day they aired and what thematic competi-

tion a given show faced, see Harrison Summers, *A Thirty-Year History of Programs Carried on National Radio Networks in the United States, 1926–1956* (New York: Arno Press/ New York Times, 1971).

197. Sanjek, *Pennies from Heaven*, 70.

198. Taylor, *Sounds of Capitalism*, 52.

199. Brooks, "High Drama in the Recording Industry," 47.

200. Sanjek, *Pennies from Heaven*, 66.

201. Sanjek, *Pennies from Heaven*, 68.

202. Alfred Chandler, *Scale and Scope: The Dynamics of Industrial Capitalism* (Cambridge, MA: Harvard University Press, 1994), 219.

Chapter 5

1. B. F. Aldridge, "Chapter Eight: 1921–1930," in *The Victor Talking Machine Company*. David Sarnoff Library, College of New Jersey (DSL-CNJ), http://www.davidsarnoff.org/vtm-chapter8.html

2. Charles McGovern, *Sold American: Consumption and Citizenship, 1890–1945*. Chapel Hill: University of North Carolina Press, 221.

3. Sanjek, *Pennies from Heaven*, 71–72.

4. Sanjek, *Pennies from Heaven*, 71–72.

5. Vincent Lopez, *Lopez Speaking: An Autobiography* (New York: Citadel Press, 1960), 251–52.

6. Eddie Cantor and David Freedman, *Yoo-Hoo, Prosperity!: The Eddie Cantor Five-Year Plan* (New York: Simon & Schuster, 1931).

7. See "Radio and Columbia Drop Plan to Unite," *New York Times*, May 8, 1928, 43; and "RCA to Acquire Columbia Company," *New York Times*, April 11, 1930, 25.

8. Jefferson Cowie, *Capital Moves: RCA's Seventy-Year Quest for Cheap Labor* (New York: The New Press, 1999), 14. Cowie's book traces RCA's quest for inexpensive labor as it moved from New Jersey to Indiana to Mexico.

9. "RCA Victor's Message for 1930," *Voice of the Victor*, January 1920, 1.

10. "RCA Victor's Message for 1930," *Voice of the Victor*, January 1920, 1.

11. "RCA Victor's Message for 1930," *Voice of the Victor*, January 1920, 1.

12. "RCA Victor's Message for 1930," *Voice of the Victor*, January 1920, 1.

13. William Faulkner, *Sanctuary* (New York: The Modern Library, 1958), 108.

14. William H. Young and Nancy K. Young, *The 1930s* (Westport, CT: Greenwood Press, 2002), 6.

15. Cohen, *Making a New Deal*, 2.

16. Cohen, *Making a New Deal*, 2.

17. Cohen, *Making a New Deal*, 9–10.

18. Kenney, *Recorded Music in American Life*, 158.

19. Gronow and Saunio, *An International History of the Recording Industry*, 57.

20. The U.S. economy, along with its recording companies, would recover by 1938 and match 1928 sales by 1942 (it would take the U.K. industry until 1959 to do the same). See Peter Martland, "Abandoning the Ship: The U.S. Recording Industry in 1931," *ARSC Journal* 47, no. 2 (2016): 205.

21. Blue Folder No. 40, Recorded Sound Archive, Edison National Historic Park, West Orange, NJ, in Allan Sutton's *Mainspring Press Blog*, "Discontinuing the Record Business": Documents from the Final Days of Edison Record Production," 78records.wordpress.com

22. Allan Sutton, *Recording the Thirties: The Evolution of the American Recording Industry, 1930–39* (Denver: Mainspring Press, 2011), 2.

23. Sutton, *Recording the Thirties*, 2.

24. Gelatt, *The Fabulous Phonograph*, 249–50.

25. Gelatt, *The Fabulous Phonograph*, 249–50.

26. Suisman, *Selling Sounds*, 269.

27. "Unbreakable Discs Made of New Resin," *New York Times*, January 4, 1930, 4.

28. Yorke, "The Rise and Fall of the Phonograph," 12.

29. Yorke, "The Rise and Fall of the Phonograph," 12.

30. Martland, "Abandoning the Ship," 205–206.

31. McGovern, *Sold American*, 221–22.

32. McGovern, *Sold American*, 232.

33. "Workers Should Profit in 1928," 76–81 (LOC).

34. "Workers Should Profit in 1928," 77 (LOC).

35. "Business Is Tired of Long Faces," *Voice of the Victor*, February 18, 1931, 2 (MHDL).

36. Jones' general motivation for the advertisement seems likely to have been to argue that advertising—the businessman's equivalent of "a sleek chestnut horse and a shiny surrey"—would benefit businesses even in the toughest times.

37. "Business Is Tired of Long Faces," 2 (MHDL).

38. "Business Is Tired of Long Faces," 2 (MHDL).

39. See the foundational text by Max Weber, *The Protestant Work Ethic and the Spirit of Capitalism* (Mineola, NY: Dover Publications, 2003).

40. "Business Is Tired of Long Faces," 2 (MHDL).

41. Clyde Haberman, "A 1930s Song Still Resonates, *New York Times*, November 27, 2007.

42. E. Y. Yip Harburg and Jay Gorney, "Brother, Can You Spare a Dime?" in *Americana: A Musical Review*, sheet music (New York: Harms, 1932).

43. Gelatt, *The Fabulous Phonograph*, 255.

44. Gelatt, *The Fabulous Phonograph*, 256.

45. Gelatt, *The Fabulous Phonograph*, 247–48.

46. Sanjek, *Pennies from Heaven*, 71.

47. Sanjek, *Pennies from Heaven*, 72.

48. Sanjek, *Pennies from Heaven*, 120.

49. Mazor, *Ralph Peer and the Making of Popular Roots Music*, 152–53.

50. Elijah Wald, "The Bluesman Who Behaved Too Well," *New York Times*, July 18, 2004, https://www.nytimes.com/2004/07/18/arts/music-the-bluesman-who-behaved-too-well.html.

51. "Radio's Dark Town Harmonizers," *Radio Digest*, November 1932, 22–23.

52. Sanjek, *Pennies from Heaven*, 120.

53. Sanjek, *Pennies from Heaven*, 72.

54. Patrick Huber, "The New York Sound: Citybilly Recording Artists and the Creation of Hillbilly Music, 1924–1932," *Journal of American Folklore* 127, no. 504 (Spring 2014): 140–58.

55. Sanjek, *Pennies from Heaven*, 118.

56. "Great Artists *Who Will SELL*," RCA Victor advertisement, *Radio Retailing*, January 1932, 16.

57. Sanjek, *Pennies from Heaven*, 117.

58. Gelatt, *The Fabulous Phonograph*, 252–53.

59. Sanjek, *Pennies from Heaven*, 253.

60. Sanjek, *Pennies from Heaven*, 118.

61. RCA Victor also experimented between 1931 and 1932 with long-play ten- to twelve-inch transcription discs geared to the classical market, with little result; few could afford the new phonographs necessary for playback. See Sutton, *Recording the Thirties*, 19–22.

62. Spring, *Saying It with Songs*, 83; and Gelatt, *The Fabulous Phonograph*, 250.

63. Sutton, *Recording the Thirties*, 3–4.

64. Sanjek, *Pennies from Heaven*, 72.

65. Brian Rust, *The American Record Label Book* (New Rochelle, NY: Arlington House, 1978), 50.

66. Rust, *The American Record Label Book*, 50.

67. "Brunswick Radio in Deal," *New York Times*, December 12, 1931, 29. ProQuest Historical Newspapers, https://search-proquest-com.libproxy.bellarmine.edu/hnpnewyorktimes/docview/99182341/9EF56CBF284145A2PQ/1?accountid=6741.

68. Gary Giddins, *Bing Crosby: A Pocketful of Dreams, The Early Years, 1903–1940* (Boston: Little, Brown, 2001), 238–39.

69. Rust, *The American Record Label Book*, 50.

70. Rust, *The American Record Label Book*, 50.

71. "Columbia Votes to Merge with Grigsby-Grunow," *Radio Retailing*, January 1932, 43 (MHDL).

72. "Brunswick Disc Sales Taken Over by New Company," *Radio Retailing*, January 1932, 43 (MHDL).

73. Sanjek, *Pennies from Heaven*, 125.

74. Sanjek, *Pennies from Heaven*, 125.

75. Sanjek, *Pennies from Heaven*, 125.

76. Sanjek, *Pennies from Heaven*, 72.

77. "Columbia Presents: A New Profit Maker!" advertisement, *Radio Merchant* September 1931, 22–23.

78. Rust, *The American Record Label Book*, 81.

79. "Refrigeration Saved My Shirt!" *Radio Retailing*, March 1932, 26–28 (MHDL).

80. "Refrigeration Saved My Shirt!" *Radio Retailing*, March 1932, 26–28 (MHDL).

81. Rust, *The American Record Label Book*, 81.

82. Gelatt, *The Fabulous Phonograph*, 268.

83. Gelatt, *The Fabulous Phonograph*, 254–55.

84. Gary Marmorstein, *The Label: The Story of Columbia Records* (New York: Thunder's Mouth Press, 2007), 62.

85. Marmorstein, *The Label*, 62. See also Eddie Condon and Thomas Sugrue, *We Called It Music: A Generation of Jazz*, (New York: Da Capo Press, 1992), 200, in Burton W. Peretti, *The Creation of Jazz: Music, Race, and Culture in Urban America* (Urbana: University of Illinois Press, 1992), 202.

86. Sutton, *Recording the Thirties*, 48–49.

87. Sutton, *Recording the Thirties*, 90–91.

88. van der Tuuk, *Paramount's Rise and Fall*, 197.

89. van der Tuuk, *Paramount's Rise and Fall*, 185–86.

90. Stephen Calt with Gayle Dean Wardlow, "Paramount's Decline and Fall (Part 5)," *78 Quarterly* 1, no. 7 (1992): 16.

91. Calt with Wardlow, "Paramount's Decline and Fall (Part 5)," Also see van der Tuuk, *Paramount's Rise and Fall*, 141.

92. van der Tuuk, *Paramount's Rise and Fall*, 186–87.

93. van der Tuuk, *Paramount's Rise and Fall*, 213–14.

94. Kennedy, *Jelly Roll, Bix and Hoagy*, 233–34.

95. Kennedy, *Jelly Roll, Bix and Hoagy*, 234.

96. Kennedy, *Jelly Roll, Bix and Hoagy*, 233.

97. Kennedy, *Jelly Roll, Bix and Hoagy*, 242.

98. Kennedy, *Jelly Roll, Bix and Hoagy*, 242–43. The Gennett brothers' strife was still part of family lore when I met a Gennett descendant by chance at Bellarmine University. Bryan Gennett, a Bellarmine University graduate who was a standout track-and-field athlete on campus, was surprised to meet someone who knew the Gennett story. I was just as surprised to meet him.

99. Kennedy, *Jelly Roll, Bix and Hoagy*, 235.

100. Mazor notes that as relatively unhappy Carmichael was with Peer, he'd regret Hollywood deals he made for "The Nearness of You" and other tunes. His son Bix (named after Carmichael's much missed friend Bix Beiderbecke) remembered, "At least Dad was with Peer when 'Georgia on My Mind' came out." Whatever business tensions did not affect Carmichael and Peer's long-term friendship. See Mazor, *Ralph Peer and the Making of Popular Roots Music*, 176–77.

101. Mazor, *Ralph Peer and the Making of Popular Roots Music*, 238–39.

102. Mazor, *Ralph Peer and the Making of Popular Roots Music*, 240.

103. For Gennett's canary recordings, see Charlie Dahan's *Gennett Records Discography*: "Jan. 10th in Gennett History, 1923: Norman Provol's Canaries Recorded 'Souvenir'"; https://gennett.wordpress.com/2015/01/10/jan-10th-in-gennett-history-1923-norm-2/. For a broader investigation of bird calls in/as sound media, see Jacob Smith, *Eco-Sonic Media* (Berkeley: University of California Press, 2015), 42–79. For the Fewkes recordings, see J. Walter Fewkes, *The Hopi and His Music on Gennett Records*, Gennett catalog and recordings, 1926. Archives of Traditional Music, Indiana University, Bloomington, IN.

104. Mazor, *Ralph Peer and the Making of Popular Roots Music*, 246.

105. The article appeared in the "Revival of Depressed Industries" issue of *An-*

nals of the American Academy of Political and Social Science. See H. S. Maraniss, "A Dog Has Nine Lives: The Story of the Phonograph," *Annals of the American Academy of Political and Social Science* 193, no. 1 (September 1937): 8–13.

106. Maraniss, "A Dog Has Nine Lives," 8.

107. Maraniss, "A Dog Has Nine Lives," 10.

108. Maraniss, "A Dog Has Nine Lives," 10.

109. Gelatt, *The Fabulous Phonograph*, 265.

110. Gelatt, *The Fabulous Phonograph*, 265.

111. Mazor. *Ralph Peer and the Making of Popular Roots Music*, 171.

112. Sanjek, *Pennies from Heaven*, 141.

113. Sanjek, *Pennies from Heaven*, 123.

114. Bob Ault, "CBS and the Columbia Phonograph Co.," *Antiques and Collecting*, November 1986, 56.

115. Gelatt, *The Fabulous Phonograph*, 266.

116. Sanjek, *Pennies from Heaven*, 141.

117. Sanjek, *Pennies from Heaven*, 141; and John Hammond, "Wallerstein: Salesman to Titan," *Billboard*, September 26, 1970, 10.

118. Sanjek, *Pennies from Heaven*, 141.

119. John Hammond, "Wallerstein: Salesman to Titan," *Billboard*, September 26, 1970, 10.

120. Hammond, "Wallerstein: Salesman to Titan," 10.

121. "The belief that John Hammond and Henry Johnson were one and the same rests on three pieces of evidence," Barry Lee Pearson and Bill McCulloch argue: "First, the writing styles were similar. Second, the initials were the same, albeit transposed. And third, Hammond's middle name was Henry, making him junior John Henry, John's son." The connection remains a tantalizing possibility, but it is hardly conclusive to date. See Barry Lee Pearson and Bill McCulloch, *Robert Johnson: Lost and Found* (Urbana: University of Illinois Press, 2008), 20. An additional bit of circumstantial evidence: Eric Bernay, publisher of *The New Masses*, funded the legendary *From Spirituals to Swing* concert at New York's Carnegie Hall. See "Jazz at Carnegie Hall," in Robert Walser, *Keeping Time: Readings in Jazz History* (New York: Oxford University Press, 2014), 93–97.

122. The author refers to ARC's Herbert Yates as "a despot of the old school, [who] disliked dealing with unions, and upon the acquisition of the Columbia factory in Bridgeport, Conn., transferred all his record manufacturing to the latter, where despite a Socialist mayor, he was promised he would have no trouble with unions." Only RCA Victor's Camden plant had been unionized at that point, by the United Electrical & Radio Workers, but Victor would decamp for Indiana, and later Mexico, in search of inexpensive labor. See Henry Johnson, "Phonograph Music," *The New Masses*, April 20, 1937, 35–36, www.marxists.org.

123. Sanjek, *Pennies from Heaven*, 139–40.

124. "CBS Buys Columbia Phonograph Company," *Heinl Radio Business Letter*, July–December 1938, 10, http://lantern.mediahist.org/catalog/heinlradiobusine-1938hein_0295. (MHDL). Paley's use of the term "home records" is especially interesting, as it leaves open the exact object of his comparison. If "home records"

were those meant for domestic play, was his comparison to radio, or jukeboxes, or simply recordings meant for public spaces?

125. Sanjek, *Pennies from Heaven*, 142.

126. Sanjek, *Pennies from Heaven*, 143.

127. Sanjek, *Pennies from Heaven*, 142.

128. Sanjek, *Pennies from Heaven*, 142.

129. Sanjek, *Pennies from Heaven*, 142.

130. Sanjek, *Pennies from Heaven*, 127.

131. Sanjek, *Pennies from Heaven*, 127.

132. Sanjek, *Pennies from Heaven*, 267.

133. Rowland Berthoff, "Independence and Enterprise: Small Business in the American Dream," in *Small Business in American Life* (New York: Columbia University Press, 1980), 36.

134. Alfred Chandler, *The Visible Hand* (1993), 233.

135. Hammond, "Wallerstein: Salesman to Titan," 10.

136. Sanjek, *Pennies from Heaven*, 266.

137. Kenney, *Recorded Music in American Life*, 165.

138. Gronow and Saunio, *An International History of the Recording Industry*, 68.

139. Gelatt, *The Fabulous Phonograph*, 268.

140. Though this conversation is focused on its cultural meanings, the *Oxford Dictionary of Music* offers the following technical description, as a "type of rhythmic interpretation in which even pulses in duple time . . . are performed in something close to, not identical with, triple time." See Tim Rutherford-Johnson, Michael Kennedy, and Joyce Bourne Kennedy, "Swing," in *The Oxford Dictionary of Music*, 6th ed. (2013), http://www.oxfordmusiconline.com/page/The-Oxford-Dictionary-of-Music.

141. Simon Frith, *Music for Pleasure: Essays on the Sociology of Pop* (Cambridge, UK: Polity Press, 1988), 11. The Library of Congress's collection is a useful introduction to both the early recording industry and the shaping of record collector culture in the 1930s and 1940s. See the Jim Walsh collection, Recorded Sound Reference Center, Library of Congress, Washington, DC (LOC).

142. Armstrong, "What Is Swing?" in Walser, *Keeping Time*, 75.

143. Armstrong, "What Is Swing?" 104, in Walser, *Keeping Time*, 75.

144. Armstrong, "What Is Swing?" 104, in Walser, *Keeping Time*, 76.

145. Duke Ellington, "Duke Ellington Explains Swing," in Walser, *Keeping Time*, 98.

146. Arnold Shaw, *Let's Dance: Popular Music in the 1930s*, ed. Bill Willard (New York: Oxford University Press, 1998), 134.

147. Tim Snelson, "'They'll Be Dancing in the Aisles!': Youth Audiences, Cinema Exhibition, and the Mid-1930s Swing Boom," *Historical Journal of Film, Radio, and Television* 37, no. 3 (2016): 455–56.

148. Hobson's study is deeply invested in the authenticity ethos already strong in jazz writing of the late 1930s. This line of thinking perhaps predictably leads to a larger condemnation of American culture in general: "It should be unnecessary to mention the American flair for the commercial adaptation of any kind of expres-

sion which offers suggestions of salability," adding that functional design leads to "all sorts of dubiously functional vogue-furniture" and "Surrealism leads to ladies' hats." See Wilder Hobson, *American Jazz Music* (New York: W. W. Norton & Co., 1939), 152, 155–56.

149. Gioia, *The History of Jazz*, 145.

150. Joel Dinerstein, *Swinging the Machine: Modernity, Technology, and African American Culture Between the World Wars* (Amherst: University of Massachusetts Press, 2003), 5–6.

151. James Lincoln Collier, *The Making of Jazz: A Comprehensive History* (New York: Dell Publishing, 1978), 263.

152. David Margolick, *Strange Fruit: The Biography of a Song* (New York: Harper-Collins, 2001), 21.

153. John Carvalho, "'Strange Fruit': Music Between Violence and Death," *The Journal of Aesthetics and Art Criticism* 71, no. 1 (Winter 2013): 111.

154. "Jazz at Carnegie Hall," in Walser, *Keeping Time*, 93.

155. "Jazz at Carnegie Hall," 94.

156. "Jazz at Carnegie Hall," 93.

157. Chris Rasmussen, "The People's Orchestra: Jukeboxes as the Measure of Popular Musical Taste in the 1930s and 1940s," in *Sound in the Age of Mechanical Reproduction*, ed. David Suisman and Susan Strasser (Philadelphia: University of Pennsylvania Press, 2010), 181.

158. Kerry Segrave, *Jukeboxes: An American Social History* (Jefferson, NC: McFarland 2002), 48–49.

159. Gert Almind, *The History of Coin Operated Phonographs, 1888–1998*. Unpublished manuscript, 2009, http://coin-o-phone.com/history.pdf.

160. Eric Weisbard, *Top 40 Democracy: The Rival Mainstreams of American Music* (Chicago: University of Chicago Press, 2014), 5.

161. Weisbard, *Top 40 Democracy*, 5.

162. Clarke, *The Rise and Fall of Popular Music*, 231.

163. Rasmussen, "The People's Orchestra," 183.

164. Gelatt, *The Fabulous Phonograph*, 274.

165. Gronow and Saunio, *An International History of the Recording Industry*, 69.

166. "Further Analysis Cues Jukebox Take Nearer $150,000,000 a Year," *Variety*, August 21, 1940, 49. Such profits had already attracted the interest of organized crime in the United States, including Meyer Lansky and his Wurlitzer distribution business, based in New York.

167. Gelatt, *The Fabulous Phonograph*, 276.

168. Kenney, *Recorded Life in American Music*, 171.

169. Rust, *The American Record Label Book*, 82.

170. Rust, *The American Record Label Book*, 82.

171. Rust, *The American Record Label Book*, 82.

172. "'Dead' Phonograph and Record Field Approaches Peak," *Advertising Age*, December 9, 1940.

173. Ryan, *The Production of Culture in the Music Industry*, 77.

174. Ryan, *The Production of Culture in the Music Industry*, 77.

175. Sanjek, *Pennies from Heaven*, 121.

176. Sanjek, *Pennies from Heaven*, 121.
177. Goldstein, *Copyright's Highway*, 57.
178. Goldstein, *Copyright's Highway*, 57.
179. Goldstein, *Copyright's Highway*, 58.
180. Goldstein, *Copyright's Highway*, 57.
181. Bruce Dennis, "He Was Always Good at Arithmetic," *Saturday Evening Post*, October 12, 1940, 13.
182. Dennis, "He Was Always Good at Arithmetic," 13.
183. Dennis, "He Was Always Good at Arithmetic," 13.
184. Goldstein, *Copyright's Highway*, 57.
185. Goldstein, *Copyright's Highway*, 57.
186. Sanjek, *Pennies from Heaven*, 140.
187. Sanjek, *Pennies from Heaven*, 140.
188. "Is AFM Acting as the Catspaw of ASCAP?" *Broadcasting and Broadcast Advertising*, August 15, 1937, 10 (MHDL).
189. "Is AFM Acting as the Catspaw of ASCAP?" 10 (MHDL).
190. "Is AFM Acting as the Catspaw of ASCAP?" 10 (MHDL).
191. Bruce Robertson, "Industry Acts to Avoid Musicians Strike," *Broadcasting and Broadcasting Advertising*, September 1, 1937, 9.
192. "Music Boost Adopted; NAB Calls Session," *Broadcasting and Broadcasting Advertising*, September 15, 1937, 11.
193. "Radio Maps Union Fight," *Variety*, November 29, 1939, 25.
194. "AFM Strike-Out," *Broadcasting*, January 15, 1940, 52.
195. Sanjek, *Pennies from Heaven*, 175.
196. Sanjek, *Pennies from Heaven*, 175.
197. "Statement of the Department of Justice on the Closing of the Anti-Trust Division's Review of ASCAP and BMI Consent Decrees," https://www.justice.gov/atr/ascap-bmi-decree-review.
198. Charles Williams, "Mr. Petrillo's Hopeless War," *The Nation*, October 3, 1942, 291.
199. "Decca Pact Covers Fees Direct to Union," *Broadcasting*, September 27, 1943, 9.
200. "FDR Telegram to Petrillo," *Broadcasting*, October 9, 1944, 11.
201. Emil Pinta, "Efforts by Record Companies, the American Federation of Musicians, and Truman's Cabinet to Resolve the 1948 Recording Ban," *ARSC Journal* 41, no. 1 (Spring 2010): 23.
202. Emil Pinta, "The First RCA Victor Recording of 1948: Petrillo, Truman and the AFM Recording Ban of 1948," *ARSC Journal* 39, no. 2 (Fall 2008): 283.
203. MacDonald, *Don't Touch That Dial*, 60–61.
204. Correspondence from Fred Gennett to George Kay, July 19, 1962, John K. MacKenzie Collection (IHS).
205. John Fahey, "American Quick Fix Religion," liner notes to *American Primitive, Vol. 1*, Revenant Records CD, 1997.
206. Barnett, "Record Men," 140.

Bibliography

1925 List of All Lateral Gennett Records—Complete Catalog No. 5. January 1, 1925. Starr-Gennett Foundation, Richmond, IN.

"104,628.74 Paid for 'Black Swan' Records." *New York Age*, January 28, 1922.

Aberjhani and Sandra L. West. "Hot Chocolates." *Encyclopedia of the Harlem Renaissance*. New York: Facts on File, 2003. 157–58.

Abbott, Lynn, and Doug Seroff. *Out of Sight: The Rise of African-American Popular Music, 1889–1895*. Jackson: University Press of Mississippi, 2002.

"AFM Strike-Out." *Broadcasting*, January 15, 1940.

"Al Jolson Signs Contract to Record Exclusively for the Brunswick Co." *Music Trade Review*, February 21, 1924.

Aldridge, Benjamin L. *The Victor Talking Machine Company*. Edited by L. Frederic Bayh. David Sarnoff Library, College of New Jersey. http://www.davidsarnoff.org/vtm.html

Alexander, Ben. "'For Posterity': The Personal Audio Recordings of Louis Armstrong." *American Archivist* 71, no. 1 (Spring-Summer 2008): 50–86.

Allison, Madeline G. "The Horizon." *The Crisis*, March 22, 1922.

Almind, Gert. *The History of Coin Operated Phonographs, 1888–1998*. Unpublished manuscript, 2009. http://coin-o-phone.com/history.pdf

"Amateur Radio Restored." In Hugo Gernsback, *The Perversity of Things: Hugo Gernsback on Media, Tinkering, and Scientification*, edited by Grant Wythoff. online edition. Minneapolis: University of Minnesota Press, 2016. https://manifold.umn.edu/projects/the-perversity-of-things

Anderson, Christopher, and Michael Curtin. "Writing Cultural History: The Challenge of Radio and Television." In *Media History: Theories, Methods, Analysis*, edited by Niels Brügger and Søren Kolstrup. Aarhus, DK: Aarhus University Press, 2002.

Anderson, Maureen. "The White Reception of Jazz in America." *African-American Review* 38, no. 1 (2004): 135–45.

"Any Woman's Blues." *Printers' Ink*, May 29, 1924.

"Appeal of Primitive Jazz." In *Jazz in Print (1856–1929): An Anthology of Collected Writings of Selected Early Readings in Jazz History*, edited by Karl Koenig, 119–20. Hillsdale, NY: Pendragon Press, 2002.

Armstrong, Louis. "Joe Oliver Is Still King." *The Record Changer*, July–August 1950.

Armstrong, Louis. "What Is Swing?" In *Keeping Time: Readings in Jazz History*, edited by Robert Walser, 73–76. 2nd ed. New York: Oxford University Press, 2015.

Ault, Bob. "CBS and the Columbia Phonograph Co." *Antiques and Collecting*, November 1986.

"Automatic Music in the Home." *The Phonogram*, June 1901.

Barlow, William. "Black Music on Radio in the Jazz Age." *African-American Review* 50, no. 4 (Winter 2017): 934–37.

Barlow, William. *"Looking Up at Down": The Emergence of Blues Culture*. Philadelphia: Temple University Press, 1989.

Barlow, William. *Voiceover: The Making of Black Radio*. Philadelphia: Temple University Press, 1999.

Barnett, Kyle. "Furniture Music: The Phonograph as Furniture, 1900–1930." *Journal of Popular Music Studies* 18, no. 3 (December 2006).

Barnett, Kyle. "Record Men: Talent Scouts in the U.S. Recording Industry, 1920–1935." In *Making Media Work: Cultures of Management in the Entertainment Industries*, edited by Derek Johnson, Derek Kompare, and Avi Santo, 113–41. New York: New York University Press, 2014.

Baumbach, Robert W. *Look for the Dog: An Illustrated Guide to Victor Talking Machines*. Woodland Hills, CA: Stationery X-Press, 1981.

Berliner, Emile. *The Gramophone: Etching the Human Voice*. Pamphlet. Berliner Gramophone Company, 1909.

Berrett, Joshua. *Louis Armstrong and Paul Whiteman: Two Kings of Jazz*. New Haven, CT: Yale University Press, 2004.

Berthoff, Rowland. *Small Business in American Life*. New York: Columbia University Press, 1980.

Biel, Michael. "The Making and Use of Recordings in Broadcast Before 1935." PhD dissertation, Northwestern University, 1977.

Bourdieu, Pierre. *Distinction: A Social Critique of the Judgement of Taste*. Cambridge, MA: Harvard University Press, 1993.

Bourdieu, Pierre. *Outline of a Theory of Practice*. New York: Cambridge University Press, UK: 1977.

Bowers, David Q., and Kathryn Fuller-Seeley. *One Thousand Nights at the Movies: An Illustrated History of Motion Pictures*. Atlanta: Whitman Publishing, 2013.

Bowker, Geoffrey C., and Susan Leigh Star. *Sorting Things Out: Classification and Its Consequences*. Cambridge, MA: MIT Press, 2000.

Brackett, David. *Categorizing Sound: Genre and Twentieth-Century Popular Music*. Berkeley, CA: University of California Press, 2016.

Bradford, Perry. *Born with the Blues*. New York: Oak Publications, 1965.

Brooks, Tim. "High Drama in the Recording Industry." *ARSC Journal* 33, no. 1 (Spring 2002): 21–76.

Brooks, Tim. *Lost Sounds: Blacks and the Birth of the Recording Industry*. Urbana: University of Illinois Press, 2004.

Brothers, Thomas, ed. *Louis Armstrong, In His Own Words*. New York: Oxford University Press, 2001.
Brügger, Niels, and Søren Kolstrup, eds. *Media History: Theories, Methods, Analysis*. Aarhus, DK: Aarhus University Press, 2002.
"Brunswick Disc Sales Taken Over by New Company." *Radio Retailing*, January 1932.
"Brunswick Radio in Deal." *New York Times*, December 12, 1931.
Bryan, Martin F. "And We Wonder Why They Are So Common?" *New Amberola Graphic*, October 1999, 8–10.
Buckner, Reginald T., and Steven Weiland, eds. *Jazz in Mind: Essays on the History and Meanings of Jazz*. Detroit: Wayne State University Press, 1991.
"Business and Business Men." *New York Evening Press*, May 13, 1924.
"Business Is Tired of Long Faces." *Voice of the Victor*, February 18, 1931.
Callen, Robert J. "The Education and Tribulations of a Precursory Disc Recording Engineer." Presented at the Fortieth American Engineering Society (AES) Convention, April 27–30, 1971, Los Angeles. http://aes.org/aeshc/docs/callen.education.and.tribulations.pdf
Calt, Stephen. *I'd Rather Be the Devil: Skip James and the Blues*. Chicago: Chicago Review Press, 1994.
Calt, Stephen. "Paramount—Part Two: Anatomy of a 'Race' Label." *78 Quarterly* 1, no. 4 (1989): 9–30.
Calt, Stephen, with Gayle Dean Wardlow. "Paramount's Decline and Fall (Part 5)." *78 Quarterly* 1, no. 7 (1992): 7–29.
Cantor, Eddie, and David Freedman. *Yoo-Hoo, Prosperity!: The Eddie Cantor Five-Year Plan*. New York: Simon & Schuster, 1931.
Carvalho, John. "'Strange Fruit': Music between Violence and Death." *Journal of Aesthetics and Art Criticism* 71, no. 1 (Winter 2013): 111.
"CBS Buys Columbia Phonograph Company." *Heinl Radio Business Letter*, July–December 1938, 10. http://lantern.mediahist.org/catalog/heinlradiobusine1938hein_0295
Chanan, Michael. *Repeated Takes: A Short History of Recording and Its Effects on Music*. London: Verso, 1995.
Chandler, Alfred. *The Essential Alfred Chandler: Essays Toward a Historical Theory of Big Business*. Boston: Harvard Business School Press, 1991.
Chandler, Alfred. *Scale and Scope: The Dynamics of Industrial Capitalism*. With the assistance of Takashi Hikino. Cambridge, MA: Harvard University Press, 1994.
Chandler, Alfred. *The Visible Hand: The Managerial Revolution in American Business*. Cambridge, MA: Harvard University Press, 1993.
Clarke, Donald. *The Rise and Fall of Popular Music*. New York: Viking/Penguin Books, 1996.
Clock, W. H. "How Columbia Sold Whiteman," *Talking Machine World*, September 9, 1928, 24.
Cohen, Lizabeth. *A Consumer's Republic: The Politics of Mass Consumption in Postwar America*. New York: Alfred A. Knopf, 2003.
Cohen, Lizabeth. *Making a New Deal: Industrial Workers in Chicago, 1919–1939*. 2nd ed. New York: Cambridge University Press, 2008.

Collier, James Lincoln. *The Making of Jazz: A Comprehensive History*. New York: Dell Publishing, 1978.

"Columbia Votes to Merge with Grigsby-Grunow." *Radio Retailing*, January 1932.

Condon, Eddie, and Thomas Sugrue. *We Called It Music: A Generation of Jazz*. New York: Da Capo Press, 1992.

Consider Fifty Million Customers: A Study of the Rural Section of the American Market. Spencer, IN: Farm Life Publishing, ca. 1920s.

Cooke, Mervyn. *The Chronicle of Jazz*. New York: Abbeville Press, 2013.

Cowan, Lester, ed. *Recording Sound for Motion Pictures*. New York: McGraw-Hill, 1931.

Cowie, Jefferson. *Capital Moves: RCA's Seventy-Year Quest for Cheap Labor*. New York: The New Press, 1999.

Cox, Ted. *Sold on Radio: Advertisers in the Golden Age of Broadcasting*. Jefferson, NC: McFarland, 2008.

Crafton, Donald. *The Talkies: American Cinema's Transition to Sound, 1926–1931*. Berkeley: University of California Press, 1997.

Craig, Steve. "'The Farmer's Friend': Radio Comes to Rural America, 1920–1927." *Journal of Radio Studies* 8, no. 2 (2001): 330–46.

"Crooner Vallee on Jazz Tour." *Voice of the Victor*, February 18, 1931.

Dahan, Charlie B., and Malcolm Rockwell. "Starr Piano/Gennett Records Discography." *ARSC Journal* 39, no. 2 (Fall 2008): 292–94.

Daniel, Wayne. "Old Time Fiddlers' Contests on Early Radio." *JEMF Quarterly* 17 (Fall 1981): 159–65.

Davis, Angela Y. *Blues Legacies and Black Feminism: Gertrude "Ma" Rainey, Bessie Smith, and Billie Holliday*. New York: Vintage Books, 1998.

"'Dead' Phonograph and Record Field Approaches Peak." *Advertising Age*, December 9, 1940.

"Decca Pact Covers Fees Direct to Union." *Broadcasting*, September 27, 1943.

DeGraaf, Leonard. *Edison and the Rise of Innovation*. New York: Sterling Signature, 2013.

Dennis, Bruce. "He Was Always Good at Arithmetic." *Saturday Evening Post*. October 12, 1940.

Dick, Bernard. *City of Dreams; The Making and Remaking of Universal Pictures*. Lexington: University Press of Kentucky, 1997.

Dinerstein, Joel. *Swinging the Machine: Modernity, Technology, and African American Culture Between the World Wars*. Amherst: University of Massachusetts Press, 2003.

"Distribution System of Black Swan Phonograph Records." *The Crisis*, March 1922, 221.

Dixon, Robert M. W., and John Godrich. "Recording the Blues," In *Yonder Comes the Blues: The Evolution of a Genre*, 2nd ed., edited by Paul Oliver, Tony Russell, Robert M. W. Dixon, John Godrich, and Howard Dye, 243–44. Cambridge, UK: Cambridge University Press, 2001.

Doerksen, Clifford. *American Babel: Rogue Radio Broadcasters of the Jazz Age*. Philadelphia: University of Pennsylvania Press, 2005.

"Don't Try to Burn a Jazz Record." *Talking Machine World*, December 15, 1920.

Drowne, Kathleen Morgan, and Patrick Huber. *The 1920s*. Westport, CT: Greenwood Press, 2004.

du Gay, Paul, Stuart Hall, Linda Janes, Anders Koed Madsen, Hugh Mackay, and Keith Negus. *Doing Cultural Studies: The Story of the Sony Walkman*. 2nd ed. Thousand Oaks, CA: Sage Publications, 2013.

Early, Gerald. "White Noise and White Knights: Some Thoughts on Race, Jazz, and the White Jazz Musician." In *Jazz: A History of America's Music*, edited by Geoffrey C. Ward and Ken Burns, 324–25. New York: Knopf, 2000.

Edison, Thomas. "The Phonograph and Its Future." *The North American Review*, May 1, 1878, 527–36.

Ewen, David. *All the Years of American Popular Music*. Englewood Cliffs, NJ: Prentice Hall, Inc., 1977.

"Exploit-O-Grams: 'Chicken A La King.'" *The Film Daily*, July 8, 1928.

"Exploit-O-Grams: 'Reckless Romance.'" *The Film Daily*, July 2, 1928.

Fahey, John. "American Quick Fix Religion." *American Primitive, Vol. 1*. Liner notes. Revenant Records CD, 1997.

Faulkner, William. *Sanctuary*. New York: The Modern Library, 1958.

"FDR Telegram to Petrillo." *Broadcasting*, October 9, 1944, 11.

Fewkes, J. Walter. *The Hopi and His Music on Gennett Records*. Richmond, IN: Gennett catalog and recordings, 1926.

"Fiddlin' John Carson OKeh Records Popular in Trade." *Talking Machine World*, April 15, 1925.

Filzen, Sarah. "The Rise and Fall of Paramount Records." *Wisconsin Magazine of History* 82, no. 2 (Winter 1998–1999): 104–27.

Fischer, Suzanne. "Friars Society Orchestra/New Orleans Rhythm Kings." *Starr-Gennett Foundation Inc. News*, Winter 2004–5, 3

Fitzgerald, F. Scott. "Echoes of the Jazz Age." In *The Crack Up*, edited by Edmund Wilson. New York: J. Laughlin, 1945.

Fitzgerald, F. Scott. *Tales of the Jazz Age*. New York: Scribner and Sons, 1922.

Fleischer, Dave, dir. *I'll Be Glad When You're Dead and Gone You Rascal You*. Paramount Pictures, 1932.

"Fletcher Making Tour." *New York Clipper*, February 7, 1923.

Foreman, Ronald Jr. "Jazz and Race Records, 1920–1932: Their Origins and Their Significance for the Record Industry and Society," PhD dissertation, University of Illinois, 1968.

"Fred Wiggins Dies Suddenly at Elks Club." *Richmond Palladium-Item & Sun-Telegram*, August 10, 1948.

Frith, Simon. *Music for Pleasure: Essays on the Sociology of Pop*. Cambridge, UK: Polity Press, 1988.

Frith, Simon. *Performing Rites: On the Value of Popular Music*. Cambridge, MA: Harvard University Press, 1996.

Frith, Simon. *Popular Music: Critical Concepts in Media and Cultural Studies: Vol. 4, Music and Identity*. London: Routledge, 2004.

"Further Analysis Cues Jukebox Take Nearer $150,000,000 a Year." *Variety*, August 21, 1940.

Gabbard, Krin. *Jammin' at the Margins: Jazz and the American Cinema*. Chicago: University of Chicago Press, 1996.

Gara, Larry. *The Baby Dodds Story*. Los Angeles: Contemporary Press, 1959.

Gates, W. Francis. "American Folk Music: A Negligible Quality." *The Etude*, January 1918.

Gaudreault, André. "The Diversity of Cinematographic Connections in the Intermedial Context of the Turn of the 20th Century." Translated by Wendy Schubring. In *Visual Delights: Essays on the Popular and Projected Image in the 19th Century*, edited by Simon Popple and Vanessa Toulmin, 8–15. Wiltshire, UK: Flick Books, 2000.

Gelatt, Roland. *The Fabulous Phonograph: 1877–1977*. 2nd rev. ed. New York: MacMillan and Company, 1977.

Giddins, Gary. *Bing Crosby: A Pocketful of Dreams, The Early Years, 1903–1940*. Boston: Little, Brown, 2001.

Gioia, Ted. *The History of Jazz*. New York: Oxford University Press, 1997.

Gitelman, Lisa. *Always Already New: Media, History and the Data of Culture*. Cambridge, MA: MIT Press, 2008.

"A Glance into the Trade Future." *Talking Machine World*, January 1928, 34b.

Goldstein, Paul. *Copyright's Highway: From Gutenberg to the Celestial Jukebox*. Rev. ed. Palo Alto, CA: Stanford University Press, 2003.

Gomery, Douglas. *The Coming of Sound*. New York: Routledge, 2005.

Goodrod, R. M. "Conquering the Jazz Craze of Young Pianists." *The Etude*, February 1934, 5, 17.

"Government Moves Against Publishers." *Musical Courier*, August 12, 1920, 5, 17.

Gray, Herman. *Producing Jazz: The Experience of an Independent Record Company*. Philadelphia: Temple University Press, 1988.

Grenfell, Michael. *Pierre Bourdieu: Key Concepts*. 2nd ed. Durham, UK: Acumen, 2012.

Gronow, Pekka. "The Record Industry: The Growth of a Mass Medium." *Popular Music* 3 (January 1983): 53–75.

Gronow, Pekka, and Ilpo Saunio. *An International History of the Recording Industry*. London: Cassell, 1999.

Grossberg, Lawrence. "On Postmodernism and Articulation: An Interview with Stuart Hall." *Journal of Communication Inquiry* 10, no. 2 (June 1986): 45–60.

Haberman, Clyde. "A 1930s Song Still Resonates." *New York Times*, November 27, 2007.

Haden, Walter Darrell. "Vernon Dalhart: His Rural Roots and the Beginnings of Commercial Country Music." *ARSC Journal* 3, no. 1 (Winter 1970/71): 3.

Hagan, Chet. *Grand Ole Opry*. New York: Henry Holt and Company, 1989.

Hagstrom Miller, Karl. *Segregating Sounds: Inventing Folk and Pop Music in the Age of Jim Crow*. Durham, NC: Duke University Press, 2010.

Hammond, John. "Wallerstein: Salesman to Titan." *Billboard*, September 26, 1970, 10.

Handbook of Texas Online, s.v. "Alexander Robertson." http://tsha.utexas.edu/handbook/online/articles/RR/froam.html

Harburg, E. Y., and Jay Gorney. "Brother, Can You Spare a Dime?" In *Americana: A Musical Review*, sheet music. New York: Harms, 1932.

Hardin-Armstrong, Lil. "Lil and Louis: Satchmo and Me (Excerpts)." In *The Louis Armstrong Companion*, edited by Joshua Berrett. New York: Schirmer Books, 1999.

Harris, Rodger. "Wars of the Confederacy, Wars with the Indians, Wars with the 'Fiddle and Bow': The Life of Henry C. Gilliland." Oklahoma Historical Society. http://www.ok-history.mus.ok.us/folk/gilliland.html

Harrison, Daphne Duval. *Black Pearls: Blues Queens of the 1920s*. New Brunswick, NJ: Rutgers University Press, 1988.

Harrison, Daphne Duval, and John Cowley. "'Classic' Blues and Women Singers." In *The New Blackwell Guide to Recorded Blues*, edited by John Cowley and Paul Oliver, 125–54. Oxford, UK: Blackwell Publishers, 1996.

Hartley; Eugene F., ed. Department of Commerce, Bureau of the Census. *Abstract of the Census of Manufactures, 1919*. Washington, DC: Government Printing Office, 1923.

"Henry Whitter, OKeh Artist, Real Hill Country Type." *Talking Machine World*, May 15, 1925, 35.

Herbert v. Shanley Co. 242 U.S. 591 (1917).

Hesmondhalgh, David. "Bourdieu, the Media, and Cultural Production." *Media, Culture & Society* 28, no. 2 (November 9, 2010): 211–31.

"Hill Billies' OKeh Records." *Talking Machine World*, April 15, 1925, 50d.

"Hill-Billy Music." *Variety*, December 29, 1926, 1.

Hilmes, Michele. "The 'Ban' That Never Was: Hollywood and the Broadcasting Industry in 1932." *The Velvet Light Trap* 23 (Spring 1989): 39–47.

Hilmes, Michele. *Hollywood and Broadcasting*. Urbana: University of Illinois Press, 1990.

Hilmes, Michele. "Nailing Mercury: The Problem of Media Industry Historiography." In *Media Industries: History, Theory and Method*, edited by Jennifer Holt and Alisa Perren, 21–33. Malden, MA: Wiley-Blackwell, 2009.

Hilmes, Michele. *NBC: America's Network*. Berkeley: University of California Press, 2007.

Hilmes, Michele. *Radio Voices: American Broadcasting, 1922–1952*. Minneapolis: University of Minnesota Press, 1997.

Hilton, A. W. "How Victor Is Forcing Sales in Cities and Trade Centres." *Printers' Ink*, March 19, 1925.

Hobson, Wilder. *American Jazz Music*. New York: W. W. Norton & Co., 1939.

Hoffmann, Franz, ed. *Jazz Advertised, 1910-1967, Vol. 1: Out of the New England Press, 1910-1934*. Berlin: Franz Hoffmann, 1997.

Horning, Susan Schmidt. "Chasing Sound: The Culture and Technology of Recording Studios in America, 1877–1977." PhD dissertation, Case Western Reserve University, 2002.

Horning, Susan Schmidt. *Chasing Sound: Technology, Culture, and the Art of Studio Recording from Edison to the LP*. Baltimore: Johns Hopkins University Press, 2015.

"How Popular Song Factories Manufacture a Hit." *New York Times*, September 18, 1910.

Huber, Patrick. "Black Hillbillies: African American Musicians on Old-Time Records, 1924–1925." In *Hidden in the Mix: The African American Presence in Country Music*, edited by Diane Pecknold, 19–81. Durham, NC: Duke University Press, 2013.

Huber, Patrick. "The New York Sound: Citybilly Recording Artists and the Creation of Hillbilly Music, 1924–1932." *Journal of American Folklore* 127, no. 504. (Spring 2014): 140–58.

"Is AFM Acting as the Catspaw of ASCAP?" *Broadcasting and Broadcast Advertising*, August 15, 1937, 10.

Israel, Paul. *Edison: A Life of Invention*. New York: Wiley & Sons, 1998.

Jackson-Brown, Irene. "Developments in Black Gospel Performance and Scholarship" *Black Music Research Journal* 10, no. 1 (Spring 1990): 36–42.

James, Arthur. "Sound? Well, Listen." *Exhibitors' Daily Review*, December 21, 1928.

Jensen, Joli. *The Nashville Sound: Authenticity, Commercialization, and Country Music*. Nashville: Vanderbilt University Press, 1998.

Johnson, Henry. "Phonograph Music." *The New Masses*, April 20, 1937, 35–36.

Jones, Isham. "American Dance Music Is Not Jazz." *The Etude*, August 1924, 526.

Jones, Leroi (Amiri Baraka). *Blues People: Negro Music in White America*. New York: Harper Perennial, 1999.

Jones, Max, and John Chilton. *Louis: The Louis Armstrong Story*. Boston, Little, Brown, 1971.

Kando, T. M. *Leisure and Popular Culture in Transition*. 2nd ed. St. Louis, MO: C. V. Mosby, 1980.

Kay, George W. "Those Fabulous Gennetts!: The Life Story of a Remarkable Label." *The Record Changer*, June 1953, 4–13.

Kelley, Norman. *R&B (Rhythm and Business): The Political Economy of Black Music*. New York: Akashic, 2002.

Kennedy, Rick. *Jelly Roll, Bix and Hoagy: Gennett Records and the Rise of America's Musical Grassroots*. 2nd ed. Bloomington and Indianapolis: Indiana University Press, 2013.

Kennedy, Rick. "The Man Who Made the Label a Legend. *Starr-Gennett Foundation Inc. News*, 2002.

Kennedy, Rick, and Randy McNutt. *Little Labels, Big Sound: Small Record Companies and the Rise of American Music*. Bloomington and Indianapolis: Indiana University Press, 1999.

Kenney, William Howland. *Chicago Jazz: A Cultural History, 1904–1930*. New York: Oxford University Press, 1993.

Kenney, William Howland. "Negotiating the Color Line: The Autobiographies of Louis Armstrong." In *Jazz in Mind: Essays on the History and Meanings of Jazz*, edited by Reginald T. Buckner and Steven Weiland, 38–59. Detroit: Wayne State University Press, 1991.

Kenney, William Howland. *Recorded Music in American Life: The Phonograph and Popular Memory*. New York: Oxford University Press, 1999.

Ketcham, C. L. "Let the Customer Play the Records." *Music Trade Review*, February 22, 1919, 36.

Kingsbury, Paul. *The Encyclopedia of Country Music*. New York, Oxford University Press, 1998.

Klein, Andrew. "A History of the Starr Piano Factory." *Talking Machine Review*, February 1983, 1787–89, 1818.

Klemm, Gustave. "The Jargon of Jazz." *The Etude*, August, 1934, 155.

Kornweibel, Theodore Jr. "An Economic Profile of Black Life in the Twenties." *Journal of Black Studies* 6, no. 4 (June 1976): 307–20.

Kraft, James P. "Musicians in Hollywood: Work and Technological Change in Entertainment Industries." *Technology and Culture* 35, no. 2 (1994): 289–314.

Kruse, Holly. "Early Audio Technology and Domestic Space." *Stanford Humanities Review* 3, no. 2 (Autumn 1993): 1–14.

Laing, Dave. "A Voice without a Face: Popular Music and the Phonograph in the 1890s." *Popular Music* 10, no. 1 (1991): 1–9.

Laird, Ross, and Brian Rust. *Discography of OKeh Records, 1918–1934*. Westport, CT: Praeger, 2004.

Lange, Jeffrey. *Smile When You Call Me Hillbilly: Country Music's Struggle for Respectability, 1939–1954*. Athens: University of Georgia Press, 2004.

"Last Week's Ten Best Sellers among the Popular Songs." *Billboard*, September 15, 1917, 14.

Latour, Bruno. *Reassembling the Social: An Introduction to Actor-Network Theory*. New York: Oxford University Press, 2007.

Layton, James, and David Pierce. *King of Jazz: Paul Whiteman's Technicolor Revue*. Severn, MD: Media History Press, 2016.

Leeds & Catlin Co. v. Victor Talking Machine Co., 213 U.S. 301 (1909).

Lena, Jennifer. *Banding Together: How Communities Create Genres in Popular Music*. Princeton, NJ: Princeton University Press, 2014.

Lescarboura, Austin. "At the Other End of the Phonograph." *Scientific American*, August 31, 1918, 164.

Library of Congress. "History of the Cylinder Phonograph." https://www.loc.gov/collections/edison-company-motion-pictures-and-sound-recordings/articles-and-essays/history-of-edison-sound-recordings/history-of-the-cylinder-phonograph/

Lilly, John. "Jimmie Rodgers and the Bristol Sessions." In *The Bristol Sessions: Writings about the Big Bang of Country Music*, edited by Charles K. Wolfe and Ted Olson, 54–65. Jefferson, NC: McFarland, 2005.

"Local Firm Takes Over New York Phonograph Company" *Port Washington Herald*, April 2, 1924, 1.

Lomax, John A. *Cowboy Songs and Other Frontier Ballads*. New York: Sturgis & Walton Company, 1911.

Lopez, Vincent. *Lopez Speaking: An Autobiography*. New York: Citadel Press, 1960.

Lott, Eric. *Love & Theft: Black Minstrelsy and the American Working Class*. New York: Oxford University Press, 1995.

Lynd, Robert S., and Helen Merrell Lynd. *Middletown: A Study in Contemporary Culture*. New York: Harcourt, Brace & World, 1929.

MacDonald, J. Fred. *Don't Touch That Dial: Radio Programming in American Life. 1920–1960*. Chicago: Nelson-Hall, 1979.

MacKenzie, John K. *Gennett Records of Old Time Tunes: A Catalog Reprint*. JEMF Special Series, no. 6. Los Angeles: John Edwards Memorial Foundation at the Folklore and Mythology Center, University of California at Los Angeles, 1975.

"'Making Record No Easy Job' says Pittsburgh Musicians by Deppe (Himself)." *Pittsburgh Courier*, October 20, 1923.

Malone, Bill C. *Country Music, U.S.A.* Rev. ed. Austin: University of Texas Press, 1997.

Maraniss, H. S. "A Dog Has Nine Lives: The Story of the Phonograph." *Annals of the American Academy of Political and Social Science* 193, no. 1 (September 1937): 8–13.

Marcus, Greil. *The Old, Weird America: The World of Bob Dylan's Basement Tapes*. New York: Picador/Henry Holt, 2001.

Marden, S. G. "Phonograph Books Have New Motif." *Printers Ink Monthly*, April 20, 1920.

Margolick, David. *Strange Fruit: The Biography of a Song*. New York: HarperCollins, 2001.

Marmorstein, Gary. *The Label: The Story of Columbia Records*. New York: Thunder's Mouth Press, 2007.

Martin, Lerone. *Preaching on Wax: The Phonograph and the Shaping of Modern African American Religion*. New York: New York University Press, 2014.

Martland, Peter. "Abandoning the Ship: The U.S. Recording Industry in 1931." *ARSC Journal* 47, no. 2 (2016): 205–14.

Mazor, Barry. *Meeting Jimmie Rodgers: How America's Original Roots Music Hero Changed the Pop Sounds of a Century*. New York: Oxford University Press, 2012.

Mazor, Barry. *Ralph Peer and the Making of Popular Roots Music*. Chicago: Chicago Review Press, 2015.

McCracken, Allison. "Real Men Don't Sing Ballads: The Radio Crooner in Hollywood 1929-1933." In *Soundtrack Available: Essays on Film and Popular Music*, edited by Pamela Robertson Wojcik and Arthur Knight, 105–33. Durham, NC: Duke University Press, 2001.

McCracken, Allison. *Real Men Don't Sing: Crooning in American Culture*. Durham, NC: Duke University Press, 2015.

McDaniel, William R., and Harold Seligman. *Grand Ole Opry*. New York: Greenburg, 1952.

McFall, Liz. "What about the Old Cultural Intermediaries?: An Historical Review of Advertising Producers." *Cultural Studies* 16, no. 4 (November 9, 2010): 532–52.

McGovern, Charles. *Sold American: Consumption and Citizenship, 1890–1945*. Chapel Hill: University of North Carolina Press, 2006.

Merriam, Alan P., and Fradley H. Garner. "Jazz—the Word." *Ethnomusicology* 12, no. 3 (September 1968): 373–96.

Messenger, Cory. "Calling the Tune: Hollywood and the Business of Music," PhD dissertation, Griffith University, 2011.

"Microphones, Part 2: The Electrical Era," Audio Engineering Society. http://www.aes-media.org/historical/html/recording.technology.history/microphones2.html

Middleton, Richard. *Pop Music and the Blues: Its Relationship and Significance*. London: Gollancz, 1972.

"Momentous Musical Meeting: Thomas A. Edison and Lt. Comm. John Philip Sousa Meet for the First Time and Talk upon Music," *The Etude*, October 1923, 663–64.

Morritt, Eric. "Early Sound Recording Technology and the Bristol Sessions." In *The Bristol Sessions: Writings about the Big Bang of Country Music* edited by Charles K. Wolfe and Ted Olson, 7–16. Jefferson, NC: McFarland, 2005.

"Movie Stars Go On Air." *St. Petersburg Times*, March 25, 1928.

"Music Boost Adopted; NAB Calls Session." *Broadcasting and Broadcasting Advertising*, September 15, 1937, 11–12, 74–75.

"Music Publishers Sued Here on Trust." *New York Times*, August 4, 1920, 24.

Nanry, Charles, ed. *American Music: From Storyville to Woodstock*. New Brunswick, NJ: Transaction Books, 1972.

Negus, Keith. *Music Genres and Corporate Cultures*. New York: Routledge, 1999.

Negus, Keith. *Popular Music in Theory: An Introduction*. Hanover, NH: Wesleyan University Press, 1996.

Negus, Keith. "The Work of Cultural Intermediaries and the Enduring Distance Between Production and Consumption." *Cultural Studies* 16, no. 4 (November 9, 2010): 501–15.

"No Victor Records for August." *Talking Machine World*, July 15, 1919, 172.

"On With the Dance." *Voice of the Victor*, July–December 1928.

"Our National Victory Should Make Business Men Optimistic." *Talking Machine World*, January 15, 1919.

"Paul Whiteman Signs to Record Exclusively for Columbia Co." *Talking Machine World*, May 1928.

Pearson, Barry Lee, and Bill McCulloch, *Robert Johnson: Lost and Found*. Urbana: University of Illinois Press, 2008

Pecknold, Diane, ed. *Hidden in the Mix: The African American Presence in Country Music*. Durham, NC: Duke University Press, 2013.

Pecknold, Diane. *The Selling Sound: The Rise of the Country Music Industry*. Durham, NC: Duke University Press, 2007.

Peretti, Burton W. *The Creation of Jazz: Music, Race, and Culture in Urban America*. Urbana: University of Illinois Press, 1992.

Peterson, Richard A. *Creating Country Music: Fabricating Authenticity*. Chicago: University of Chicago Press, 1997.

Peterson, Richard, and Paul DiMaggio. "The Early Opry: Its Hillbilly Image in Fact and Fancy." *Journal of Country Music* 4 (1974): 39–40.

Phonogram abcedary series, May 1900–June 1901.

"Phonograph and Radio Industries Pool Their Interests." *The Magazine of Wall Street*, January 2, 1926.

"Phonograph Sales Have Declined." *Wall Street Journal*, May 16, 1924.

Pinta, Emil. "Efforts by Record Companies, the American Federation of Musicians, and Truman's Cabinet to Resolve the 1948 Recording Ban." *ARSC Journal* 41, no. 1 (Spring 2010): 24–42.

Pinta, Emil. "The First RCA Victor Recording of 1948: Petrillo, Truman and the AFM Recording Ban of 1948." *ARSC Journal* 39, no. 2 (Fall 2008): 284–91.

Pitman, Isaac. *Phonography; or Writing by Sound*. London: S. Bagster & Sons, 1942.

Popple, Simon, and Vanessa Toulmin, eds. *Visual Delights: Essays on the Popular and Projected Image in the 19th Century*. Translated by Wendy Schubring. Trowbridge, UK: Flick Books, 2000.

Porterfield, Nolan. *Jimmie Rodgers: The Life and Times of America's Blue Yodeler*. Urbana: University of Illinois Press, 1992.

"Radio and Columbia Drop Plan to Unite." *New York Times*, May 8, 1928, 43.

"Radio Helps to Sell Your Victor Records." *Voice of the Victor*, February 1924, 24.

"Radio Maps Union Fight." *Variety*, November 29, 1939, 25.

"Radio Review of 1927 and Hopes for the New Year." *New York Times*, January 1, 1928, 12.

"Radio's Dark Town Harmonizers." *Radio Digest*, November 1932, 22–23.

Radio Service Bulletin, August 31, 1931. https://www.fcc.gov/sites/default/files/radio-service-bulletin-FRC-general-orders-list.pdf

"Radio v. Phonograph." *Time*, October 4, 1926, 38.

Ramsey, Frederick Jr., and Charles Edward Smith, eds. *Jazzmen*. New York: Harcourt, Brace & Company, 1939.

Randle, William M. Jr. "A History of Radio Broadcasting and Its Social and Economic Effect on the Entertainment Industry." PhD dissertation, Case Western Reserve University, Cleveland, OH, 1966.

Rasmussen, Chris. "The People's Orchestra: Jukeboxes as the Measure of Popular Musical Taste in the 1930s and 1940s." In *Sound in the Age of Mechanical Reproduction*, edited by David Suisman and Susan Strasser, 181–98. Philadelphia: University of Pennsylvania Press, 2010.

Rayno, Don. *Paul Whiteman: Pioneer in American Music, Vol. 1: 1890–1930*. Lanham, MD: Scarecrow Press, 2012.

Razoglova, Elena. *The Listener's Voice: Early Radio and the American Public*. Philadelphia: University of Pennsylvania Press, 2011.

"RCA to Acquire Columbia Company." *New York Times*, April 11, 1930.

"RCA Victor's Message for 1930." *Voice of the Victor*, January 1920.

Read, Oliver, and Walter Welch. *From Tin Foil to Stereo: Evolution of the Phonograph*. Indianapolis: Howard W. Sams/Bobbs Merrill Company, Inc., 1959.

"The Real Publicity Value of the Dodge Radio Hour." *Variety*, January 11, 1928.

"Refrigeration Saved My Shirt!" *Radio Retailing*, March 1932.

Reich, Howard, and William Gaines. *Jelly's Blues: The Life, Music, and Redemption of Jelly Roll Morton*. Cambridge, MA: Da Capo Press, 2003.

Robertson, Bruce. "Industry Acts to Avoid Musicians Strike." *Broadcasting and Broadcasting Advertising*, September 1, 1937, 9.

Robertson Wojcik, Pamela, and Arthur Knight. *Soundtrack Available: Essays on Film and Popular Music*. Durham, NC: Duke University Press, 2001.

"R. S. Peer on Extended Trip." *Talking Machine World*, April 15, 1925, 50d.

"Rural Residents Average Well As Music Lovers." *Talking Machine World*, December 15, 1920.

Russell, Tony. "Blacks, Whites, and the Blues." In *Yonder Comes the Blues: The Evolution of a Genre*, 2nd ed., edited by Paul Oliver, Tony Russell, Robert M. W. Dixon, John Godrich, and Howard Dye, 143–44. Cambridge, UK: Cambridge University Press, 2001.

Russo, Alexander. *Points on the Dial: Golden Age Radio Beyond the Networks*. Durham, NC: Duke University Press, 2010.

Rust, Brian. *The American Record Label Book*. New Rochelle, NY: Arlington House, 1978.

Rutherford-Johnson, Tim, Michael Kennedy, and Joyce Bourne Kennedy. "Swing." *The Oxford Dictionary of Music*. 6th ed. (2013). http://www.oxfordmusiconline.com/page/The-Oxford-Dictionary-of-Music

Ryan, John. *The Production of Culture in the Music Industry: The ASCAP-BMI Controversy*. Lanham, MD: University Press of America, 1985.

Rypinski, M. C. "Will the Jobber Survive in Radio?" *Electrical World*, January 20, 1925.

Samuels, Edward. *The Illustrated Story of Copyright*. New York: Thomas Dunne Books, 2000.

Sanjek, Russell. *Pennies from Heaven: The American Popular Music Business in the Twentieth Century*. Updated by David Sanjek. New York: Da Capo Press, 1996.

Schatz, Thomas. "Film Studies, Cultural Studies, and Media Industries Studies." *Media Industries Journal* 1, no. 1 (2014). mediaindustriesjournal.org

Schiffer, Michael Brian. *The Portable Radio in American Life*. Tucson: University of Arizona Press, 1992.

Schnirring, F. W. "Does the Phonograph Need a New Copy Appeal?" *Printer's Ink*, October 28, 1926.

Schwartz, Laszlo. "The Phonograph Industry Has Become an Ideal Musical-Commercial Power." *Music Trade Review*, June 14, 1919.

Scotto, Aubrey, dir. *A Rhapsody in Black and Blue*. Paramount Pictures, 1932.

Seeley, Ward. "Will the Great Artists Continue?" *The Wireless Age*, June 25, 1923.

Seeley, Ward. "Yes, the Radiofan Buys Records!" *The Wireless Age*, August 1923.

Segrave, Kerry. *Jukeboxes: An American Social History*. Jefferson, NC: McFarland 2002.

Seldes, Gilbert. *The Seven Lively Arts*. New York: Harper & Brothers, 1924.

Settel, Irving. *A Pictorial History of Radio*. New York: Grosset and Dunlap, 1967.

Shapiro, Nat, and Nat Hentoff. *Hear Me Talkin' to Ya: The Story of Jazz as Told by the Men Who Made It*. New York: Dover Publications, 1966.

Sharpe, Cecil. *Folksongs from the Southern Appalachians*. Oxford, UK: Oxford University Press, 1974.

Shaw, Arnold. *Let's Dance: Popular Music in the 1930s*. Edited by Bill Willard. New York: Oxford University Press, 1998.

Sherman, Mike, and Kurt Nauck. *Note the Notes: An Illustrated History of the Columbia Record Label, 1901–1958*. New Orleans: Monarch Record Enterprises, 1998.

Shilkret, Nathaniel. *Nathaniel Shilkret: Sixty Years in the Music Business*. Edited by Niels Shell and Barbara Shilkret. Lanham, MD: Scarecrow Press, 2005.

Smith, Basil, dir. *Jailhouse Blues*. Columbia Pictures, 1929.

Smith, Jacob. *Eco-Sonic Media*. Berkeley: University of California Press, 2015.

Smulyan, Susan. *Selling Radio: The Commercialization of American Broadcasting, 1920–1934*. Washington, DC: Smithsonian Institution Press, 1996.

Snelson, Tim. "'They'll Be Dancing in the Aisles!': Youth Audiences, Cinema Exhibition, and the Mid-1930s Swing Boom." *Historical Journal of Film, Radio, and Television* 37, no. 3 (2016): 455–56.

Sousa, John Philip. "The Menace of Mechanical Music." *Appleton's*, September 1906, 278–84.

Spring, Katherine. *Saying It with Songs: Popular Music and the Coming of Sound to American Cinema*. New York: Oxford University Press, 2013.

Starr, Paul. *The Creation of Media: Political Origins of Modern Communications*. New York: Basic Books, 2004.

"Statement of the Department of Justice on the Closing of the Anti-Trust Division's Review of ASCAP and BMI Consent Decrees." https://www.justice.gov/atr/ascap-bmi-decree-review

Stearns, Marshall. *The Story of Jazz*. London: Oxford University Press, 1972.

Sterne, Jonathan. *The Audible Past: Cultural Origins of Sound Reproduction*. Durham, NC: Duke University Press, 2003.

Sterne, Jonathan. "There Is No Music Industry." *Media Industries* 1, no. 1 (2014). mediaindustriesjournal.org

Stewart, Jacqueline Najuma. *Migrating to the Movies: Cinema and Black Modernity*. Berkeley: University of California Press, 2005.

Sudhalter, Richard M. *Lost Chords: White Musicians and Their Contributions to Jazz, 1915–1945*. New York: Oxford University Press, 1999.

Suisman, David. *Selling Sounds: The Commercial Revolution in American Music*. Cambridge, MA: Harvard University Press, 2009.

Suisman, David, and Susan Strasser, eds. *Sound in the Age of Mechanical Reproduction*. Philadelphia: University of Pennsylvania Press, 2010.

Summers, Harrison. *A Thirty-Year History of Programs Carried on National Radio Networks in the United States, 1926–1956*. New York: Arno Press/New York Times, 1971.

Sutton, Allan. *A Phonograph in Every Home: The Evolution of the American Recording Industry, 1900–1919*. Denver: Mainspring Press, 2010.

Sutton, Allan. *Race Records and the American Recording Industry, 1919–1945*. Denver: Mainspring Press, 2016.

Sutton, Allan. *Recording the Thirties: The Evolution of the American Recording Industry, 1930–39*. Denver: Mainspring Press, 2011.

Sutton, Allan. *Recording the Twenties: The Evolution of the American Recording Industry, 1920–29*, Denver: Mainspring Press, 2008.

Sutton, Allan. "Sorting Out Puritan." *Victrola and 78 Journal*, Issue 13 (Autumn 1998).

Sutton, Allan, and Kurt Nauck. *American Record Labels and Companies, 1891–1943*. Denver: Mainspring Press, 2000.

Swift, Gertrude. "What Do You Think of Jazz?" *Voice of the Victor*, December 1924.

Swinton, Paul. "A Twist of Lemon" *Blues and Rhythm* 121. www.bluesandrhythm.co.uk/documents/121.pdf

Szwed, John. "Negro Music: Urban Renewal." In *Our Living Traditions: An Introduction to American Folklore*, edited by Tristram Potter Coffin, 272–82. New York: Basic Books, 1968.

Taylor, Timothy. *The Sounds of Capitalism: Advertising, Music, and the Conquest of Culture*. Philadelphia: University of Pennsylvania Press, 2012.

Thompson, Emily. "Machines, Music, and the Quest for Fidelity: Marketing the

Edison Phonograph in America, 1877–1925." *The Musical Quarterly* 79, no. 1 (Spring 1995): 131–71.

Thompson, Emily. *The Soundscape of Modernity: Architectural Acoustics and the Culture of Listening in America, 1900–1933*. Cambridge, MA: M.I.T. Press, 2004.

"Thos. A. Edison, Inc., Not to Enter Radio Field." *Talking Machine World*, January 15, 1925.

Titon, Jeff Todd. *Early Downhome Blues: A Musical and Cultural Analysis*. 2nd ed. Chapel Hill: University of North Carolina Press, 1994.

Toll, Robert. *The Entertainment Machine: American Show Business in the Twentieth Century*. New York: Oxford University Press, 1982.

"Tommy's Only Foolin'." *Radio Digest*, October 15, 1926.

Tosches, Nick. *Country: Living Legends and Dying Metaphors in America's Biggest Music*. Rev. ed. New York: Charles Scribner's Sons, 1985.

Traver, Jerome D., and Joel M. Maring. "*Stand By*: Journalistic Response to a Country Music Radio Audience." *JEMF Quarterly* 14, no. 71 (Autumn 1983): 149–61.

Tschmuck, Peter. *Creativity and Innovation in the Music Industry*. Berlin: Springer Verlag, 2012.

Tschudi, Ernst Felix. "The Immortals Protest against Jazz." *The Etude*, August 1924, 670.

Tsotsi, Tom. "Gennett-Champion Blues." *78 Quarterly* 1, no. 3 (1988).

Tyler, Paul L. "Hillbilly Music Re-imagined: Folks and Country Music in the Midwest." *Journal of American Folklore* 127, no. 504 (Spring 2014): 159–90.

"Unbreakable Discs Made of New Resin." *New York Times*, January 4, 1930, 4.

United States v. Motion Picture Patents Co., 225 F. 800 (D.N.J. 1915).

"Utilizing Heart Throbs as Trade Ticklers." *Talking Machine World*, March 13, 1919.

VanCour, Shawn. *Making Radio: Early Radio Production and the Rise of Modern Sound Culture*. New York: Oxford University Press, 2018.

van der Tuuk, Alex. *Paramount's Rise and Fall: The Roots and History of Paramount Records*. Denver: Mainspring Press, 2012.

"Victor Co. Loses Suit." *Variety*, April 7, 1922.

"Victor Self-Service Record Cabinet a Novel Method Adopted by Dealers." *Music Trade Review*, January 18, 1919.

Wald, Elijah. *Escaping the Delta: Robert Johnson and the Invention of the Blues*. New York: Amistad/Harper Collins, 2004.

Wald, Elijah. "The Bluesman Who Behaved Too Well." *New York Times*, July 18, 2004. https://www.nytimes.com/2004/07/18/arts/music-the-bluesman-who-behaved-too-well.html

Walser, Robert. *Keeping Time: Readings in Jazz History*. New York: Oxford University Press, 2014.

Ward, Geoffrey C., and Ken Burns. *Jazz: A History of America's Music*. New York: Knopf, 2000.

Wardlow, Gayle Dean. "Godfather of Delta Blues: H. C. Speir." In *Chasin' That Devil Music: Searching for the Blues*, edited by Edward Komara, 131–51. San Francisco: Miller Freeman Books, 1998.

Wardlow, Gayle Dean. "Initial Inquiries and Encounters." In *Chasin' That Devil Music: Searching for the Blues*, edited by Edward Komara, 10–73. San Francisco: Miller Freeman Books, 1998.

Wardlow, Gayle Dean. "Legends of the Lost (The Story of Henry Speir)." *Blues Unlimited* 31 (March 1966).

Wardlow, Gayle Dean. "The Talent Scouts: H. C. Speir (1895–1972)." *78 Quarterly* 1, no. 8 (1993).

Weber, Max. *The Protestant Work Ethic and the Spirit of Capitalism*. Mineola, NY: Dover Publications, 2003.

Weisbard, Eric. *Top 40 Democracy: The Rival Mainstreams of American Music*. Chicago: University of Chicago Press, 2014.

"West Virginia Hills Are in the Bronx, Says Barn Barnum." *Billboard*, March 4, 1939.

"Where Is Jazz Leading America?" *The Etude*, August and September 1924.

"Where the Etude Stands on Jazz." *The Etude*, August 1924.

Whiteman, Paul, and Mary Margaret McBride. *Jazz*. New York: J. H. Sears & Company, 1926.

White-Smith Music Publishing Company, Appt., v. Apollo Company, 209 U.S. 1 (1908).

Wiggins, Gene. "John Carson: Early Road, Radio, and Records," *Journal of Country Music* 3, no. 1 (May 1979): 30–38.

Wile, Raymond. "The American Graphophone Company and the Columbia Phonograph Company Enter the Disc Record Business, 1897–1903." *ARSC Journal* 22, no. 2 (Fall 1991): 207–9.

Wile, Raymond, ed. *Proceedings of the 1890 Convention of Local Phonograph Companies*. Nashville, TN: Country Music Foundation Press, 1974.

Wilgus, D. K. "The Hillbilly Movement." In *Our Living Traditions: An Introduction to Folklore*, edited by Tristram Potter Coffin, 263–71. New York: Basic Books, 1968.

Willems, Joseph. *All of Me: The Complete Discography of Louis Armstrong*. Oxford, UK: Scarecrow Press, 2006.

Williams, Charles. "Mr. Petrillo's Hopeless War." *The Nation*, October 3, 1942.

Williamson, John, and Martin Cloonan. "Rethinking the Music Industry." *Popular Music* 26, no. 2 (May 2007): 305–22.

"Will the Phonograph Industry Stage a Sensational Comeback?" *Sales Management*, November 14, 1925, 639.

Wilson, Carl. *Let's Talk About Love: Why Other People Have Such Bad Taste*. New and expanded ed. New York: Bloomsbury 2014.

Winchell, Walter. "On Broadway." *The New York Mirror*, March 14, 1930.

Winters, S. R. "An Interview with Secretary H. C. Hoover." *Radio News* 6 (October 1924): 568.

Witt, Doris. "Soul Food: Where the Chitterling Hits the (Primal) Pan." In *Eating Culture*, edited by Ron Scapp and Brian Seitz, 258–87. Albany: State University of New York Press, 1998.

Wolfe, Charles K. *A Good Natured Riot: The Birth of the Grand Ole Opry*. Nashville: Country Music Foundation/Vanderbilt University Press, 1999.

Wolfe, Charles. "Nashville and Country Music, 1926–1930: Notes on Early Nashville Media and Its Response to Old-Time Music." *Journal of Country Music* IV, no. 1: 2–16.

Wolfe, Charles, and Ted Olson, eds. *The Bristol Sessions: Writings about the Big Bang of Country Music.* Jefferson, NC: McFarland, 2005.

Wondrich, David. *Stomp 'n' Swerve: American Music Gets Hot, 1843–1924.* Chicago: Chicago Review Press, 2003.

Woodward, C. Vann. *The Strange Career of Jim Crow.* 3rd ed. New York: Oxford University Press, 1974.

"Workers Should Profit in 1928." *Talking Machine World,* January 1928, 76.

Wurtzler Steve. *Electric Sounds: Technological Change and the Rise of Corporate Mass Media.* New York: Columbia University Press, 2007.

Yorke, Dane. "The Rise and Fall of the Phonograph." *The American Mercury*, September 1932, 1–12.

Young, William H., and Nancy K. Young. *The 1930s.* Westport, CT: Greenwood Press, 2002.

Zwonitzer, Mark, and Charles Hirshberg. *Will You Miss Me When I'm Gone?: The Carter Family and Their Legacy in Music.* New York: Simon & Schuster, 2002.

Index

***Note**: Page numbers in italics indicate photographs or illustrations.*